P9-CSG-089

Air Apparent

Air

Apparent

HOW METEOROLOGISTS

LEARNED TO MAP, PREDICT,

AND DRAMATIZE WEATHER

Mark Monmonier

THE UNIVERSITY OF CHICAGO PRESS ❈ CHICAGO AND LONDON

3 1923 00372219 2

MARK MONMONIER is Distinguished Professor of Geography at Syracuse University. He is the author of *Maps, Distortion, and Meaning; Computer-Assisted Cartography; Technological Transition in Cartography; Maps with the News; Mapping It Out; Drawing the Line; How to Lie with Maps;* and *Cartographies of Danger.*

The University of Chicago Press, Chicago 60637
The University of Chicago Press, Ltd., London
© 1999 by Mark Monmonier
All rights reserved. Published 1999

09 08 07 06 05 04 03 02 01 00   1 2 3 4 5

ISBN: 0-226-53422-7 (cloth)

Library of Congress Cataloging-in-Publication Data

Monmonier, Mark S.
    Air apparent : how meteorologists learned to map, predict, and
dramatize weather / Mark Monmonier.
        p.   cm.
    Includes bibliographical references and index.
    ISBN 0-226-53422-7 (cloth : alk. paper)
    1. Meteorology—Charts, diagrams, etc. 2. Weather forecasting—
Technique. I. Title.
QC878.M59 1999
551.63'022'3—dc21   98-25797
                                                     CIP

♾ The paper used in this publication meets the minimum requirements of the American National Standard for Information Sciences—Permanence of Paper for Printed Library Materials, ANSI Z39.48–1992.

For George A. Schnell,
effective educator,
respected realist, and
amiable advisor

Contents

Preface

I am fascinated with the Weather Channel. Minutes after checking into a hotel far from home, I'm either viewing its narrated sequence of colorful, timely weather maps or cursing the stinginess of the local cable system. These maps are important to me: I want to know what to wear the next day, whether snow or strong wind is likely to delay my return flight, and what conditions will be like back in Syracuse. Short-range cartographic forecasts are informative and reliable, much more so than the terse pronouncements of radio news and dial-up recordings, rife with vague slogans like "partly cloudy" and "30 percent chance of thundershowers." The difference is profound: no word-bound forecast can capture the uncertainty of an approaching low-pressure cell that might with equal likelihood pass to the west or the east, with drastically different effects on flying and driving. The weather map is like a chessboard, with which the knowledgeable observer can speculate intelligently about the atmosphere's next few moves.

At home we don't have cable. Marge and I enjoy books and magazines more than made-for-TV movies, and reception in our neighborhood is adequate, if not perfect, for newscasts and public affairs programming. Besides, at least one of the local stations has a lively team of competent meteorologists who succeed at being both informative and personable. I know where to find them, around supper and again before bedtime.

Until the mid-1980s, I knew little about the origin of weather maps. I had taught cartography for several years, using textbooks that discussed landforms but largely ignored weather and climate.

And my growing interest in the history of cartography was nourished largely by scholars preoccupied with the maps of monarchs, explorers, and colonial administrators. In 1983, more on whim than a realistic expectation of winning, I asked the Guggenheim Foundation to sponsor a sabbatical year of looking at and thinking about another neglected cartographic genre, the news map. The fellowship—the biggest and most pleasant surprise of my life—bought me time to visit libraries, talk with journalists, dig up facts, digest ideas, and spend countless days looking at old newspapers on microfilm. A nuisance at first, weather maps in newspapers had their own story: an intriguing blend of weather science, electronic telecommunications, newspaper design, and media competition. In addition to describing temperature and air masses, the lowly weather map had become a journalistic fashion statement.

News publishers are not the only organizations to exploit the weather map as propaganda—in the annual battle to protect their slice of the federal budget, Washington's weather bureaucrats update a running cartographic campaign that links weather disasters to the need for reliable forecasts based on accurate data. Their appeals have a vivid history, fueled by the atmosphere's sporadic assaults on unsuspecting communities and by technological innovations promising richer insights and earlier warnings.

In the late nineteenth century, when wannabe forecasters studied maps rather than physics, the weather map was both a symbol of scientific achievement and a metaphor for the Weather Bureau itself. Meteorologists relied on timely data, and the synchronous weather map, made possible by telegraphic reports from a far-flung network of paid observers, was an object of awe. By the end of twentieth century, weather maps had changed markedly in look and role. Timely data are as important as ever, but meteorologists now have the theoretical foundation for projecting their measurements farther forward in time with greater reliability. Mathematical equations and computer software have replaced pen and paper, and their projections are numerical, not graphic. Although visual maps display the results, the map inside the computer is more a tool than a product. Forecasters employ several of these "numerical models," and their skill lies not in drawing up and interpreting a single map but in choosing among a range of cartographic views.

In what follows I trace the evolution of the weather map from simple hand-drawn plots of temperature, pressure, and wind direction to a varied menu of customized displays. The marvel of contemporary meteorological cartography is not the increased reliability of its forecasts but its growing relevance to a broad array of human concerns, from tornadoes and flash flooding to air pollution and global warming. This is a cartographic story, best told by the maps themselves: exploratory maps, which continue to refine our knowledge of storms, atmospheric motion, and climatic change; weather charts, numerical models, and other everyday tools of forecasters, pilots, and farmers; radar and satellite monitoring systems essential to emergency managers, physicians, power companies, and commodities traders; hazard maps, which guide building contractors and regional planners; and media maps, which report weather as a soap opera of powerful forces that affect us all.

The story of weather maps is very much a story of electronic technology, which provided efficient mechanisms for recording measurements, disseminating data, and drawing maps. Atmospheric cartography is indebted to many electronic innovations, including digital computing, wireless broadcasting, radar imaging, and satellite remote sensing. Yet no single tool or technique was as catalytic as the electric telegraph, which shrunk space literally and figuratively through timely miniature snapshots of surface weather: *synchronous* weather maps with which forecasters could track storms and warn of danger.

By today's standards the early history of the weather map might seem painfully slow. The invention of the thermometer and the barometer preceded the systematic measurement of temperature and pressure by at least a century and a half, the first synchronous weather map described data collected several decades earlier, and the isolines that made the air visible and its spatial patterns apparent were a much earlier cartographic innovation. What's more, "natural philosophers" in Europe and America recognized the feasibility of forecasting weather a half century before the first government-run weather bureau. But a closer look reveals that implementation is seldom an immediate consequence of discovery, even today. Numerical forecasting was first proposed in 1922—a half century before even vaguely reliable computer models—and workable strategies for forecasting climatic change have yet to be tried. Not surprisingly, the weather map's slow early development holds valuable insights for its continued evolution.

Many of these insights reflect the interaction of ideas and controversy, people and institutions, competition and cooperation. In the first half of the nineteenth century, for instance, maps of the atmosphere led to competing theories about storms, which further mapping helped resolve. Informal networks for data collection identified the value of federal coordination and international cooperation. Progress depended on persuasive champions, pivotal events that dramatized need, and the evolving structure of science and government agencies. The current situation is much the same: although we know a lot more about the atmosphere, the continuing saga of the evolving weather map remains in large part a people story.

Telling this story to a mixed audience of cartographic historians, weather professionals, and intelligent lay readers is a daunting task—please one group and alienate another with an account that's either too complex or too watered down. Be warned, then, that what follows is a broad synthesis intended for amateur weather enthusiasts likely to appreciate an excursion through meteorology's cartographic history as well as for cartographic historians and map enthusiasts curious about a neglected but momentous area of map use. In particular, I emphasize historical development and the roles of technology and institutions, introduce the book by looking at the use of weather maps in the late nineteenth century, focus on map content rather than aesthetics, and avoid textbook explanations of atmospheric processes as well as didactic deconstructions of cartographic design. To serve a largely American audience, I report lengths in miles and temperatures in Fahrenheit degrees and draw largely on examples from the National Weather Service and its predecessors. Limited to a single volume, I touch marginally on a few developments important to contemporary meteorology and climatology, namely, isentropic analysis, polar-orbiting satellites, GOES Sounder imagery, wind-profilers, general circulation models, and the AWIPS workstation (still under development as I write). Extensive treatments of contouring algorithms and critiques of Web-site maps I leave to the cartography-for-meteorologists text that, I hope, a cartographer and an atmospheric scientist will soon see the need to collaborate on. Also neglected are vertical cross-sections, meteorographs (time-series records), and other nonmap graphics important to the development and practice of weather science. And the examples excerpted in my later chapters offer at best a selective introduction to the many dozens of distinct map types that pervade contemporary meteorology—arguably today's single most map-intensive scientific enterprise. Even so, the more than a hundred illustrations discussed here afford valuable insights to the weather map's evolution, varied roles, explosive growth, and increased foresight.

Acknowledgments

For help in uncovering the story, I am indebted to numerous people who willingly shared insights or discussed their work. Several staff at the National Oceanic and Atmospheric Administration's National Centers for Environmental Prediction were especially helpful: Kenneth Comba, executive officer; Jerry Delaney of the Technical Support Group; Kim Comba, in Public Affairs; Jim Hoke, director of the Hydrometeorological Prediction Center and the Marine Prediction Center; Stephen Lord and Steve Tracton, in the Environmental Modeling Center; and James Laver, Jamie Kousky and Ed O'Lenic, in the Climate Prediction Center. I also appreciate the assistance of Jeff Waldstreicher and Pete Ahnert, at the Weather Forecast Office in Binghamton, New York; John Paquette, at the National Environmental Satellite, Data, and Information Service; Vico Baer, at the National Weather Service's Office of Systems Operation; Dan O'Brien and William Peat, Jr., at the New York State Emergency Management Office; Dave Eichorn, at Syracuse's WIXT-TV; the late Fred Ludwick, at the Niagara Mohawk Power Corporation; and Rick Booth, at the Weather Channel.

For assistance with reference materials, I am grateful to Dottie Anderson and Carla Wallace, at the NOAA Central Library; Lee Murray, Tom Keays, and Lockhart Russell, at Syracuse University's Science and Technology Library; and Karen Ingeman and Wayne Stevens, at the E. S. Bird Library. Ralph Jewell, at the University of Bergen, graciously provided reproducible copies of Tor Bergeron's and Halvor Solberg's historic correspondence. Geographers who shared examples and experiences include Jim Carter, at Western Illinois University; Alan MacEachren, at the Pennsylvania State University; and Mike Peterson, at the University of Nebraska.

At Syracuse University, the Maxwell School's Appleby-Mosher fund supported research travel, and Geography 155 (the Natural Environment), a section of which I have taught every year since I started the project, enhanced my perceptions of what's important and how to explain it. Mike Kirchoff, at the Syracuse University Cartography Laboratory, and Jeff Bittner, at the Faculty Academic Computing Support Services, provided invaluable assistance with image-processing software. At the University of Chicago Press, I am indebted to Penelope Kaiserlian, my editor, and to Carol Saller and Mike Brehm. In addition, Edward Aguado and Julie A. Winkler provided useful suggestions on content and presentation. On the home front, my wife Marge and daughter Jo offered encouragement and tolerant understanding.

Seeing and Forecasting

*L*ike the flashing red signals at railroad crossings, weather maps regularly announce the imminent arrival of an unstoppable threat with severe consequences for those who ignore the warning. But while train tracks are rigid, well defined, and narrow, cyclonic storms are fluid, free ranging, and several hundred miles across. What's more, because parents and schoolteachers instill respect for traffic signals at an early age, crossing lights are intelligible to everyone, whereas weather charts are meaningful only to those who understand atmospheric processes and have learned the maps' graphic codes. As a result, we rely largely on professional meteorologists, working around the clock, who scrutinize their charts for ominous signs of change and issue warnings when storms approach. With skill and luck, forecasters can look ahead one to five days—long enough to prepare for, if not avoid, disaster.

The value of weather maps is most apparent when a warning is ignored by someone who should know better. Someone like Hollis Blanchard, captain of the *Portland*, a 320-foot-long wooden sidewheel steamship that foundered in Massachusetts Bay, several miles west of Cape Cod, on Sunday morning, November 27, 1898, around ten o'clock—the time frozen on wristwatches strapped to bodies that washed ashore later that day. Everyone aboard, approximately 120 passengers and crew, perished.[1] (Counts vary because the passenger list went down with the ship, and many bodies were washed out to sea or buried in the Cape's reworked beaches.) In describing the disaster as "one so far from accidental, one in which the fixing of direct responsibility is so easy," the *New York Times* condemned Blanchard for taking "chances which no man in his position had a right to take."[2] Because of machismo, prior reprimands for being too cautious, or a direct order from his employer, Captain Blanchard sailed on schedule, at 7 P.M., Saturday evening, eight hours after the Weather Bureau hoisted storm signals along the New England coast. As predicted, the snow began shortly after the *Portland* left Boston's India Wharf for its namesake city in Maine. Within two hours the light snow had become a raging blizzard, and through the night the wind's velocity increased steadily to 40, 50, and 60 miles per hour. On late Sunday morning, about the time the *Portland* went down, the storm centered just east of Cape Cod was lashing Boston with winds of 72 miles per hour.[3]

Had other skippers been as foolish as Captain Blanchard, the storm might have eclipsed the Blizzard of 1888, which killed 400 people.[4] Instead, what history books call the *Portland* storm ravaged the coast from Virginia to eastern Canada, claimed more than 50 additional lives in smaller tragedies on both land and water, injured thousands, and trashed property worth several billion of today's dollars. People trapped outside by heavy snows died from exposure; trains and trolley cars were stalled for days; urban grocers could not restock; and strong sustained winds damaged or destroyed many ships, especially small fishing vessels, on Cape Cod and Long Island. The Cape's outermost village, Provincetown, abruptly lost its fleet, its wharves, and its role as New England's largest fishing port. Even so, the toll would have been far higher without the accurate, insightful reading of a timely weather chart.

Figure 1.1 describes the weather map for 8 A.M., November 26, 1898—the map that kept Captain Blanchard's peers in port. I've long detested the drab, barely legible (if at all) full-view halftone photos of historic maps reduced to fit a book page, and prefer instead to scan the best representation available, extract the relevant portion, retrace important features, eradicate whatever is distracting or illegible, and add type and symbols that preserve much if not all of the original cartographic flavor. For this example, I've enlarged an excerpt from a countrywide map printed in black and light orange, darkened otherwise faint coastlines and state boundaries, and added a light-gray tint to better differentiate land and water.[5] The result captures the map's frantic clutter as well as the positions and design of its symbols for temperature, pressure, wind direction, and local weather. By omitting the western half of the country and much of the south, I can focus on the map's most frightening feature: the huge area of low pressure centered on southeastern Michigan.

Several symbols reveal the threat. The solid line labeled 29.90 en-

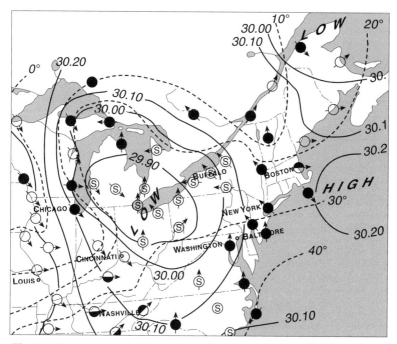

Fig. 1.1. Weather map for 8 A.M., Saturday, November 26, 1898. Excerpt redrawn from *Maryland Weather Service* (Baltimore: Johns Hopkins Press, 1899), 1: 244.

closes an area extending from southwestern Pennsylvania to central Michigan within which barometric pressure is less than 29.9 inches of mercury. Called an *isobar* (meaning equal weight or pressure), this line separates areas with slightly higher pressure, just outside its perimeter, from areas with slightly lower pressure, just within. Near Detroit, which registered the region's lowest pressure, 29.8 inches, the word "LOW" marks the center of the storm. At representative cities, tiny circles reflect sky conditions: black for cloudy, white for clear, and half black–half white for partly cloudy. Circles inscribed with the letter "S" identify cities where snow is falling, and arrows showing wind direction reveal air moving toward, as well as in a counterclockwise direction around, the center of low pressure. Additional isobars labeled 30.00 and 30.10 reveal the extent of the storm, which is responsible for snow in Virginia and cloudy skies in upper Michigan. By contrast, Boston, well east of the storm center and still under the influence of high pressure off the coast, reported partly cloudy skies at 8 A.M. and winds from the west.

Meteorologists discover storms by plotting surface pressure and wind on the same map. Nested sets of closed isobars, like the lines labeled 29.90, 30.00, and 30.10, signify pressure cells, described as either high or low depending upon whether pressure increases or decreases toward the center. Wind direction is part of the pattern: air not only flows outward from a surface high and inward toward a surface low but in the northern hemisphere circulates in a clockwise direction around a high and in a counterclockwise direction around a low. This circular movement reflects the *Coriolis effect*, whereby the rotation of the earth deflects moving objects—air included—to the right in areas north of the equator and to the left in the southern hemisphere.[6] The map thus shows a storm extending well beyond the large, closed 30.10 isobar, with winds veering to the north in eastern New York and to the south in western Illinois.

Counterclockwise circulation also affects air temperature, portrayed on the map by dashed lines called *isotherms:* lines of equal temperature. The isotherms for Saturday morning, November 26, show a predictable south-to-north decline in temperature from 40°F along the Carolina coast to 10°F in Quebec and northern New Hampshire. Note, though, how the 10° and 20° isotherms bulge northward around the storm, which is pumping warmer air from the south into western New York, southern Ontario, and northern Michigan. The threat of a broad low-pressure cell lies not only in its strong winds but also in its counterclockwise, or *cyclonic*, flow, which can import moist air from the ocean and convert a gale into a blizzard. Equally ominous for New England is the bitter-cold air directly behind the storm. Pulled from the north by this massive counterclockwise vortex, frigid polar air behind the low guaranteed that the heavy snow would not soon melt.

As early as 1830, natural philosophers had recognized the basic relationship between wind and atmospheric pressure as well as the tendency of winter storms to travel from west to east. For New England these phenomena present the apparent contradiction of the *northeaster:* a severe storm that enters the region from the southwest, as in figure 1.2, but attacks the coast with strong winds from the northeast.[7] Only when the low-pressure center moves past and away from an area, do its winds mimic the storm's southwest-to-northeast advance.

Recognizing an approaching northeaster on the morning weather map for November 26, forecasters telegraphed gale warnings to Weather Bureau stations from Virginia northward through Maine.[8] The warning posted from Newport, Rhode Island, to Eastport, Maine, read:

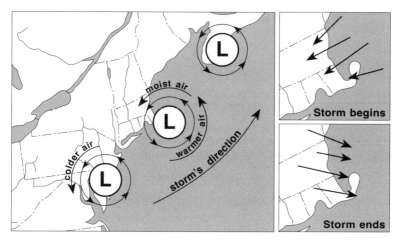

Fig. 1.2. Generalized pattern of cyclonic winds in a northeaster. Although a northeaster approaches New England from the southwest, the storm begins with strong winds from the east or northeast. Rain or snow occurs as counterclockwise circulation around the storm pulls moist air inward from the ocean. This circular air flow is partly responsible for cold polar air behind the storm, which typically is much broader than the symbols in this diagram.

> Storm central near Detroit moving east. East to northeast gales with heavy snow to-night. Wind will shift to west and northwest with much colder Sunday.

Connecticut and Long Island, which the storm would hit a few hours earlier, received a slightly different warning:

> Storm central near Detroit moving east. Wind will increase to south and southeast gales this afternoon and shift to west to-night, with snow. Decidedly colder Sunday.

The Weather Bureau also advised the Baltimore and Ohio and the Pennsylvania railroads, which crossed the Appalachians, to plan for heavy snow and stranded passengers.

The broad cyclone apparent on the November 26 morning weather map did not surprise Edward Garriott, the Weather Bureau's chief forecaster. Plotting and studying pressure measurements and other observations telegraphed to Washington twice each morning and evening, forecasters had recognized the storm two days earlier and tracked its movement from North Dakota across northern Minnesota into Michigan's northern peninsula, and then south toward Detroit. Experience had taught them where to look for new storms, how storms evolve and move, and what areas would be most

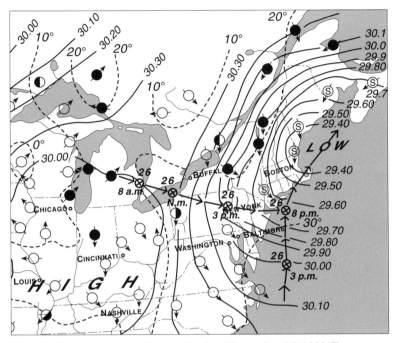

Fig. 1.3. Weather map for 8 A.M., Sunday, November 27, 1898. Excerpt re-drawn from *Maryland Weather Service* (Baltimore: Johns Hopkins Press, 1899), 1:248.

vulnerable. Meteorologists knew, for instance, that winds are strongest where the most closely spaced isobars reflect the greatest pressure difference, and they also knew that falling pressure at the center typically signals a growing storm, with more intense winds. Using additional weather observations collected to monitor the storm, Garriott's staff tracked the low-pressure center's rapid move-ment into western Pennsylvania by noon and across the Appalachi-ans into southeastern Pennsylvania by 3 P.M., when another low-pres-sure center appeared off the coast of Virginia. But because these new revelations merely confirmed the terse warnings issued at 10:30 A.M., modifications were not needed.

Storm tracks on figure 1.3, the weather chart for the following morning, describe what happened next. Around 8 P.M. on the 26th, an hour after the *Portland* left Boston harbor, the two low-pressure centers fused into a more intense storm, which drifted northward, off the coast. Pressure at the storm center fell from 29.6 inches on Saturday evening to 29.3 inches on Sunday morning, when the storm was centered about 50 miles east of Boston. The steep pressure gradient reflected in the closely spaced isobars produced vicious gales that toppled piers and whipped up 16-foot snowdrifts. That evening, after Boston had accumulated more than a foot of snow, pressure at the center dropped to 28.38 inches. By then, the wind had shifted, skies were clearing, and the storm was advancing northward toward amply forewarned Canadians along the Gulf of St. Lawrence.

Maps that could warn of storms and cold waves were a triumphant collaboration of science, technology, bureaucracy, and cartography. Science provided the instruments for measuring temperature and pressure as well as broad but useful theories of atmospheric behavior. Technology furnished the electric telegraph, which allowed the rapid collection of perishable data from widely separated weather observers and the equally efficient dissemination of forecasts. Bureaucracy afforded the institutional framework for hiring, training, paying, and supervising several hundred weather observers as well as for building and maintaining a vast network of weather stations. And cartography contributed the base maps and graphic codes with which nineteenth-century meteorologists organized their data, visualized the atmosphere, and made educated guesses about the next day's weather.

Meteorological cartography ran like clockwork. At 8 A.M. and 8 P.M. each day, eastern standard time, observers at more than 150 weather stations used standardized thermometers, barometers, and other instruments to sample temperature, pressure, wind direction and velocity, and cloudiness.[9] After correcting for instrumentation error and adjusting barometric pressure to sea level, they converted their figures to an alphabetic cipher that reduced transmission errors as well as telegraph charges.[10] Although most weather stations were near or directly connected to a local telegraph office, nearly 500 miles of the Weather Bureau's own lines linked remote coastal stations with the commercial network. Cooperating stations in Canada and the West Indies sent additional data, for a broader, more complete atmospheric snapshot. Long-distance communication was slow and labor-intensive by modern standards, but because weather telegrams and cablegrams had priority over other messages, most observations reached Washington within the hour.

The mechanical, military efficiency of the central forecast office betrayed the civilian Weather Bureau's origin in the Army Signal

Service. As telegraphers decoded incoming messages, a platoon of clerks entered the readings in statistical registers and distributed data to a squad of experienced atmospheric cartographers. One clerk drew up a composite weather chart similar to the generalized atmospheric snapshots in figures 1.1 and 1.3. On a 22-by-16-inch blank outline map showing state boundaries, coastlines, important rivers, and weather stations, he neatly recorded each station's air temperature, sea-level pressure, wind speed and direction, degree of cloudiness, and amount of rain or snow since the last report. After transcribing the data, the clerk outlined broad areas with current precipitation, drew isotherms and isobars, and identified centers of low and high pressure. A second worker compared current temperatures with readings taken 24 hours earlier and used an identical blank chart to construct a map highlighting invasions of markedly cooler or warmer air. A third clerk prepared a similar map showing the 24-hour change in barometric pressure, while a fourth drafted two maps, one portraying regional variations in humidity and the other describing the type, height, and movement of clouds. All five charts helped the forecaster in charge project the atmosphere's recent past into the next 36 hours.

Drawing isolines on a weather chart was so routine that a search of turn-of-the-century meteorological texts and weather service publications failed to turn up a single step-by-step illustration.[11] Even so, readers unfamiliar with the process of threading contours though a set of data points should appreciate the example in figure 1.4, which compares logical and illogical renderings of a hypothetical temperature contour representing 30°. As the upper diagram illustrates, an acceptable isotherm passes directly through a reading of 30° but midway between 31° and 29°. In threading the line between 31° and 26°, for example, the conscientious mapmaker shifts the line toward the closer temperature—in this case by mentally dividing an invisible line between the points into five equal parts (one for each degree of difference) and placing the isotherm only one unit away from 31° but four units out from 26°.[12] Because not all readings on the cool side of the isotherm are less than 30°, a small closed contour isolates locally high temperatures of 30° and 34°. By contrast, the unacceptable isotherm in the lower diagram runs through 28°, but not 30°, and passes between 22° and 28°, rather than between 22° and 39°—glaring mistakes rarely made.

The Weather Bureau trusted its clerks' judgment in positioning isolines but proscribed the values portrayed with standardized intervals, which promoted rapid, mechanical mapmaking as well as read-

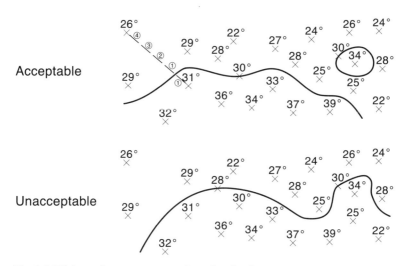

Fig. 1.4. Right and wrong ways to thread an isotherm.

ily comparable maps. Isotherms representing multiples of 10° yielded an informative, rarely cluttered temperature surface with neither too many nor too few contours. (More problematic was the Fahrenheit scale, for which a 30° line could only approximately depict the extent of freezing temperatures.) Isobars plotted every tenth of an inch afforded an equally lucid portrait of barometric pressure, except when an extraordinarily intense depression darkened the map with tightly spaced contours.[13] After drawing and labeling isolines, the clerk inspected the pressure map for small closed contours, which he quickly labeled "high" or "low."

Weather Bureau offices in more than a hundred cities produced their own weather snapshots. Smaller than the Washington weather map and based on fewer observations, these station weather maps let forecasters throughout the country relate the national pattern to local conditions.[14] A meteorologist in Oswego, New York, for instance, knew that west-northwest winds across Lake Ontario in winter usually meant "lake-effect" snow south and east of the lake. Although the central forecast office tracked storms and coordinated warnings, cities and their hinterlands required prompt customized forecasts based on experience with regional peculiarities. Because the Washington map could not be faxed out to the provinces, detailed specifications in the bureau's telegraph contract assured each forecast office a timely sampling of observations sufficient to draw its own national weather chart.[15]

Despite the local importance of regional anomalies, turn-of-the-century storm forecasting was very much a process of early recognition, informed guessing, and careful tracking. The weatherman vigilantly awaited the birth of each new low-pressure center, studied its early movement as well as whatever terrain and high pressure it faced, made an educated prediction of its path, and monitored its progress. A storm's origin and subsequent movement could reveal its destination, but because weather is quirky, the forecaster remained alert for a sudden acceleration or shift in direction. This strategy worked much of the time, and what he learned from one storm, he filed away in mind and map to help predict others.

Not surprisingly, forecasters developed a deep respect for weather charts. As Weather Bureau meteorologist George Bliss told *Scientific American* readers in 1917,

> The most profound students of atmospheric physics . . . were far from being the best forecasters. . . . In order to excel in the profession one must possess a special faculty for intuitively and quickly weighing the forces indicated on the weather map and calculating the resultant. This special faculty is developed by long and continued study and association with the maps, rather than by a profound study of atmospheric physics.[16]

Bliss was not exaggerating. Officials responsible for daily forecasts owed much of their understanding of storms and cold waves to the 8 A.M. Washington daily weather map, lithographed copies of which the bureau dutifully mailed to all weather stations as well as to colleges with a meteorology course. As a detailed history of atmospheric behavior, bound collections of weather maps served as both reference and textbook.

Eager like all scientists for formal theory, meteorologists recognized the practical and pedagogic value of summarizing thousands of individual cartographic snapshots in a single map: a storm-track chart that not only pointed out where to look for infant storms but suggested likely trajectories. In *Storms, Storm Tracks, and Weather Forecasting*, published in 1897 as Weather Bureau Bulletin no. 20, Frank Bigelow (1851–1924) discussed the derivation of storm-track maps from an examination of 1,133 storms observed between 1884 and 1893. One of several senior research employees with the title of professor, Bigelow simplified the analysis by delineating the nine districts in figure 1.5 and counting the number of low-pressure centers

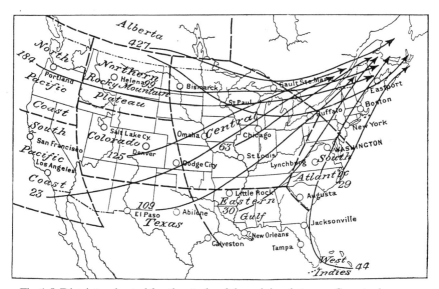

Fig. 1.5. Districts adopted for the study of the origin of storms. Counts show total numbers of storms originating in each region over the period 1884–93, and curved arrows depict average trajectories. From Frank H. Bigelow, *Storms, Storm Tracks, and Weather Forecasting*, U.S. Weather Bureau Bulletin no. 20 (Washington, D.C., 1897), 13.

originating in each. As his map illustrates, more than a third of the storms originated north of Montana, in a broad district centered on Alberta. Lows originating here typically took one of two general routes: a comparatively direct course eastward across Canada, or a warped path diverging southward, through Minnesota and Michigan, before turning northeastward, toward Maine. Storm tracks from the other districts also converged on New England. Not content with the map's concise explanation of the Northeast's harsh and fluky weather, Professor Bigelow could not resist observing that "if all roads lead to Rome, all storms and weather tracks lead to New England."[17] A similar analysis revealed that centers of high pressure, sometimes associated with destructive cold waves, originated in only two districts: Alberta and the North Pacific Coast.

Winter weather is not like autumn weather, of course, nor do December's lows heed October's storm tracks. To address seasonal and intraseasonal variations, Bigelow derived separate storm-track maps for each month—an understandable approach for a government agency accustomed to describing weather in monthly chunks. As his November map (figure 1.6) shows, each district typically originated

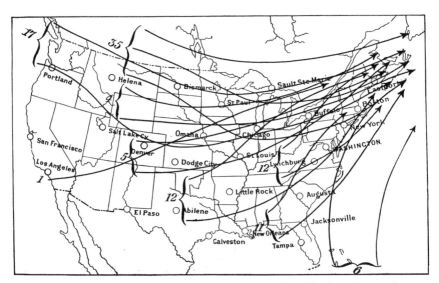

Fig. 1.6. Storm tracks for November. From Frank H. Bigelow, *Storms, Storm Tracks, and Weather Forecasting*, U.S. Weather Bureau Bulletin no. 20 (Washington, D.C., 1897), 27.

two tracks. The exceptions here are the South Pacific Coast, which generated only one storm, and Alberta, for which three tracks summarized 35 lows. In general, the number of storm tracks emanating from a district reflected the number of storms generated as well as their geographic diversity, which could be highly varied. For example, although Alberta's 35 November storms spawned only three tracks, the same district's 45 August storms produced six.

Bigelow never described how he (or someone else) plotted the storm tracks, but I can make an educated guess about his strategy.[18] The *Monthly Weather Review*, initiated in 1872, when the weather service was still part of the Army Signal Office, included monthly plots of storm tracks and probably provided his basic data. I can picture Bigelow or an assistant sitting at a light table—a glass table top with a light source underneath—tracing storm tracks from each year's November map onto nine composite November maps, one for each originating district.[19] (Although the South Pacific Coast district originated just a single November storm in the ten-year period, he probably treated it separately, like the other eight districts, for consistency.) Then, after examining each district's composite November map for several minutes, he used a French curve (or a similar drafting template) to approximate the storm tracks' trend with one, two,

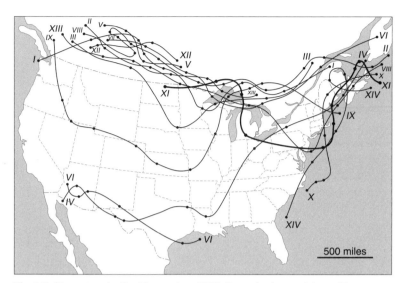

Fig. 1.7. Storm tracks for November 1898. Dots depict position of low-pressure centers at 8 A.M. and 8 P.M., and the heavier line represents the *Portland* storm. Redrawn from *Monthly Weather Review* 26 (November 1898), n.p., chart 2.

or three comparatively smooth curves. Smoothing was essential because actual storm tracks, like those for November 1898 (figure 1.7), were markedly more contorted than his published generalizations. A modern cartographer might invoke a computer to pick and smooth the more representative curves, but the nineteenth-century meteorologist had to rely on his intuition, experience, and graphic dexterity. In short, he eyeballed it.

However vague about the construction of Bigelow's maps, Bulletin no. 20 left no doubt about their role and significance. Its second page, opposite the table of contents, reproduced Weather Bureau chief Willis Moore's letter of transmittal to the secretary of agriculture, the cabinet officer responsible for the weather service. In words carefully chosen for congressmen and the news media, Moore praised the report's intention

> to present the latest phases of the science of meteorology as regards practical forecasting of weather conditions. It is believed that the bulletin will be found of particular value to those engaged in teaching meteorology, and also to students of the subject. Furthermore, it is intended to be of assistance to the observers and other officials of the Weather Bureau, and, in so far as it treats of weather forecasting by

months and generalizations on the same, it is recommended for their careful study, as the most lucid exposition so far presented for their information.[20]

Two years later, in a general essay on weather forecasting, another Weather Bureau professor, Cleveland Abbe, underscored Moore's confidence in storm tracks with a concise prescription for their use:

> although the forecaster knows the various types of paths that storms have taken when passing though any given region, ... in making specific forecasts he always waits a few hours until he has observed the path of any particular storm long enough to know along which of these tracks and at what speed the present storm is travelling.[21]

Informative to lay readers as well as professional forecasters, generalized storm tracks can greatly oversimplify real storms, as figure 1.7 illustrates for the *Portland* storm and other centers of low pressure identified during November 1898. In condensing a much larger *Monthly Weather Review* map, I've retained the roman numerals showing the sequence of individual storms but replaced with mere dots the more complex symbols showing the 8 A.M. and 8 P.M. positions and barometric pressure of low-pressure centers. These dots are important because radical variation in their spacing along individual paths reveals considerable uncertainty in the speed (and predictability) of November storms. Especially deviant is storm IV, which toddled around west Texas at less than eight miles an hour but a day and a half later sprinted from Louisiana to Ohio in a mere 12 hours. More troubling, though, are storm IX, which accords poorly with the two generalized tracks originating in North Pacific Coast district, and storm XIV, which executed a puzzling 270° turn over northern Maine before taking a more predictably eastward route.

Equally atypical is the *Portland* storm of November 26–27, labeled XI and highlighted by a thick line that describes its plunge southward through Michigan, its rapid dash to the Atlantic, and its meandering exit through eastern Canada—a markedly more sinuous course than any of Bigelow's November exemplars. Despite a southerly swing more characteristic of an Alberta low, the storm made its cartographic debut in the Central district but probably originated several hours earlier in the Northern Rocky Mountain Plateau. Lacking a clear storm-track analog, the Washington office relied on meteorology's broadest, most durable principles—storms move eastward; falling pressure signifies a growing threat—and chose a common-sense double strategy: warning the coast and monitoring the storm. Special maps based on observations taken at noon and 3 P.M. con-

firmed the forecasters' judgment and compensated for the imprecision and only marginal relevance of streamlined storm tracks.[22]

Their appetites whetted by Bigelow's rhetorically impressive maps,
Washington's weather bureaucrats took another step along the empirical path they hoped would lead to precise, reliable predictions.
Their new strategy, called "weather typing," was based on the com-
monsense notion that the atmosphere repeats its own past. To exploit
atmospheric history as a forecasting model required a comprehensive
catalog of storms and weather maps, similar in function to the "knowledge base" that provides the memory, if not the brain, of contemporary computer-based decision-making strategies known as expert systems and artificial intelligence. Simply put, the forecaster would
peruse the catalog for a series of weather maps identical in pattern
and evolution to those of the past few days.[23] If he found a similar set
of maps: Bingo!—he knew the next day's (or next several days')
weather. If not, well, at least the experience would enrich the catalog.

A staunch champion of weather typing was chief Washington forecaster Edward Garriott. Writing in the December 1901 issue of the
Monthly Weather Review, Garriott argued that "the increased accuracy of the forecasts and the lengthening of the period for which they
are made, which will surely follow a vigorous prosecution of this line
of work, will, however, justify the great expenditure of time, labor,
and skill which the preparation of reference charts and their classification will demand."[24]

The key, of course, was an efficient scheme for classifying, indexing, and retrieving weather maps: a need addressed in the same issue
by Denver forecaster Frederick Brandenburg, who developed his
system during the summer of 1900, while on temporary assignment
to the Washington office.[25] Figure 1.8 is a reduced facsimile of Brandenburg's solution: a printed form that juxtaposed a small outline
map of the United States with columns labeled "LOW" and "HIGH" for
listing pressure centers. A clerk classified each day's weather by
sketching "principal isobars" from the much larger morning weather
map and marking pressure cells with prominent labels. He then
added vertical shading identifying areas with precipitation during the
next 24 hours, and equally prominent plus or minus signs showing
where temperature the following morning was higher or lower by at
least 20°. Other annotations included a small "T" (as in southeastern
Colorado) to denote significant, widespread thunderstorms.

Carbon paper helped the clerk create separate catalog entries for

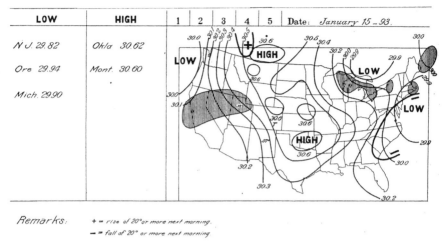

LOW	HIGH	1	2	3	4	5	Date: *January 15 _ 93.*
N J. 29.82	*Okla 30.62*						
Ore 29.94	*Mont. 30.60*						
Mich. 29.90							

Remarks: + = *rise of 20° or more next morning.*

             − = *fall of 20° or more next morning*

             Figures represent Rainfall in this Forecast District during 12 hours ending 8 p.m to morrow

Fig. 1.8. Sample entry in Brandenburg's file of classified weather. From F. H. Brandenburg, "Facilities for Systematic Study of Corresponding Weather Types," *Monthly Weather Review* 29 (1901), 549.

each high- and low-pressure center. Pressing firmly on several blank forms interleaved with graphite-coated tissue paper, he made multiple copies of the map. He then filled out each form separately, listing pressure centers by state and intensity in the appropriate columns and rotating their order so that each center appeared once at the top of the leftmost column—a permutation promoted by two otherwise identical forms, one as shown in figure 1.8 and the other with the column labels reversed. Although identified by the state listed at the top of the left-hand column, each entry was filed under one of "ten or eleven" multistate districts.[26] Each district's entries were stored in two boxes, one for highs and one for lows, and to promote seasonally coherent matching, entries were sorted further by month and day. Brandenburg's schema was no mere off-the-cuff proposal: in devising a practicable method for classifying weather, he compiled a cartographic knowledge base with more than 15,000 separate entries covering ten years.[27]

Despite Garriott's enthusiasm and Brandenburg's impressive (if not intimidating) data bank, weather typing gained few adherents.[28] Although its implicit recognition of the interaction of distant highs and lows addressed a significant limitation of Bigelow's myopically simplistic storm trajectories, the three-dimensional atmosphere is much too complex and ornery to betray its long-term intentions by

wind might prove revealing. Interested in describing climate, not in forecasting weather, they had little sense of the atmosphere as a geographic phenomenon.[6] Like eighteenth-century meteorologists elsewhere, they focused instead on celestial influences as well as on local links among weather, magnetism, insects, and health.

Brandes, by contrast, viewed weather as a spatial problem and was confident his maps would be revealing:

> If one could collect somewhat more precise reports of the weather, even if only for the whole of Europe, it would surely yield very instructive results. If one could draw maps of Europe according to the weather for all the 365 days of the year, then it would of course show, for instance, where the boundary of the great rain-bearing clouds, which in July covered the whole of Germany and France, lay; it would show whether this limit gradually shifted farther towards the north or whether fresh thunderstorms suddenly formed over several degrees of longitudes and latitudes and spread over entire countries.[7]

Plotting graphic symbols on a map base, he argued, would be more informative than merely listing the data, especially if the maps were viewed as a series. Even so, Brandes was concerned about the areal extent and density of the data:

> In order to initiate a representation according to this idea, one must have the observations of 40 to 50 places scattered from the Pyrenees to the Urals. Although this would still leave very many points uncertain, yet by this procedure, something would be achieved, which up to now is completely new.[8]

Without a precedent, though, the task seemed daunting:

> These bold ideas cannot of course be carried out in practice so easily.[9]

In announcing his results three years later, Brandes focused on two severe storms, identified as centers of low pressure, ringed by isobars.

> On the 6th of March 1783 in Amsterdam and Franekar the barometer stood nearly 17 lines below the average height. . . . The low pressure was accompanied by storms, which can well be due to the mass of air rushing into this depression. . . .
>
> Still more remarkable is the 12th of March 1783, when the barometer stood lowest in Switzerland. Here the lines, in which the barometric pressures were equally low, may be traced more or less completely around Switzerland, and the southeasterly storm which raged over Italy at that time and which was simultaneous with the strong

northwesterly wind in France, the northerly wind in Germany and the easterly wind in Ofen looks entirely like air rushing in from all sides into this region of low pressure.[10]

Maps of isobars and wind directions had rewarded Brandes's confidence with an insightful spatial hypothesis (rotational storms with winds converging on a center of low pressure) and that gave meteorologists a reason to make further maps.

Ironically, neither Brandes's short article in *Annalen der Physik* nor his 1820 book on atmospheric conditions throughout the year 1783, *Beiträge zur Witterungskunde*, (Contributions to meteorology) contained a single map.[11] This lack of cartographic evidence seems not to have bothered historians of science. Karl Schneider-Carius, in observing that the maps had never been published, conceded that "in all probability he [Brandes] himself plotted the maps suggested by him earlier for this year [1783]."[12] Less trusting, if not blatantly cynical, was Aleksandr Khrgian, who remarked that "sometimes it has even been thought that [the charts] may never have been drawn up at all."[13] Concealing the source of this doubt, Khrgian further noted, perhaps too pointedly, that "Brandes himself uses the conditional when he refers to these charts: '. . . if we were to prepare weather charts of Europe for each of the 365 days of the year, then it would be possible to determine . . .'"[14]—an inappropriate reference to a wholly appropriate if-then proposition in Brandes's 1816 letter, which was clearly just a proposal.[15]

Eager for visual authentication, at least one scholar settled for a map from Brandes's later work, *Dissertatio physica de repentinis variationibus in pressione atmosphaerae observatis* (Scientific discussion of rapid changes in atmospheric pressure), published in Latin in 1826.[16] Hans Scultetus, a meteorologist interested in the history of his discipline, focused his 1943 article "Die erste Wetterkarte" (The first weather map) on the rather disappointing map in figure 2.1.[17] A simple grid of meridians and parallels on which Brandes plotted deviations from mean air pressure recorded at 6 P.M. on Christmas Eve 1821 at various European cities, this chart says nothing about winds and pressure centers. Its significance as a "first," I'm convinced, rests largely on its position as the earliest of four cartographic snapshots describing a two-day storm. For me, Brandes's words are sufficient evidence that he mapped the 1783 storms.

Other authors eagerly accepted an apparent reconstruction of Brandes's weather map for March 6, 1783. In my largely fruitless search for an image of the first weather map, I encountered two coun-

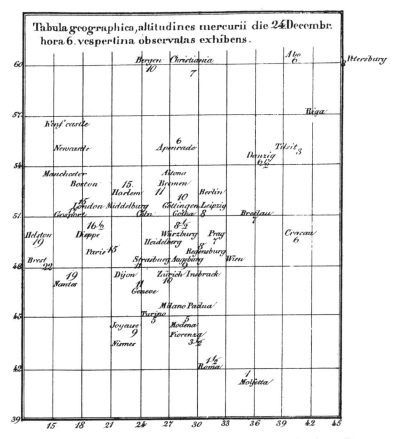

Fig. 2.1. Chart showing deviations from mean air pressure for 6 P.M., December 24, 1821. From H. W. Brandes, *Dissertatio physica de repentinis variationibus in pressione atmosphaerae observatis* (Leipzig, 1826), n.p.

terfeits, one German and the other French. The German version, in figure 2.2, is from Wilhelm Trabert's textbook *Meteorologie und Klimatologie*, published in Leipzig in 1905.[18] With loosely concentric isolines, the map describes the broad low-pressure center synthesized and interpreted in Brandes's *Beiträge*, but its narrow typeface and parallel-line shading reflect the graphic style of other illustrations in Trabert's book. More revealing is the title's telltale "nach Brandes" (after Brandes), which implies that someone (Trabert? a predecessor?) either retraced the earlier, original drawing or, more plausibly, independently plotted and contoured the Palatine data. The French version, with noticeable differences in the curvature of its isolines

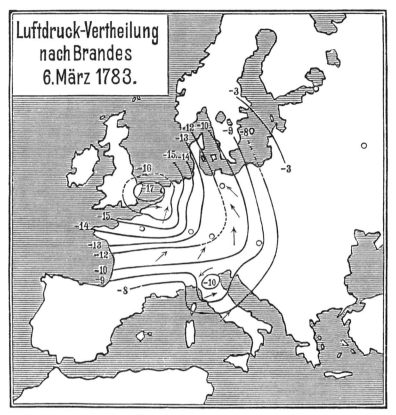

Fig. 2.2. Reconstruction of Brandes's weather map for March 6, 1783. From Wilhelm Trabert, *Meteorologie und Klimatologie* (Leipzig: Franz Deuticke, 1905), 65.

and the number and placement of its arrows, appeared (source not cited) in the October/November 1994 issue of *Weatherwise*, a magazine for weather enthusiasts.[19] The map's title "H. W. Brandes: Carte synoptique du 6 Mars 1783" suggests with equal conviction that its isobars too, whether redrafted or reconstructed from the same data, portray original delineations. I traced the French version first to *Wetter und Wettervorhersage* (Weather and weather forecasting), a 1918 textbook by Albert Defant, who obtained the image from *Les bases de la météorologie dynamique* (The basis of dynamic meteorology), a 1907 text by Hildebrandsson and de Bort, who admitted they reconstructed isobars from the original data.[20] Nice to look at, I suppose, if Brandes's words seem insufficient.

Whether verbal or visual, these descriptions are important to historians of cartography, for whom Brandes's map signifies two distinct innovations: the portrayal of atmospheric pressure with isolines, and the juxtaposition of wind and air pressure on the same map. Like most important inventions and discoveries, these contributions were equally consequential as breakthroughs upon which others could build. In *Cartographical Innovations: An International Handbook of Mapping Terms to 1900,* Helen Wallis and Arthur Robinson described several key extensions.[21] As a snapshot of deviations from the average pressure at each weather station, Brandes's isolines invited two crucial enhancements: the reduction of barometric pressure measurements to sea level (which had become common practice by 1870) and the mapping of average barometric pressure (the extension of the isobar from weather to climate). And in linking wind and pressure, Brandes's weather map was a precursor of cartographic composites that included temperature, precipitation, cloudiness, fronts, and air masses.

Of Brandes's two innovations, the weather map is arguably the more fundamental, not only because of its seminal impact on meteorology (which I'll get to shortly) but because the isobar was a clever but somewhat obvious extension of an older, more general cartographic symbol, the isoline. Arthur Robinson, who developed a genealogy of equal-value line symbols, noted the appearance of the isobar's meteorological cousin, the isotherm, in 1817, more than a hundred years after English astronomer-geophysicist Edmond Halley (1656–1742) mapped the earth's magnetic field over the Atlantic Ocean with isogons, or lines of equal magnetic attraction.[22] Introduced in prominent French scientific journals the year that Brandes announced his intention to map weather, the isotherm invited a parallel, contourlike strategy for describing geographic variation in barometric pressure. But lines of equal temperature were not the only plausible intellectual catalyst—Brandes could have gained inspiration from at least three earlier exemplars: the elevation contour (proposed in 1777), the isogon (used as early as the 1630s), and the isobath, or line of equal depth (used on a marine chart published in 1584). Or he could have thought it up spontaneously, which seems unlikely.

Were I forced to chose a single isoline for Brandes to adapt, I'd pick the isogon, made prominent by Halley, an influential scientist credited with several noteworthy maps as well as numerous important discoveries, including the comet that bears his name.[23] Halley probably inherited the idea from Christoforo Borri, an Italian Jesuit

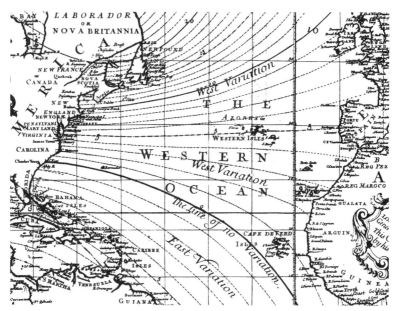

Fig. 2.3. Portion of Edmond Halley's Atlantic chart, ca. 1701. Excerpt from L. A. Bauer, "Halley's Earliest Equal Variation Chart," *Terrestrial Magnetism* 1 (1896): facing p. 33.

who plotted isogons on a map that survived only verbally, as a description in Athanasius Kircher's 1641 book on magnetic variation, which Halley cited.[24] But even though Halley's concept was not wholly original, his *New and Correct Chart Showing the Variations of the Compass in the Western and Southern Oceans*, printed in 1701, was a cartographic example others could imitate.[25]

Halley's Atlantic chart, which cartographic scholar Norman Thrower called one of the most important maps in the history of cartography, is worth at least a glance. As illustrated by the much reduced excerpt in figure 2.3, smoothly curved lines describe the deviation of magnetic north from true north, or in Halley's words, "shew at one View all places where the Variation of the Compass is the same."[26] The interval between adjacent isogons is one degree, and a noticeably thicker "line of no variation," along which the compass needle accurately reflects true north, runs eastward from the Carolina coast in a gentle arc passing just south of Bermuda and the Cape Verde Isles before turning more sharply southward between Africa and South America. Above and to the right of this *agonic line*, the earth's magnetic field deflects the needle to the west, whereas below and to the left

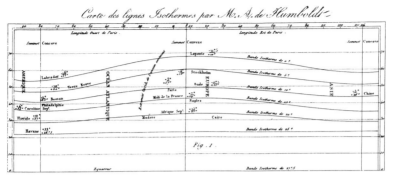

Fig. 2.4. Alexander von Humboldt's map of average temperatures. The area portrayed ranges from the Equator to 70°N and from North America to China. From Alexandre de Humboldt, "Sur les lignes isothermes" (On isothermal lines), *Annales de chemie et de physique* 5 (1817): map between pp. 112 and 113.

of the line, declination is to the east. Halley based his isogons on approximately 150 observations collected between 1698 and 1700, when he crossed the Atlantic several times on the royal navy ship *Paramour*, which he commanded with the temporary rank of captain; the irregular dashed line describes his route.[27] Although the Atlantic chart had become so rare by 1895 that discovery of a second copy was noted in the British scientific journal *Nature*, Halley assured the isogon a wider, more influential audience by extending isogons across the Indian Ocean (but not the Pacific) and publishing this newly engraved compilation in 1702 as the *Sea-Chart of the Whole World, Shewing the Variations of the Compass*.[28] According to Thrower, several atlases incorporated smaller versions of the *Whole World* chart and at least one publisher sold a revised separate-sheet version as late as 1794.[29]

However uncertain their influence on isobars, Halley's lines had an undisputed catalytic effect on isotherms.[30] At least that's what the father of temperature mapping, German naturalist Alexander von Humboldt (1769–1859), insinuated in a single sentence that described his map, named its innovative symbols, and noted their similarity with isogons: "Thus we see that the lines of equal annual heat, or to coin a new word, the isotherms, do not run parallel to the equator but like the magnetic lines, they also cut the geographical latitudes at a different angle."[31] A similarity in purpose is also noteworthy.[32] In the same way that Halley's smoothly curving Atlantic isogons in figure 2.3 depict the departure of magnetic north from geographic north, Humboldt's gently warped isotherms in figure 2.4 describe the deviation of average temperature from the purely solar effects of geographic latitude. In western Europe, for instance, the isotherms bend northward to reflect generally warmer temperatures than in Asia or over the Atlantic Ocean—clear testimony to the efficacy of the Gulf Stream, which supports palm trees in western Ireland and agriculture in Norway. Based on 58 mean temperatures and Humboldt's observation of a milder climate on the west coast of North America than in the interior, his highly generalized map successfully challenged the prevailing scientific notion that climate depended solely on latitude.[33]

Best known for his contributions to climatology, Humboldt was also a pioneer of well-focused single-theme graphic design.[34] Notice, in particular, how he directed the viewer's attention to the broad pattern of average temperatures with a minimally distracting geographic frame of reference: parallels of latitude every 10 degrees, grid ticks instead of lines for all but three meridians, vertical labels instead of boundary lines to mark the ocean and continents, and a mere handful of place names, principally in Europe and North America. According to Arthur Robinson and Helen Wallis, this visual effect—with a strikingly convincing simplicity that cried out for elaboration —accounted for much of the map's success.[35]

Edmond Halley anticipated another component of Brandes's weather map: directional arrows representing winds. Although earlier world maps had portrayed winds and currents of interest to navigators, Halley's map of trade winds, published in 1686, is significant to historians of cartography and meteorology as the first meteorological map.[36] But as with Humboldt's map of average temperature, its focus was climate, not weather.

Too wide for a standard book page, Halley's map of trade winds and monsoons separates conveniently into two parts in figure 2.5, a substantial reduction from the original size (19.2 by 5.7 inches) that nonetheless captures the map's design and content. Similar at first glance to Halley's world chart of magnetic declination, the winds map omits most of the Pacific Ocean but focuses on the tropics, extending beyond 30° north and south only far enough to identify zones of "Variable Winds" dominated by cyclonic storms. As with his map of the earth's magnetic field, Halley delineates wind direction only over water, a reflection of his own marine observations as well as the records of other English navigators. Norman Thrower, who compared the winds chart to modern maps, found Halley's portrayal of continents impressively accurate for its time, except in the vicinity of

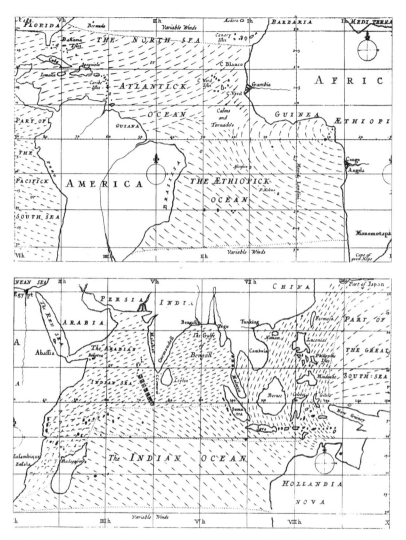

Fig. 2.5. Halley's map of trade winds and monsoons. From E. Halley, "An Historical Account of the Trade Winds, and Monsoons, Observable in the Seas between and near the Tropicks, with an Attempt to Assign the Phisical Cause of the Said Winds," *Philosophical Transactions* 16 (1686): map opposite p. 151.

New Guinea and Australia, where English explorers were well behind the Spanish, Dutch, and Portuguese.[37]

Halley included the map with a paper published in *Philosophical Transactions*, a journal he edited for the Royal Society of London between 1685 and 1692. In this article he attempted a general explanation of the east-to-west trade winds, which, except in monsoon areas, provided dependable year-round conveyance for sailing ships. The map followed a detailed description of wind variations throughout the tropics.[38] His words, however quaint, reflect a profound awareness of cartography's role in both communication and discovery:

> To help the conception of the reader in a matter of so much difficulty, I believed it necessary to adjoyne a scheme, shewing at one view all the various Tracts and Courses of these Winds; whereby 'tis possible the thing may be better understood, than by any verbal description whatsoever.[39]

After stating the reason for the map, he described its symbols:

> I could think of no better way to design the course of the Winds on the Mapp, than by drawing rows of stroaks in the same line that a Ship would move going alwaies before it; the sharp end of each little stroak pointing out that part of the Horizon, from whence the Wind continually comes; and where there are Monsoons the rows of the stroaks run alternatingly backwards and forwards, by which means they are thicker there than elsewhere.[40]

Halley's streamlined, cometlike directional symbol, subtle even when enlarged, never gained the wide acceptance of the more blatantly graphic arrow, but his winds chart afforded the revolutionary concept of a general pattern of atmospheric circulation, which physical laws might explain. Rejecting the notion that the trade winds were merely the result of air lagging behind the rotating planet, he attributed the convergence of winds at the equator to thermal convection: rising air caused by intense equatorial heating created a vacuum at the surface, which cooler air (the trade winds) rushed in to fill.[41] By noting the seasonal shift in this belt of peak solar radiation, his theory also accounted for the tropical monsoon, which in India brings wet winds from the south in June and July and dry winds from the north in December and January.

This interpretation of the winds chart prevailed until 1735, when London barrister George Hadley (1685–1758) modified Halley's explanation by balancing ascending air along the equator with twin belts of descending air at roughly 30°N and 30°S—a discovery commemorated in present-day climatology textbooks as "Hadley cells."[42] In addition to hypothesizing the poleward movement of tropical air at a higher altitude, Hadley also observed (but failed to understand) the deflecting force of the earth's rotation, which turns winds to the right in the northern hemisphere and to the left in the southern hemisphere.[43] By exciting Hadley's interest, Halley's primitive winds chart contributed to the discovery of the Coriolis effect as well as to recognition a century and a half later of the need to map the upper atmosphere.

Often described as boring by American and European visitors, tropical weather (hot, with either daily thunderstorms or monsoon winds and seasonal rains) is mostly a matter of climate, for which a single map of average winds proved highly revealing. By contrast, only a series of cartographic snapshots could unravel the "variable winds" of the middle latitudes, largely ignored on Halley's chart. There's no guarantee, though, that mapped patterns will suggest a single, universally obvious explanation. In 1828, for instance, a student of Brandes, Heinrich Wilhelm Dove (1803–1879) drew a markedly different conclusion from data his mentor had explained just a few years earlier, in *Dissertatio physica de repentinis variationibus*.[44]

Influenced by Hadley's explanation of tropical climate, Dove viewed midlatitude storms as a contest between radically different air currents: a warm, moist flow from the southwest and a cooler, drier current from the northeast.[45] (Geography textbooks call these winds the westerlies and polar easterlies, respectively.) Meeting along a line of low pressure, the opposing equatorial and polar currents alternately advance and retreat, much like the polar front, recognition of which revolutionized weather forecasting 90 years later. As the line of low pressure approaches, barometric pressure drops and the wind direction begins to shift, or as Dove described it, "turn clockwise through the complete wind rose."[46] This succession of winds, first from the south, and then from the west, north, east, and south again

> was most conspicuous in winter. When, at least, the southwest wind blew more and more violently and alone prevailed, it raised the temperature . . . and therefore it could not snow any more, but it rained when the barometer reached its lowest reading. Now the wind turns toward the west and the dense flaky snow indicates the incident cold wind just as well as does the rapidly rising barometer.

With this "law of turning," Dove dismissed the notion of rotary storms in which winds converge on centers of low pressure.

Dove expounded his hypothesis in an article with nine abstract diagrams, only two of which were remotely cartographic.[47] Figure 2.6, the most revealing and relevant of his two maps, describes pressure and wind for a huge storm that moved across Europe in late Decem-

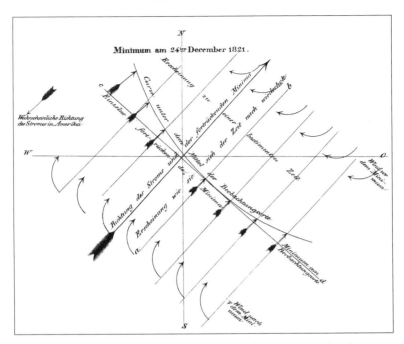

Fig. 2.6. Dove's explanation of the "turning winds" accompanying the storm of December 24, 1821. From H. W. Dove, "Ueber barometrische Minima" (On barometric minima) *Annalen der Physik und Chemie* 13 (1828): fig. 8 on plate 7.

ber 1821—the storm depicted in Brandes's first published weather maps. A large arrow indicates the storm's movement from the southeast, as the equatorial current temporarily displaces the polar current, and line *c–d*, representing an advancing line of minimum pressure, depicts the storm's extent. The lowest pressure in this encroaching trough follows line *a–b*, and the perpendicular curved line shows a slight but steady increase in pressure to both the left and the right of this "greatest minimum." At the upper right, where the minimum-pressure line has not yet arrived, clockwise arrows describe winds blowing toward the south but veering to the west. At the lower left, through which the minimum-pressure line has recently passed, clockwise arrows reflect a shift in wind direction toward the north and east. At the far left, an arrow pointing to the southwest describes the equatorial current's simultaneous retreat in America—a countermovement that balances the warm current advancing across Europe.

Although Dove's theory accounts for local sequences of pressure and wind, his maps reflect a geographically restricted view of midlatitude storms, the centers of which typically pass to the north of central Europe, so that winds accompanying the storm appear to be blowing in one direction. It's understandable, then, that other central European meteorologists, trained by Dove and exposed largely to the southeastern side of vast cyclonic storms, interpreted changing weather as turning winds as late as the 1870s.[48] According to Khrgian, an entrenched belief in opposing currents also explains the apparent reluctance of Dove and his students to make weather maps.[49]

American contributions to nineteenth-century meteorology reflect the country's vast size as well as a population sufficiently dispersed to foster frequent observations on all sides of most large storms.[50] Other key ingredients were present as well: scientific curiosity, academies eager to organize networks of volunteer weather observers, and an efficient postal system. The result, though, was less a steady refinement of knowledge than a prolonged debate among feisty rivals with conflicting theories—a controversy in which maps served both analysis and rhetoric.[51]

The earliest significant contributor, William Redfield (1798–1857) didn't need an observer network, at least not at first. In 1821, while working as a saddle- and harness-maker and living in Connecticut, Redfield experienced an epiphany of sorts while returning from a temporary job in western Massachusetts.[52] Several months earlier a violent storm had toppled trees throughout the region. At the beginning of his journey, he noticed an alignment of fallen trees indicating winds from the northeast. But closer to home the pattern was reversed, suggesting a circular storm, or "whirlwind." Intrigued by this discovery, Redfield collected additional information from mariners and other record keepers. In 1831, in the *American Journal of Science*, he described the damage pattern of the 1821 storm, traced its movement from the West Indies, related these observations to several other large storms, and presented a theory of cyclonic whirlwinds advancing from southwest to northeast in a parabolic path.[53]

Between 1831 and 1856, Redfield published additional evidence for the circular motion of storms, including tropical hurricanes and tornadoes. His most widely cited paper is "Observations on the Storm of December 15, 1839," which he read before the American Philosophical Society in 1841 and published in the Society's *Transactions* two years later.[54] With only four pages of text, a table, and a single map, Redfield presented a concise summary of 48 observations compiled from meteorological journals, newspapers, personal correspondents, and ships' logs. A snapshot of wind direction at noon, when the storm was centered southeast of Rhode Island, his map (figure 2.7 portrays wind direction with arrows and identifies observations with numbers. Although the vortex is clear, Redfield added concentric circles, spaced 30 miles apart, "not as precisely indicating the true course of the wind, but to afford better means of comparison for the several observations."[55] Even so, this persuasive enhancement promotes as well as helps confirm his hypothesis.

Addressing an audience well aware of opposing theories, Redfield had no need to mention his rivals by name. Referring to the map, he reconciled his whirlwind theory with Dove's alternating air currents by pointing out how a storm's position might bias the pattern of winds:[56]

> As relates to the whirling or rotary action in the case before us, it may be remarked, that had we obtained no observations from the north-western side of the axis of this gale, it would have been easy, in the absence of more strictly consecutive observations than are usually attainable, to have viewed the initial south-easterly wind of the gale, and the strong north-westerly wind which soon followed, as two distinct sheets, or currents of wind, blowing in strictly opposing directions: and if we could so far lose sight of the conservation of spaces and areas, the laws of momentum and gravitation, together with a continually depressed barometer within the storm, we might then have supposed one of these great winds, if not both, to have been turned upward by an unseen deflection, and doubled back upon itself in the higher atmosphere. But the case neither calls for nor admits these speculations.[57]

By contrast, Redfield was not the least conciliatory toward his principal rival, James Pollard Espy (1785–1860). Like Brandes, Espy interpreted storms as the inward flow of air toward a center of low pressure—a phenomenon faintly apparent on Redfield's map. Although his measurements of the angles between arrows and circles reveal "an average convergence, or inward inclination, of about six degrees," Redfield dismissed this slight but largely inward flow as "errors of observations and the deflecting influences of the great valleys [as well as] errors of approximation which often arise from referring all winds to eight, or, at most, to sixteen points of the compass."[58] In response, Espy calculated the centrifugal forces of opposing winds 40 miles apart, and concluded that Redfield's whirlwind "would soon be destroyed by its outward motion, unless some mighty cause exists, of which we have no knowledge."[59] Debate intensified as the two competitors attracted disciples.

Espy was a formidable rival.[60] A successful popular lecturer

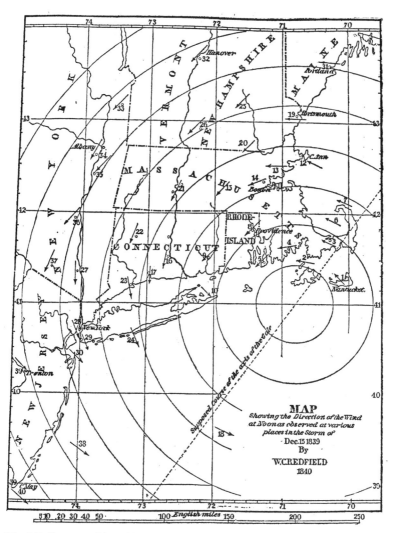

Fig. 2.7. Cartographic evidence of Redfield's whirlwind theory. From William C. Redfield, "Observations on the Storm of December 15, 1839," *Transactions of the American Philosophical Society* 8 (1843): map following p. 80.

trained in law and mathematics, he worked at the Franklin Institute, in Philadelphia, where in 1831 he organized a committee to collect meteorological data. In 1834, the Franklin Institute and the American Philosophical Society, also in Philadelphia, formed the Joint Committee on Meteorology, with Espy as its chairman. He published numerous papers in American scientific journals, and on a visit to Eu-

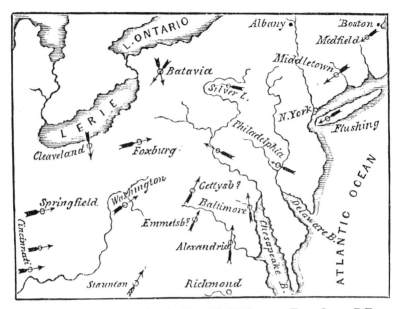

Fig. 2.8. Espy's winds map for the June 20, 1836, storm. From James P. Espy, *The Philosophy of Storms* (Boston: Charles C. Little and James Brown, 1841), 105.

rope in 1840–41, he stimulated lively, enthusiastic discussions at meetings of the British Association for the Advancement of Science and the French Academy. In 1841, his *Philosophy of Storms*, a nontechnical introduction for lay readers, caught the attention of newspaper editors, who supported his campaign for federally funded meteorological studies. In 1842, Espy became the de facto national meteorologist when Congress granted him franking privileges, and the War Department put him on the payroll as a professor.

Redfield and Espy attributed storms to radically different processes.[61] For Redfield, the driving force was gravity and the earth's rotation, which caused the winds, which in turn accounted for differences in pressure, temperature, and moisture. By contrast, Espy's theory focused on heat, which caused air to rise, winds to rush in, and clouds to form. Aware of John Dalton's and Joseph Louis Gay-Lussac's experiments with the evaporation and condensation of gases, Espy recognized the condensation of water vapor as a crucial source of the heat fueling intense vertical convection near a storm's center. But ignorant of Coriolis's recent (1836) description of the earth's rotational force, he neither looked for nor accepted the counterclockwise winds described by Redfield.

Espy and Redfield focused their maps on wind direction, but instead of a circular pattern, Espy's arrows described an inward flow toward a point or line of minimum pressure. Typical of his early meteorological cartography is figure 2.8, which depicts a storm thought to be centered over Silver Lake, New York, where nearly 3 inches of rain fell on the night of June 20, 1836. A wood-block engraving printed in the *Journal of the Franklin Institute* to illustrate an 1837 report of the Joint Committee, the map is significant as the first published American weather map based on widely spaced observations.[62] Impressed with its persuasive convergence of arrows, Espy reproduced the map in his *Philosophy of Storms*, and coaxed readers to examine its convincing patterns: "By casting the eye on the wood cut, it will be seen in a glance that the wind blew on all sides toward the point of greatest rain."[63]

Well supplied with data from the Joint Committee, military observers, and meteorological journals, Espy constructed many maps over the next two decades. His *First Report to the Surgeon General*, released in 1845, included engravings of 29 of 92 maps summarizing reports from more than 50 observers with barometers and 60 others without.[64] Typical of his later maps, these included advancing lines of high and low pressure, represented by prominent black and red lines, respectively. Espy employed similar symbols in the *Fourth Report on Meteorology*, published in 1857 with 70 maps.[65] Figure 2.9, a monochrome excerpt describing a storm advancing from the southwest, illustrates his use of longer arrows to depict stronger winds, numbers to represent inches of rainfall (R), and highly generalized pictorial escarpments to point out areas above 2,000 and 4,000 feet.[66] Conspicuously absent are isobars, which would have afforded a less constrained view of barometric pressure.

Were Espy and Redfield both wrong, or did each contribute worthwhile insight? The latter possibility intrigued Elias Loomis (1811–1889), a young professor of mathematics and natural philosophy at Western Reserve College, in Cleveland, Ohio. Having followed the exchanges among Redfield, Espy, and their supporters since the early 1830s, he sought an answer by selecting a large, presumably significant storm and requesting data from observers at widely separated locations. In a paper titled "On the Storm Which Was Experienced throughout the United States about the 20th of December, 1836," which he read to the American Philosophical Society in 1840, Loomis described variable winds blowing toward a line of minimum pressure, in accord with Espy's theory.[67] Reluctant to either confirm Espy or challenge Redfield with results he regarded as

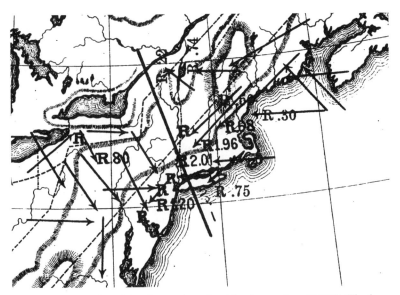

Fig. 2.9. Excerpt from Espy's weather chart for 3 P.M., June 7, 1851. The long black line, which was red on the original map, indicates the "locality of minimum barometer." From James P. Espy, *Fourth Report on Meteorology* (Washington, D.C., 1857), n.p.

preliminary, he focused on an explanation of the storm's heaviest precipitation.[68] His diagram (figure 2.10) showing moist air from the southeast forced upward by advancing cold air from the northwest was a remarkably insightful anticipation of fronts and air masses, which transformed weather forecasting in the 1920s.

Three years later, in a paper illustrated with 13 maps, lithographed and hand-colored, Loomis examined the development and demise of two additional winter storms.[69] Based on reports from 131 observers, his charts depict winds that are neither wholly inward nor entirely rotational. But as the chart for the morning of February 16, 1842 (see plate 1), demonstrates, a "prevalent tendency inward, with a disposition to circulate around the centre" confirms that Espy and Redfield were each partly right.[70]

In addition to an enlightened understanding of storms, Loomis provided the first printed examples of the *synoptic weather map*, which integrates information for pressure, temperature, wind, sky conditions, and precipitation.[71] His description of their symbols reveals acute awareness of the need for contrast, cartographic license, and graphic logic:

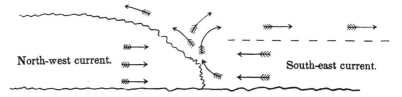

North-west current.

South-east current.

Fig. 2.10. Loomis's explanation of heavy rain along a squall front. From Elias Loomis, "On the Storm Which Was Experienced throughout the United States about the 20th of December, 1836," *Transactions of the American Philosophical Society* 7 (1841): 159.

I have represented the principal phenomena of [the February 15–17] storm on the accompanying charts, one to five. Those regions, where the sky was unclouded, or where the cloudiness was less than one half, are coloured blue; those where the sky was entirely overcast, or the cloudiness exceeded one half, but without rain or snow, are coloured brown; the fall of snow is indicated by the green colour, and rain by the yellow. The direction of the wind is represented by the arrows, and its force is, to a certain extent, indicated by their length. When, however, the winds were faint, the arrows, if drawn of a strictly proportionate length, would have been too short to attract notice, and are therefore somewhat magnified. A calm is represented by a cipher, (0).[72]

Like Brandes, Loomis compensated for geographic variations in elevation and other irrelevant influences by mapping deviations from the local mean. His system of isobars and isotherms avoids confusion and is easily decoded:

I have also represented the barometric, and thermometric observations in the following manner: Having determined, as well as I was able, the mean height of the barometer at each station, I compared each observation with the mean. I then drew a line passing through all the places, where the barometer stands at its mean height. This line is marked — — 0, and may be called the line of mean pressure. I then drew a line through all the places where the barometer stands two inches above the mean. This line is marked — — — + .2, and so of the others. In like manner, a line joining all places where the thermometer stands at its mean height for that hour and month, is marked . . . . . . 0°, and may be called the line of mean temperature. Another line joins all those places where the thermometer is 10° above the mean, and is marked . . . . . . 10°, and so of the others. Nearly every circumstance essential to a correct understanding of the phenomena of the storm is thus presented to the eye at a single glance.[73]

Loomis did not abruptly end the storms controversy, but his charts reinforced the promise of good data. A dense network of well-organized observers, he argued, could reap the advantage of America's size and common language. In comparing previous efforts to "guerrilla warfare ... with indifferent success," he called for a "general meteorological crusade." With twice-daily weather maps as the key weapon, "the war might soon be ended, and men would cease to ridicule the ideal of our being able to predict an approaching storm."[74]

Weather by Wire

Meteorologists quickly grasped the significance of the electric telegraph. In 1846, two years after Samuel Morse linked Washington with Baltimore, and a year after the wires reached Boston, William Redfield proposed using the rapidly growing network to warn Atlantic ports of storms developing in the South and Midwest.[1] The following year, Elias Loomis echoed Redfield's idea in a letter to Joseph Henry (1797–1878), the secretary and director of the newly established Smithsonian Institution.[2] Loomis recommended that Henry establish a meteorological department to study and forecast storms, and in late 1847, following a similar suggestion from James Espy, Henry pitched the idea to the Smithsonian's Board of Regents. His plan was to "organize a system of observations which shall extend as far as possible over the North American continent." The time for such an enterprise was "particularly auspicious," he argued. Because of widespread settlement in the South and the West, "the extended lines of the telegraph will furnish a ready means of warning the more northern and eastern observers to be on the watch for the first appearance of an advancing storm."[3]

A physicist noted for groundbreaking experiments on electromagnetism, Henry understood the potential of both the telegraph and the weather map.[4] Early in his career, he had taught at the Albany Academy, which participated in a regional weather network established in 1825 by Simeon DeWitt, vice chancellor of the University of the State of New York.[5] Each of the 62 academies under the state's jurisdiction received instructions for taking and reporting observations as well as a standardized thermometer and rain gauge—

equipment with which many New York observers contributed to the storm studies of Redfield, Espy, and Loomis. Between 1829 and 1833, Henry helped compile annual meteorological reports for the state.[6] In 1833, he accepted a position at Princeton University, where his studies occasionally involved electromagnetic phenomena in the atmosphere.

Impressed by endorsements from Espy and Loomis, the Smithsonian's board approved Henry's plan on December 15, 1847, and appropriated a thousand dollars for instruments.[7] Early efforts focused on establishing a broad network of volunteer observers, who would keep journals and mail monthly reports to Washington. Because instruments were expensive, the Smithsonian sorted observers into three categories: the first without instruments, the second with only a thermometer, and the third with a full set: thermometer, barometer, rain gauge, and psychrometer (for measuring humidity). In addition to calibrated instruments, Henry's staff furnished blank reporting forms as well as detailed instructions on when and how to take measurements thrice daily, at 7:00 A.M., 2:00 P.M., and 9:00 P.M.[8] Edward Foreman, the professor in charge of meteorological correspondence, constructed an outline map "for presenting the successive phases of the sky over the whole country, at different points of time."[9]

How thoroughly the Smithsonian mapped its weather data is unclear. The network, which fluctuated between two and five hundred observers, generated too much data for timely manual calculations and efficient mapping.[10] The report for 1857 describes a clerical staff of "twelve to fifteen persons, many of them female" inundated by "the records of upwards of half a million separate observations, each requiring a reduction involving an arithmetical calculation."[11] Although Henry's report for 1858 notes that "a large number of maps have been constructed for the investigation of storms," he apparently had little interest in the labor-intensive compilation of three weather maps a day—more than a thousand maps a year—based on stale data.[12] Even so, Smithsonian weather data enriched the research of James Espy, who was a professor of meteorology at the War Department until his retirement in 1857.

While Henry's telegraphic weather data were unquestionably fresh, the network was sparse and its reports comparatively vague. Once a day, around 10 A.M., concise descriptions of early morning weather (clear, cloudy, rain, snow) and wind direction arrived from points along cooperating telegraph lines. At Henry's request, telegraph operators reporting for work replaced their usual "O.K." with a "fair" or "cloudy," which the company sent on to Washington. Although the Smithsonian received electronic dispatches as early as 1856, at the end of 1858 its telegraphic weather network consisted of only 32 stations from New York to Cincinnati and New Orleans, along the circuits of the National Telegraph Company.[13] But by 1860, Henry had enlisted the cooperation of other firms, increased the number of telegraphic observations to 45, and extended coverage as far north as Burlington, Vermont, and as far west as Cedar Rapids, Iowa; and St. Louis, Missouri.[14]

To promote public interest in its meteorological work, the Smithsonian erected a large map, on which a clerk hung small cards on iron pins inserted at places providing telegraphic reports.[15] The colors indicated current weather conditions, with black for rain, gray for cloudy, and white for a clear sky. Officials later replaced the cards with small colored disks with eight holes, evenly spaced around the edge, and an arrow to represent wind direction. Like many museum attractions, the display was entertaining as well as informative. As Henry noted in his 1858 report, "This map is not only of interest to visitors in exhibiting the kind of weather which their friends at a distance are experiencing, but is also of importance in determining at a glance the probable changes which may soon be expected."[16] The map also helped Henry win the support of members of Congress, several of whom regularly consulted him for weather predictions.[17] In addition to displaying weather warnings on its tower, the Smithsonian supplied the data to the *Washington Evening Star*, which in 1857 became one of the first newspapers to publish a telegraphic weather report.[18]

Although Henry was the first to demonstrate the efficacy of forecasting weather with telegraphic data, his project never recovered from two disasters.[19] The Civil War (1861–65) severed links with the South, diverted the telegraph network in the North to military work, and cut the number of observers reporting by mail from more than 500 to less than 300. And in early 1865, when the war's end was imminent, a fire destroyed much of the Smithsonian building and stifled Henry's plans to revive weather telegraphy. Telegraph companies, he found, were no longer willing to donate their service, and funds for transmitting weather data were needed to recover from the fire. In 1870, Henry invoked the Smithsonian's policy of "do[ing] nothing which can be done as well or better by other means"—in this case, the War Department's telegraphic weather network, which Congress had established the previous year.[20] It was a graceful withdrawal: after reminding readers of the value of "simultaneous observations . . . over large portions of the earth," he called for a "still larger appropriation" for the government's new weather service.

Like Redfield and Henry, European meteorologists and scientific en-trepreneurs were enthusiastic about telegraphic weather reports. In 1848, for instance, John Ball addressed the British Association for the Advancement of Science with a plan for telegraphic storm warn-ings.[21] Noting that "atmospheric disturbances" move no more rapidly than 20 miles an hour, Ball suggested that "with a circle of stations ex-tending about 500 miles in each direction, we should in almost all cases be enabled to calculate on the state of the weather for twenty-four hours in advance." Daily weather reports commenced the fol-lowing year, after James Glaisher (1809–1903), superintendent of the Department of Magnetism and Meteorology at the Royal Observa-tory, in Greenwich, arranged for London's *Daily News* to carry a bul-letin of observations taken the previous morning at 9 A.M. at 50 rail-way stations throughout Britain and delivered to London by train and telegraph.[22] Although the newspaper never included a map, Glaisher used the reports to prepare daily weather maps for his own use.

London's Great Exhibition, held in Hyde Park in 1851, was the occasion of another meteorological milestone: the first same-day weather map. From August 8 through October 11, the Electric Tele-graph Company charged a penny for copies of a map based on ob-servations telegraphed earlier in the day to the company's exhibit, where a printing press had been set up.[23] Around 1863, the "Daily Weather-Map Company (Limited)," presumably organized by James Glaisher, further plumbed the commercial potential of weather cartography by offering subscriptions for 4 shillings a month (or "£2 : 12 s. per annum") to the timely tableau of pictorial symbols shown in figure 3.1.[24] Clever and intriguing, the symbols are similar in content but not design to geometric point symbols de-scribed two years earlier by Francis Galton (1822–1911), who had solicited data from correspondents throughout Europe.[25] A faint credit line at the lower right, "movable symbols by G. Barclay," in-dicates that the map was printed with an ordinary letterpress using specially designed type. According to its prospectus and specimen map, the company hoped to sell as many as 5,000 copies of a map based on data telegraphed from 64 places in England and Ireland. Its history is obscure, apparently because the firm attained neither its projected break-even target of 3,000 subscriptions nor the "ex-tensive display of high-class advertisements" needed to sustain a "much more moderate circulation."

While commerce can be an effective patron of atmospheric car-

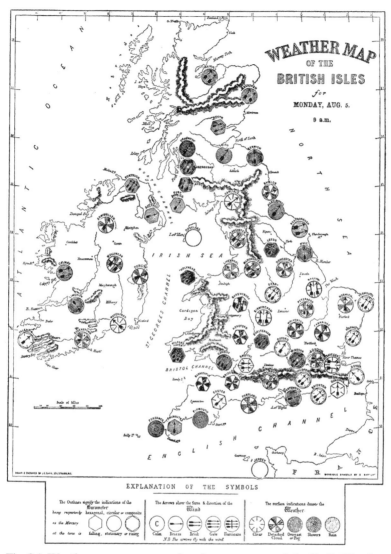

Fig. 3.1. Weather map accompanying the prospectus of the Daily Weather Map Company, ca. 1863. From Napier Shaw, *Manual of Meteorology* (Cambridge: Cambridge University Press, 1926), 1: 309.

tography, nothing was ever as effective as national defense in jumpstarting a weather service. Among the earliest meteorological entrepreneurs to benefit from military needs was French astronomer Urbain Jean Joseph Le Verrier (1811–1877), whom Napoleon III's minister of war asked to study the sudden storm that on November 14, 1854, during the Crimean War, devastated a joint British-French naval expedition near Balaklava, in the Black Sea.[26] Le Verrier reconstructed the storm's path from meteorological registers, and demonstrated that telegraphed observations and synoptic weather maps could have given naval commanders a day to prepare.[27] Using his position as director of the Paris Observatory, Le Verrier persuaded the government to contribute the services of its telegraph department. On January 1, 1858, he initiated a daily weather bulletin with observations from 18 cities, all but 4 in France, and from September 1863, his *Bulletin International de l'Observatoire de Paris* included a daily weather map. As the example in figure 3.2 illustrates, the maps focused on western Europe and the relationship between barometric pressure and wind, portrayed with isobars and arrows. Its isobars represent millimeters of mercury, and barbs on the arrows show wind velocity by category, with more barbs representing stronger winds. Historians of meteorology credit Le Verrier with the first use of isobars on a telegraphic weather map.[28] On April 24, 1878, the bulletin added a second map, with isobars as well as shading showing cloudy skies and rain.[29]

Le Verrier was not the first to offer regular telegraphic weather forecasts. On May 21, 1860, at the request of the Dutch government, Christoph Buys Ballot (1817–1890) established a system of storm warnings for port cities based on reports from six stations.[30] Although his telegraphic network was sparse, Buys Ballot had discovered a helpful rule years earlier, while making daily weather maps with wind and pressure measurements from a larger network of volunteer observers. Commemorated as Buys Ballot's law, the rule holds that the lowest pressure lies to the left (in the northern hemisphere) of an observer with his back to the wind—a reflection of counterclockwise circulation around a low and clockwise circulation around a high.[31] With observations taken several times a day, meteorologists applying this law can detect and track large storms centered outside their monitoring network.

Like France, England recognized that weather telegraphy might have averted the November 1854 disaster at Balaklava. Parliament's response was to order Capt. Robert FitzRoy (1805–1865), an experienced hydrographer, to organize a meteorological department in the Board of Trade—under an appointment back-dated to August.[32] Although his official focus was the meteorology of the sea, FitzRoy set about planning simultaneous weather observations with standardized instruments at telegraph offices widely spaced throughout

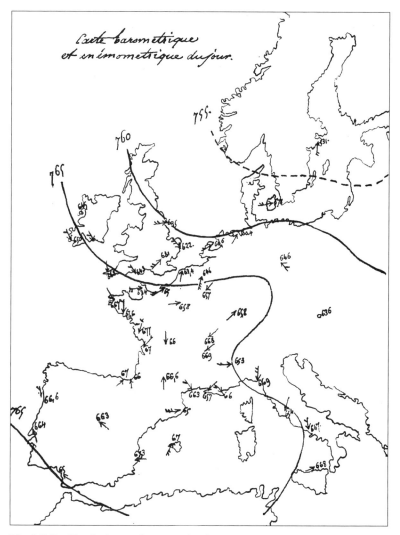

Fig. 3.2. Le Verrier's weather map for September 16, 1863. From H. Hilde-brand Hildebrandsson and Léon Teisserenc de Bort, *Les bases de la météorologie dynamique: Historique-état de nos connaissances* (Paris: Gau-thier-Villars et Fils, 1907), plate 4 in vol. 1, between pp. 66 and 67.

Britain. On September 3, 1860, a network of 15 stations began re-porting 8 A.M. observations to London, where the Meteorological Of-fice compiled a daily weather report for distribution at 11 A.M. to the Admiralty, the Horse Guards, Lloyd's, the Humane Society, and sev-eral newspapers.[33] On February 5, 1861, the office issued its first order to raise storm signals. Several days later, a severe gale struck the east coast of England, sank numerous ships, and drowned hundreds of people.[34] Eager to make a point, FitzRoy told readers of London's *Times* newspaper, "All the much-frequented parts of our coasts might have been warned—a very few places were actually warned—three days before this storm."[35] Within a year the number of ports able to receive telegraphic storm warnings rose from 50 to 130.

On August 1, 1861, the Meteorological Office inaugurated a general weather forecast: a prediction of the next day's weather and wind direction for each of four districts comprising the United Kingdom. Although the public enthusiastically approved this practical use of meteorology, many members of the Royal Academy and the British Association for the Advancement of Science considered these highly empirical weather predictions unsound and premature. Depressed by stinging criticism about his lack of theory as well as the quality of data collected by mere telegraph operators, FitzRoy killed himself on April 30, 1865.[36]

With FitzRoy gone, the theorists took over. In 1866, the government transferred the Meteorological Office to the Royal Society, which accepted the recommendations of a "Committee of Inquiry" and suspended storm forecasts altogether.[37] In 1868, after numerous complaints to Parliament, officials resumed warnings for already-evident storms likely to travel but declined to notify the public about possible storms not fully apparent. Despite its members' skepticism about empirical analysis, the Royal Society approved a *Quarterly Weather Report*, with maps inspired by the meteorological cartography of the polymath Francis Galton. Published for 12 years, starting in 1867, the *Report* included daily weather maps, on which closed isobars identified cyclones and "anticyclones," a term Galton coined for centers of high pressure.[38] On March 23, 1872, the Meteorological Office inaugurated a series of four same-day weather maps: one each for pressure (with isobars); temperature (with isotherms); wind and sea; and clouds and rain.[39] By 1879, officials were sufficiently confident to resume general forecasts.[40]

While their British colleagues whined about the dearth of theory, American meteorologists were eager to recover from the Civil War and resume forecasts based on principles discovered by Espy and Loomis. Although events in Europe had demonstrated government's role in weather telegraphy, Congress and the White House were too busy with postwar reconstruction and westward expansion to notice

Joseph Henry's difficulty in reestablishing the Smithsonian forecast network. What caught their attention was an influential Wisconsin congressman who had been skillfully lobbied by a constituent eager to reduce weather disasters on the Great Lakes by extending a successful forecast system based in Cincinnati.

Small compared to the Smithsonian's pre–Civil War telegraphic network, the Cincinnati weather service largely reflected the initiative of Cleveland Abbe (1838–1916), director of the Cincinnati Astronomical Observatory.[41] An astronomer who had worked at the National Observatory, in Washington, Abbe moved to Cincinnati in 1868 to revive a small, privately financed observatory, which had deteriorated during the war. To stimulate public interest as well as collect data on atmospheric refraction (a personal research interest), he proposed a cooperative effort involving the Western Union Telegraph Company, the Associated Press, and Smithsonian weather observers in selected cities. Observers would telegraph reports to Cincinnati, Abbe would prepare a summary, Western Union would broadcast his synopsis to the Smithsonian and newspapers throughout the country, and the AP would defray telegraph charges.

A nice idea but not easy to organize, even by someone with Abbe's perseverance—only two distant observers (one in Leavenworth, Kansas, and the other in St. Louis, Missouri) contributed to his first summary, written by hand but posted prominently at the Cincinnati Chamber of Commerce on September 1, 1869. Within a week, though, the number of stations had increased to five, including Cincinnati, and on September 8, an optimistic Abbe printed the first issue of the *Weather Bulletin of the Cincinnati Astronomical Observatory*.[42] Abbe had hoped for reports from 22 stations, but his network grew erratically, with reports often delayed or missing, and many forecasts based on fewer than a dozen observations. Even so, a confident Abbe wrote his father that "I have started that which the country will not willingly let die."[43]

Had the idea perished, the underlying cause of death would have been low birth weight. Although the Cincinnati Chamber of Commerce published Abbe's weather bulletin for the first three months, when the trial period ended his financially strapped observatory had to pick up the printing bill. Although newspapers in several cities published Abbe's summary, the Associated Press never assumed its intended role. And despite free telegraph service, the agreement with Western Union was never permanent, and the firm's participation no doubt reflected a hope of added profit more than scientific curiosity or a spirit of public service.

Whatever its motives, the company provided a 14-by-21-inch outline map boldly titled "Western Union Telegraph Co.'s Weather Report."[44] Printed on tissue paper, the map allowed Abbe or a telegraph clerk to add point symbols showing local weather and wind direction to a stack of copies interleaved with carbon paper. Abbe apparently designed the symbols: an arrow superimposed on a quarter-inch circle darkened to indicate cloudy conditions or filled with an "R" or "S" to indicate rain or snow. That the legacy of his Cincinnati project was cartographic as well as telegraphic is readily apparent in the November 1898 weather maps (figures 1.1 and 1.3) showing the *Portland* storm.

Abbe's impact on weather telegraphy owes much to the enthusiasm of his Milwaukee observer, Increase A. Lapham (1811–1875).[45] A successful civil engineer and natural scientist, Lapham had lobbied the Wisconsin legislature in the 1850s to set up a state weather service, with telegraphic warnings for ships in Lake Michigan. Impressed by the Cincinnati business community's support for Abbe's weather bulletins and alarmed by a list of 1,914 shipwrecks on the Great Lakes in 1869, Lapham petitioned Congress to follow the lead of Espy, FitzRoy, and Le Verrier:

> Now it is quite clear that if we could have the services of a competent meteorologist at some suitable place on the Lakes with the aid of a sufficient corps of observers with compared instruments at stations located every 200 or 300 miles toward the west, and the cooperation of the telegraph companies, the origin and progress of these great storms could be fully traced, their velocity and direction of motion ascertained, their destructive force and other characteristics noted—all in time to give warning of their probable effects upon the lakes.[46]

Lapham enlisted the support of his congressman, Halbert E. Paine, a retired Union Army general and a former student of Elias Loomis at Western Reserve College. On December 16, 1869, after consulting with Loomis and Joseph Henry, Paine introduced a bill calling for a telegraphic weather service administered by the War Department. Although Abbe and Henry wanted a civilian agency under the control of meteorologists, Paine argued the need for military discipline.[47]

Paine had another reason for placing the weather service in the War Department: the zeal and political connections of Gen. Albert J. Myer (1829–1880), chief of the army's Signal Office and a personal friend of President Grant.[48] As soon as he heard of the congressman's intentions, Myer asked to organize and run the new agency. As Paine recalled several decades later, "He was greatly excited and expressed a most intense desire that the execution of the new law be entrusted to him."[49] Myer clearly needed a mission—despite distinguished service in organizing and commanding the Signal Corps during the war, he had seen his postwar office shrink to just himself, a lieutenant, and two clerks. No political hack, Myer quickly impressed Paine with his enthusiasm, knowledge of telegraphy, and detailed plans for supplementing the existing telegraph network.

On February 9, 1870, President Grant signed into law a concise joint congressional resolution, approved a few days earlier, that settled the issue of control and sketched the new agency's responsibilities:

> Be it resolved by the Senate and House of Representatives of the United States of America in Congress assembled that the Secretary of War be, and hereby is, authorized and required to provide for taking meteorological observations at the military stations in the interior of the continent and at other points in the States and Territories of the United States for giving notice on the northern lakes and seacoasts, by magnetic telegraph and marine signals, of the approach and force of storms.[50]

The new law addressed only storm warnings on the Great Lakes and coasts, but Congress added river observations and warnings in 1871 and made the forecast mandate nationwide in 1872.

Myer understood military communications, but his infant Division of Telegrams and Reports for the Benefit of Commerce needed a few civilian meteorologists to make forecasts and train soldiers. In November 1870, with arrangements complete for weather telegraphy, he hired Increase Lapham to supervise storm warnings for the Great Lakes.[51] Lapham lacked forecasting experience but had learned how to read a weather map from the writings of Redfield, Espy, and Loomis, and his weather maps, unlike Abbe's Cincinnati charts, contained isobars and isotherms. Well aware of the threat implied by counterclockwise winds around a center of low pressure, he issued his first storm warning the first day on the job. But life as Myer's man on Lake Michigan was not easy for Lapham. Storms often interfered with telegraphic connections, and on one occasion severe weather reduced the number of stations reporting from 25 to 4.[52] In addition, his Signal Office duties in Chicago conflicted with his business interests in Milwaukee, which he visited two days a week. Caught between a stingy Congress and a public eager for forecasts, the Signal Office terminated its arrangement with Lapham at the end of May 1872.

Myer's next significant civilian hire stayed a bit longer—45 years, to be exact.[53] Although Cleveland Abbe had opposed military control of the weather service, Myer valued his experience as a forecaster and his skill as a teacher. Fortunately for the Signal Office, Abbe was eager to leave Cincinnati, where local contributions to the observatory would never support the research program he envisioned. When he joined the weather service in January 1871, as a professor in the Washington office, Abbe not only prepared forecasts himself but trained army officers in forecasting. As the Signal Office expanded, he assumed the roles of senior scientific advisor and editor of the *Monthly Weather Review*.

Congress quickly discovered that storm warnings were not cheap. According to annual reports and the records of congressional hearings, the weather service budget increased from a puny $15,000 for fiscal year 1871 to $250,000 for 1873 and $993,000 for 1883, while the number of full-time employees rose from 233 in 1871 to 485 in 1880.[54] More relevant to the accuracy and dissemination of Signal Office forecasts, the number of places reporting daily by telegraph to Washington jumped from 24 in 1870 to 55 in 1871 and 181 by 1880.[55]

As Henry and Abbe had demonstrated, forecasting focused on the weather map. In Washington the central forecasting office prepared maps three times a day, from synchronized observations taken at 7:35 A.M., 4:35 P.M., and 11:35 P.M., Washington time.[56] Most stations outside Washington also prepared weather charts, but not all stations received three reports a day or data from all locations. Morning and afternoon weather bulletins were posted in public places, where maps too were displayed.[57]

Guidelines advised local forecasters to prepare a map from every telegraphic bulletin, or at least once a day, and to bind these charts in books of a hundred of so, with individual sheets aligned to promote the ready comparison of successive maps, drawn on translucent paper.[58] Two maps were required: a chart with arrows and circles showing wind direction and weather at individual stations, and an auxiliary chart with isobars and isotherms.[59] Isobars were plotted with a red pencil, with the isobars from the previous map copied in blue or dotted red for ready comparison. For soldiers unfamiliar with contours, an 1871 Signal Office pamphlet titled *The Practical Use of Meteorological Reports and Weather-Maps* included a diagram (figure 3.3) illustrating how to interpolate an isobar between stations. To show what weather charts can reveal, the pamphlet reproduced the Washington map (figure 3.4) for the morning of August 26, 1871, when a hurricane centered near the Alabama-Georgia border threatened the East Coast. Although point symbols on the map depict observations from Cheyenne, Wyoming; Corinne, Utah; and San Francisco, spotty coverage of the West halted the isobars at the 100th meridian.

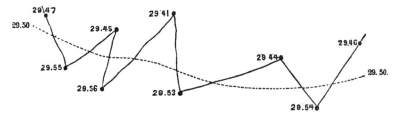

Fig. 3.3. Dotted line illustrating the interpolation of an isobar representing a barometric pressure of 29.50 inches. From Office of the Chief Signal Officer, *The Practical Use of Meteorological Reports and Weather Maps* (Washington, D.C., 1871), 12.

Fig. 3.4. Synoptic weather map, with isobars and station symbols, for 7:35 A.M., August 26, 1871. From Office of the Chief Signal Officer, *The Practical Use of Meteorological Reports and Weather Maps* (Washington, D.C., 1871), facing p. 12.

Why Corinne, Utah? Its position on the Union Pacific Railroad, of course. When the Golden Spike connected Union Pacific and Central Pacific tracks at Promontory, Utah, in 1869, Corinne became an important railway center on the first transcontinental railway and enjoyed a decade or so of notoriety as the region's first non-Mormon settlement.[60] Additional east-west rail lines—three more were completed by 1883—fostered a broader, denser telegraph network, which in turn allowed the gradual introduction of fully transcontinental telegraphic weather maps around 1881.[61]

Full coast-to-coast coverage evolved slowly. According to an 1890 study of the accuracy of cold-wave warnings, drafting clerks in Washington plotted isobars and isotherms only when they received sufficient data. Facsimiles of telegraphic maps used by forecasters show that the isolines stopped at the 100th meridian in 1880, but usually covered the entire country in 1882 and 1883. Although the validation study examined forecasts as far back as 1880, its author observed that "the material in 1880 and 1881 west of the Mississippi is not, however, as well adapted [to forecasting] as that of the later years [because] observing stations in the western country were very few at that time."[62] A similar caveat in the annual report for 1882 noted that maps published in the *Tri-Daily Meteorology Record* had been corrected whereas maps in the *Daily Bulletin of Weather Reports*—based on observations taken just an hour or so earlier—had not. For the *Tri-Daily Record*, which the Signal Office prepared for study and research, "the isobars and isotherms have been specially drawn anew . . . and are presumably free from the errors and omissions incident to telegraphic work."[63]

Although weather stations were farther apart in the West than in the East, the United States exchanged telegraphic reports with Canada, where many cold waves and winter cyclones originated. With timely reports from approximately 175 places in North America as well as nearly two decades of forecasting experience, the Signal Corps (so renamed around 1880) was sufficiently satisfied with its geographic coverage to drop the 11 P.M. observation in July 1888.[64] In cutting the number of daily observations from three to two, the weather service adopted a 12-hour sampling interval, with readings at 8 A.M. and 8 P.M., eastern standard time—a victory of sorts for Abbe, who was a strong advocate of official time zones.[65]

Despite an increasingly broader and denser telegraphic network, military stewardship of the weather service drew harsh criticism from a range of former supporters: scientists distressed by the agency's weak commitment to research, congressmen concerned with rising costs and inefficiency, soldiers troubled by nonmilitary needs draining the army budget, journalists outraged when a finance officer embezzled more than a quarter of a million dollars, and farmers and boatmen madder than hell about inaccurate forecasts or ineffective dissemination.[66] Although poor data, bad calls, or unusual weather accounted for most forecast failures, accurate storm and cold-wave warnings transmitted from Washington were occasionally lost or delayed, as when a telegraph operator at Madison, Wisconsin, ignored a frost warning that could have saved the crops of local tobacco farmers.[67] Seeing military mismanagement as a threat to their livelihood, farmers were especially eager for civilian control. Congress responded by relieving the Signal Corps of its meteorological duties and establishing the Weather Bureau within the Department of Agriculture, effective July 1, 1891.

With a clear mandate to serve localities, the Weather Bureau stepped up the modest decentralization initiated when the Signal Corps authorized local forecasts at its New York station in 1881 and at St. Paul, Minnesota; and San Francisco in 1890.[68] Chicago became a district forecast center in 1894, and by 1903, district forecasters in Boston, Chicago, Denver, New Orleans, Portland, San Francisco, and Washington were issuing daily forecasts and warnings to local offices in their multistate districts. In addition, observers at most of the 180 regular weather stations prepared local forecasts, and about a hundred weather stations issued their own daily weather maps. Mail and rail service were sufficiently efficient that maps printed at the weather station in midmorning were tacked up at post offices, schools, and other public places in nearby cities and towns later in the morning or early afternoon. Daily circulation of these locally produced "station weather maps" rose dramatically from around 150 in 1887 to 3,000 in 1891, 8,800 in 1893, and 25,000 in 1903.[69]

Much of this growth reflects a commitment to public communication by Willis Moore (1856–1927), chief of the Weather Bureau from 1895 to 1913. In his annual report for 1903, Moore appealed for funds to distribute still more maps:

> As the weather maps afford the only effective means possessed by the Weather Bureau for promptly placing before the public its daily observations and summaries, the improvement and extension of the maps . . . is urgently recommended.[70]

Moore's advocacy of meteorological cartography mirrored the appreciation of his predecessor, Mark Harrington (1848–1926). Before his appointment in 1891 as the Weather Bureau's first civilian chief, Harrington had been a university professor of astronomy as well as the founder and editor of the *American Meteorological Journal*. In 1893, at the International Meteorological Congress, in Chicago, he presented a short paper entitled "History of the Weather Map."[71] After acknowledging the contributions of Brandes, Espy, and Le Verrier, he focused on the synchronous weather maps issued by the government weather services in 18 countries, mostly in Europe. His impressive collection of facts about the maps' size, symbols, and content includes two tables with a subtle aroma of chauvinism: a list of countries noted that the United States was the second nation, after France, to issue a map of current weather, and a list of 73 U.S. Weather Bureau stations issuing weather maps documented the impressive achievement of a relatively young country eager to prove its prowess in science and technology.

Two decades later the Weather Bureau scored another technological triumph: the first regularly published current weather map of the northern hemisphere.[72] There were prominent predecessors, of course: around 1860, for instance, Robert FitzRoy had used ships' logs to construct a synoptic weather chart describing a large North Atlantic storm, and in 1869, Scottish meteorologist Alexander Buchan (1829–1907) published a series of synchronous weather maps, with isobars and isotherms, tracking storms from North America to Europe.[73] In 1868 and 1869, Le Verrier had published daily weather charts of western Europe and the North Atlantic for a 19-month period starting June 1, 1864, and in the 1880s, England, the United States, Germany, and Denmark produced weather maps of the northern hemisphere based on observations taken months earlier and collected by mail.[74] By contrast, the U.S. Weather Bureau's series of daily northern hemisphere maps, initiated on January 1, 1914, portrayed fresh observations collected within a few hours by submarine cable and "wireless telegraph."

Lithographed onto the reverse side of the 8 A.M. Washington weather map, the hemispherical map was impressive in design and content. Its polar projection (figure 3.5) provided a convenient framework for isotherms that circled the globe, and placement of North America in the center of the map's lower half afforded United States users a conveniently familiar north-is-up view. By contrast, isobars reporting pressure in millibars and isotherms labeled in absolute centigrade degrees (now called Kelvin degrees), rather than Fahrenheit degrees, expressed solidarity with the international scientific community. A bold isotherm representing the freezing point of water (273°) described the general pattern of frigid weather, and isobars at a rounded interval of 5 mb defined individual cyclones and anticyclones. Although coverage was inherently vague within the Arctic Circle and isolines usually stopped at 25°N, the new map presented a dynamic overview of the highs and lows affecting the world's weather.

Publication of the map was largely a public service to the international meteorological community, which contributed much of the

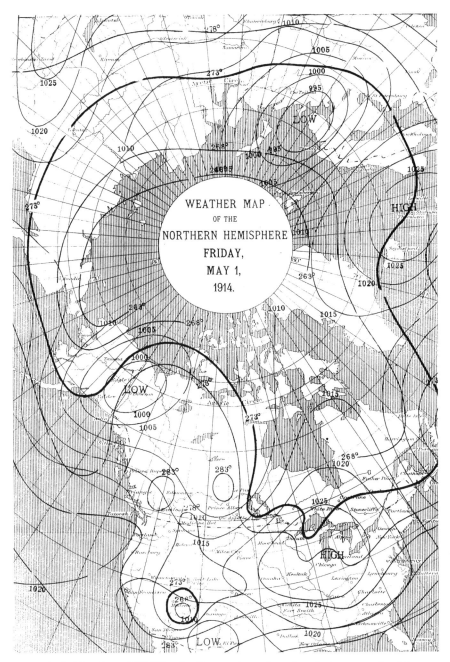

Fig. 3.5. Portion of U.S. Weather Bureau's northern hemisphere map for May 1, 1914. Excerpt from Vandevyver, "Les nouvelles cartes synoptiques du 'Weather Bureau' de Washington," *Ciel et Terre* 35 (June 1914): folded map bound at front.

data. The Washington forecast office had produced a similar map in manuscript for several years, and found it useful in extending forecasts a bit further and more reliably into the future. Although the printed version arrived too late to inform forecasts in Europe, it offered valuable insights to students and researchers, many of whom received complimentary copies of the first edition. Several acknowledgments appeared in the *Monthly Weather Review* for January 1914. Particularly gratifying were comments from Wladimir Köppen, a prominent German climatologist, and Vilhelm Bjerknes, an influential Norwegian meteorologist. Köppen expressed "great interest and delight" in the map but suggested drawing isotherms for "268°, 273°, 278°, etc., instead of for 270°, 275°, etc." and making areas with frost more prominent with a thicker isotherm for 273°.[75] Bjerknes was pleased with the map and praised the use of metric units as "a very great progress indeed."[76] But none of these kudos were as strong as a short essay by Cleveland Abbe, who was convinced that the new map "forces us as meteorologists to give our prime attention to the fundamental mechanics of the atmosphere."[77]

The printed northern hemisphere map's short life underscores the importance of international cooperation. On August 6, 1914, the Weather Bureau suspended publication "owing to the state of war involving the great nations of Europe."[78] Military censorship and disrupted communications had cut off observations from Europe and Asia, and even a credible manuscript map was impossible. Resumption of the northern hemisphere map in 1921, with twice-daily views in 1922, attests to the hemispherical chart's value as a forecasting tool.[79] By that time, though, the map had lost its novelty—forecast offices around the hemisphere were drafting their own charts, and publication was unnecessary.

4

Looking Up

Remembered for prosperity and prohibition, the 1920s were anything but roaring for American meteorology. Smugly confident their successful forecasts were not likely to get much better, Weather Bureau officials stubbornly resisted the promising theoretical and practical advances of a small but energetic band of Norwegian meteorologists.[1] The leader of this group, known as the Bergen school, was Vilhelm Bjerknes (1862–1951), a physicist eager to devise graphic methods for applying hydrodynamics to the atmosphere.[2] Had weather service bureaucrats been more receptive to radical ideas from tiny Norway, our daily weather map might have incorporated at least a decade earlier a duo of features indispensable to present-day forecasters: warm fronts and cold fronts.

Like many scientific discoveries, weather fronts were not without precedent. In 1828, for instance, Heinrich Dove described invasive, frontlike advances of warm and cold currents (figure 2.6), and in 1840, Elias Loomis presented a profile diagram (figure 2.10) indistinguishable in principle if not appearance from the cold fronts in modern meteorological textbooks.[3] Like many innovations, fronts were as much a reinterpretation as a discovery. People could feel their passage, but before Bjerknes, meteorologists gave little thought to mapping their locations.

Although the Weather Bureau would not acknowledge fronts on its maps for another two decades, the *Monthly Weather Review* for February 1919 carried one of the classic Bergen papers, "On the Structure of Moving Cyclones," in which Vilhelm's son, Jacob Bjerknes (1897–1975), summarized a detailed study based on highly pre-

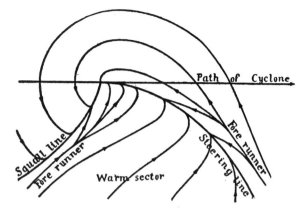

Fig. 4.1. Lines of flow in a moving cyclone. From J. Bjerknes, "On the Structure of Moving Cyclones," *Monthly Weather Review* 49 (1919): 95.

cise weather maps and sky reports from a dense network of observation stations in western Norway.[4] His goal was a conceptual model of summer cyclones with which forecasters could overcome the limitations of conventional weather maps based on more widely spaced stations sampled only every 6 to 12 hours. An elaboration of earlier work by his father and others, Jacob Bjerknes's graphic model described a moving low-pressure center with a counterclockwise flow of warm and cold air, which converged without mixing. Incorporated into the traditional weather chart, frontal analysis promised more reliable and locally precise forecasts.

Figure 4.1 is the key to Bjerknes's model: a skeleton of hydrodynamic lines of flow moving from west to east across a hypothetical weather map. Curved, asymmetric streamlines describe the inward movement of air around the cyclone. Rather than intersect at the storm center like the fins of a giant pinwheel, the streamlines merge into two lines of convergence: a "squall line" representing the cold front and a "steering line" representing the warm front. Seldom used today, the terms "squall line" and "steering line" distinguish the two-dimensional surface front portrayed on a map from the three-dimensional front separating cold and warm air masses. Just ahead of each line of convergence, air diverges from a "fore runner." Later versions of the Bergen model recognized a single line of advance, or front, for each air mass.[5] In the aftermath of World War I, the battleground term "front" was a compelling metaphor that enhanced the cyclone model's appeal.[6]

Three years later Bjerknes and colleague Halvor Solberg

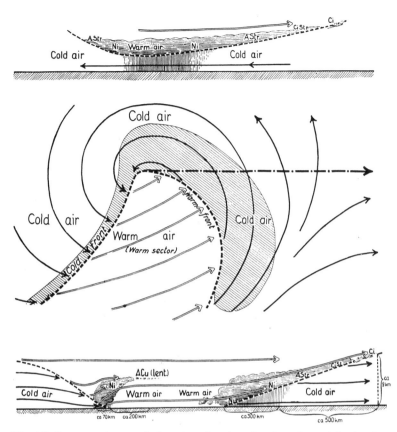

Fig. 4.2. Cross-sections and plan view showing clouds and precipitation in a model cyclone. The appreciably greater vertical scale of the cross-sections exaggerates the slope of both fronts. From J. Bjerknes and H. Solberg, "Life Cycle of Cyclones and the Polar Front Theory of Atmospheric Circulation," *Geofysiske Publikationer* 3, no. 1 (1922): 3.

(1895–1974) published a fuller treatment entitled "Life Cycle of Cyclones and the Polar Front Theory of Atmospheric Circulation."[7] On the first page of their short but richly illustrated paper, the three-part diagram in figure 4.2 recapitulated the main features of the Bergen cyclone. The plan view in the center relates the distribution of cloud cover and precipitation (shaded gray) to the counterclockwise movement of air around the cyclone's center. The labels "warm" and "cold" identify air masses differing in density and moisture, and light and dark arrows contrast the flow of warm air and cold air. Bounding the noticeably smaller "warm sector" in the cyclone's southeast quadrant, an advancing cold front chases a warm front, which defines the trailing edge of a broad band of rain and clouds. By contrast, the cold front marks the leading edge of a much narrower strip of precipitation.

Below the map in figure 4.2 an east-west cross section describes the storm's vertical structure 50 to 100 miles south of the cyclone's center. The long, comparatively light arrow across the top of the profile represents the current of warm air, which is more extensive aloft than the ground distance between surface fronts implies. Pictograms indicate the warm air's significance as a source of the rain and clouds. Directly west of the warm sector, a mass of cold, heavy air pushes into and under the relatively warm, lighter, and more humid air to produce a three-dimensional cold front, which extends sharply upward from the ground and quickly reaches an altitude of a thousand feet to a couple miles—although the map's dimensions are proportionately accurate, the profile's exaggerated angles and heights reflect considerable artistic license. Nimbus (Ni) and thunderhead clouds along the front bring moderate-to-heavy rain to a belt roughly 50 miles wide, while a high-level current of warm air carries lenticular altocumulus (ACu) clouds more than a hundred miles ahead of the front. This diagram is a dramatic demonstration that the surface front on a weather map provides a very limited view of a three-dimensional boundary between air masses.

Precipitation is common along cold fronts, especially when the warm air is relatively humid. Forced upward by an invasion of heavier cold air, the warm air cools and starts to surrender its moisture. Because condensation releases heat, humid air often turns buoyant and rises higher, releasing still more moisture and forming tall cumulonimbus clouds, called thunderheads. Although the physics is a bit more complicated, advancing cold air can trigger heavy rain, accompanied even in winter by thunder and lightning. By contrast, the isolated clouds ahead of, or far behind, the front bring little if any precipitation.

Ahead of the cold front, warmer air overriding a mass of cooler air forms a warm front, which Bjerknes also called a "steering surface." As with the cold front, the air masses don't mix and precipitation can occur where the advancing warm, moist air rises and cools. Warm fronts are markedly less steep than cold fronts—the diagram grossly exaggerates an angle of inclination of 1° or less—so that rain or snow falls over a much broader area, perhaps several hundred miles wide, in a band that precedes, rather than follows, the surface front Bjerknes originally called a steering line. And because the upward movement of warm air is less abrupt, precipitation accompanying a warm front tends to be gentler and of longer duration than along a cold front. Bjerknes's profile also describes how the gently rising low-angle warm front announces its arrival well ahead of the rain clouds and steering line with altostratus (AStr), cirrostratus (CiStr), and cirrus (Ci) clouds, which mark the discontinuity aloft between warm and cold air masses.

The upper cross section in figure 4.2 depicts the sequence of weather to the north of the storm track, where the front brings rain but never touches the ground. A dashed line portrays the discontinuity separating cold air near the surface from warm air lifted by cyclonic circulation, and ominous gray pictograms show rain clouds near the storm center, stratus clouds at higher altitudes, and cirrus clouds farther ahead. Tor Bergeron (1891–1977), another member of the Bergen team, described this elevated trough of warm air as an "occlusion" resulting when the cold front overtakes the warm front so that the warm air no longer touches the ground.[8] Bergeron's observation led to the concept of an occluded front, often portrayed cartographically as an amalgam of intersecting warm and cold fronts.

The idealized cyclone was not a rigid set of fronts, flow lines, and precipitation zones that merely drift across the map. Cyclones arose, developed, and died, Jacob Bjerknes argued, "as a combined effect of the barometric depression and the deflecting force of the earth's rotation" on warm air from the south and cold air from the north, so that "both currents turned about the cyclonic center." Because cold air is heavier, "the cold current is screwed underneath the warm one, and the warm current [is] screwed up above the cold one." Although the warm current "will be able to keep to the ground only in the warm sector, southeast of the center, ... it will flow [upward] over the cold current, joining the generally western drift in the higher strata."[9] Bjerknes and Solberg described these moving spirals of cold and warm air with the eight-step cartographic narrative in figure 4.3.[10] Formed from a wrinkle along the border between two air masses (a), the cyclone bulges toward the cold side and swells in size and strength as a counterclockwise vortex until its fronts merge (e) and the storm occludes (f). Because the cyclone ends much like it began—as a cold easterly current adjacent to a warm westerly current—the life cycle soon begins anew.

Bjerknes viewed cyclones as part of a global circulation system that avoided a disastrous buildup of solar energy near the equator by exchanging cold polar air for warm tropical air. The notion of a polar front had prominent antecedents, including Heinrich Dove's "op-

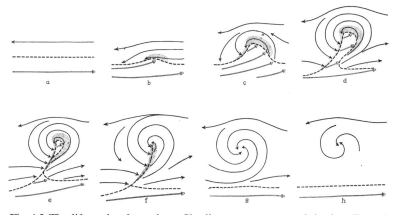

Fig. 4.3. The life cycle of a cyclone. Shading represents precipitation. From J. Bjerknes and H. Solberg, "Life Cycle of Cyclones and the Polar Front Theory of Atmospheric Circulation," *Geofysiske Publikationer* 3, no. 1 (1922): 5.

posing currents" and the three-cell convection model (figure 4.4) proposed in 1856 by the American William Ferrel (1817–1891), who extended Hadley's model by proposing descending cold air at the poles and ascending warm air at about 60°N and S—similar in effect, if not process, to the ascent of warm air in a cyclone.[11] According to Bjerknes, cyclonic circulation complemented and confirmed Ferrel's theory of convection:

> The general effect of the motion described is that cold air is conveyed to regions previously covered with warm, and there spread along the ground; and that in compensation, warm air is conveyed to previously cold air regions, and there distributed in the higher strata. Generally speaking, therefore, the cyclones may be said to be links in the interchange of air between the polar regions and the equatorial zone. This interchange, which is effected continuously in the zone of the trade winds, takes the irregular and intermittent character of cyclonic motions in the latitudes outside the high-pressure belts limiting the trade winds.[12]

Given the visual suggestiveness of cyclones as "links" in a common polar front, it was not surprising that in March 1920 Halvor Solberg found evidence for a continuous circumpolar discontinuity. In a letter to the director of the Norwegian Meteorological Institute, which collaborated with the Bergen group, Solberg described his serendipitous examination of a set of northern hemisphere weather maps (the Hoffmeyer charts) for 1907:

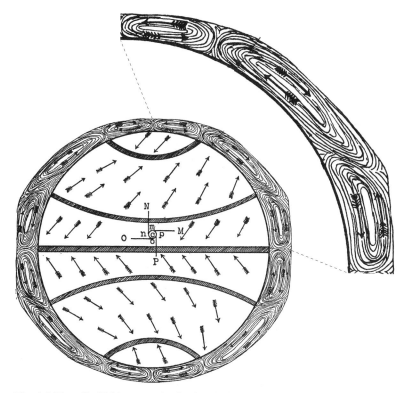

Fig. 4.4. Ferrel's 1856 atmospheric circulation model, with section enlarged
to show convection cells. From Henry Helm Clayton, *World Weather* (New
York: Macmillan, 1923), 42.

During Jack's (Jacob's) absence I reflected on the Hoffmeyer
charts. I managed to find something which we had suspected for a long
time, namely a line of discontinuity in temperature, wind etc. which
stretched right across the map from one side to the other. There is no
doubt at all that this line continues all around the Pole, even if the
Hoffmeyer charts do not extend so far. To the North of the line we
have "cold Polar air," to the South we have "warm Equatorial air." It
is only along this line that rain falls, in most cases at least. It marks the
front battle-line between two bodies of air and that is why it has such
a contorted course, since now the warm and then the cold has domi-
nance over the other.[13]

A facsimile of Solberg's sketch (plate 3) shows how he constructed
the polar front by threading a continuous line through the cyclones

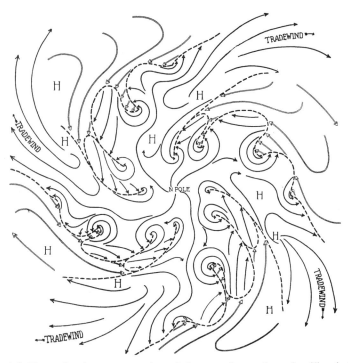

Fig. 4.5. The polar front as a series of circumpolar cyclone families, interspersed with anticyclones. From J. Bjerknes and H. Solberg, "Life Cycle of Cyclones and the Polar Front Theory of Atmospheric Circulation," *Geofysiske Publikationer* 3, no. 1 (1922): 15.

of January 3–4, 1907.[14] In his joint article with Jacob Bjerknes, Solberg pictured the polar front as a "string of families of cyclones," described by the dashed lines linking successive warm and cold fronts in figure 4.5.[15] A typical family is a chronological chain, with the oldest member, a fully occluded cyclone, on the eastern end, just ahead of a newly occluding cyclone, followed in turn by a less mature, not-yet-occluded cyclone, attached to an emergent, "secondary" cyclone on the western end. Interspersed with anticyclones (H), these cyclone families drifted and twisted their way eastward around the pole in an elaborately choreographed pageant of developing and dying swirls.

The Bergen school gave meteorology another tradition: the emphatically efficient map symbols that portray cold fronts as lines of sharp, triangular barbs; warm fronts as lines of rounded, semicircular pips; and occluded fronts as lines of alternating barbs and pips. With its tiny elements pointing in the direction of advance, the front sym-

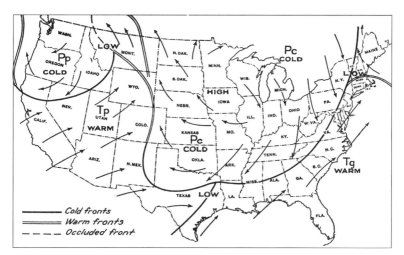

Fig. 4.6. Fronts and air masses for 8 A.M., eastern time, November 1, 1934. The map shows an occluded front (dashed line) in western Montana, where a cold front (solid line) of Polar Pacific (Pp) air has overtaken a warm front ("white" line) of Tropical Pacific (Tp) air. From H. R. Byers, "Weather Forecasting by the Air-Mass System," *Air Commerce Bulletin* 6 (1935): 161.

bol has become deeply embedded in present-day weather maps as an omen of rain and changing temperature. Tor Bergeron conceived this graphic shorthand to eliminate the expense of printing cold fronts in blue and warm fronts in red.[16] On an extended visit to Leipzig, he sketched his solution on the back of a postcard (plate 4) to his friend Jacob Bjerknes, back in Bergen.[17] In addition to differentiating among the three principal types of air-mass boundary, Bergeron's code had two additional graphic dimensions: open triangles and half-circles differentiated upper-air fronts from surface fronts, and more widely separated triangles and half-circles distinguished secondary fronts from their more mature cyclonic kin. These enhancements reflected the eagerness of Bergen researchers to account for a wider range of atmospheric phenomena with more complex notions of air masses and revolving storms.

In addition to delineating fronts, Vilhelm Bjerknes and a growing number of scientific converts began to classify air masses according to temperature, humidity, and area of origin. Because the conditions under which an air mass forms affect its interaction with other air masses, its region of origin became an important descriptor. As the assemblage of fronts, air masses, and winds in figure 4.6 illustrates, cartographic characterizations of air masses are usually alphabetic, rather than symbolic, with principal distinctions between polar (P) and tropical (T) air as well as between the high humidity of maritime origins in the Atlantic (a) or Pacific (p) ocean or the Gulf of Mexico (g) and the comparatively low humidity of a continental (c) origin.[18] This example suggests a polar front along the southern boundary of three adjacent air masses: a massive outbreak of Polar Continental (Pc) air that has pushed south into Texas and less extensive tongues of Polar Pacific (Pp) air in the Pacific Northwest and Polar Atlantic (Pa) air in northern New England. In addition to mnemonic efficiency, alphabetic labels on larger, more detailed maps avoid the graphic clutter of patterns or shaded area symbols forced to share space with complex point symbols describing wind direction, wind speed, and sky conditions at individual weather stations.

Neither Norwegian shyness nor American isolationism accounts for the U.S. Weather Bureau's resistance to Bergen principles. Vilhelm Bjerknes was an ambitious and energetic entrepreneur who sought international prestige by publishing and proselytizing, and the American weather service obliged by reprinting Norwegian articles in its *Monthly Weather Review*.[19] The February 1919 issue, for instance, not only published Jacob Bjerknes's classic paper on the structure of cyclones but included two articles by Vilhelm Bjerknes. The first was a positivist manifesto of sorts, in which the Bergen patriarch described his colleagues' successful forecasts.[20] In a fit of scientific ebullience, he asserted that because "all atmospheric processes obey the laws of . . . mechanics and thermodynamics, we are . . . really in possession of all the theoretical knowledge necessary to determine future weather."[21] In the second paper, a short essay on "Possible Improvements in Weather Forecasting, with Special Reference to the United States," he linked improved prognostication to "good charts representing the lines of wind flow."[22] The only other impediments in the United States, Bjerknes argued, were too few weather stations along the Pacific and Gulf coasts and the need to collect data by telegraph from stations currently reporting only by mail.

The need for additional stations was a convenient excuse for rejecting Bergen methods as grossly expensive and impracticable. In 1922, for instance, *Monthly Weather Review* editor Alfred Henry (1858–1931) concluded a critique of air-mass techniques with a cleverly crafted quantitative put-down:

It may not be amiss to here consider very briefly the suggestions of Professor Bjerknes in a previous article that the number of telegraphic stations in the United States be increased by about 4,500. A little calculation will uncover the difficulties which lie in the way of carrying out the suggestion. The number of telegraphic stations at present is slightly more than 200. Under the most favorable conditions the data from these stations can be charted in 35 minutes and, allowing 15 minutes additional for generalizing the data, the forecaster is able to begin issuing forecasts within an hour from the time of observation. If the number of telegraphic stations should be increased upward of *twenty-fold* [emphasis added] it would be physically impossible to chart and generalize the data within a reasonable time after the observing hour.[23]

However convincing Henry's objection, Bjerknes had in fact suggested a far more modest increase:

Besides about 300 telegraphic stations, the United States has a great number of climatological stations, about 3,000 if my memory is correct. If all these were made telegraphic we should get about the same number in proportion to areas as are used in western Norway. As desirable as an expansion on this scale would be, considered as an experiment, it would probably meet with difficulties from the point of view of the telegraph service; and quite likely it would not be necessary. The close network of stations in western Norway is necessary, partly on account of the complicated topography and partly on account of the exceptional difficulties during the war, when practically no weather telegrams from abroad are received. With the simpler topographic conditions in the United States, and the comprehensive view obtained from the great area of observations, it is probable that much could be accomplished by *doubling or tripling* [emphasis added] the number of stations.[24]

That's not all that editor Henry didn't like. The occasion of his complaint was a concise but insightful review of Bergen theories and forecasting methods by Anne Louise Beck, a geography graduate student at the University of California at Berkeley. Awarded a one-year fellowship from the American-Scandinavian Foundation, Beck had studied with Bjerknes and his colleagues, who guided her analysis of the 31 American weather maps for January 1921. Although the effort had "shown the placing of [fronts] to be more or less satisfactory," she was frustrated not only by the "diversity of topography from west to east" and the sparse telegraphic coverage but also by the

"inadequate data" on American weather charts, including barometric pressure reported only to the nearest tenth of an inch, temperature recorded only to the nearest even Fahrenheit degree, and wind
direction registered only to the closest of eight compass points.[25]

To illustrate the practicality of air-mass analysis, Beck's article included (with Henry's cooperation) a three-color weather map for
January 1st. The map showed a single warm front, a highly contorted—and thus questionable—cold front, a prominent cyclone, and
several principal air streams. That the first of an entire month of analyzed weather maps proved only mildly informative was convenient
for editor Henry, who appropriated Beck's analysis to confirm the fu-
tility of the Bergen approach:

> The remainder of the maps for January, 1921, considered by Miss
> Beck, have not been reproduced for want of space, and other reasons.
> Many of the maps were unsatisfactory for the reasons so clearly stated
> by Miss Beck on a previous page.[26]

Henry's skepticism reflected a firm belief that the moisture content and stability of the air were not, as the Norwegians claimed,
more important than the distribution of barometric pressure. While
Bjerknes and his disciples sought truth in hydrodynamics and thermodynamics, Henry embraced the simpler, more mechanistic forecasting canon of Espy, Loomis, and Bigelow: principles he and several
colleagues had summarized in *Weather Forecasting in the United
States*, a textbook published in 1916 at the direction of Weather Bureau chief Charles Marvin (1858–1943). An exemplar of cartographic
fundamentalism, Henry's text marshaled more than a hundred small-
scale weather maps—mostly printed on foldout pages in groups of
four with isobars in black and isotherms in red—to describe a litany
of cold waves, spring frosts, autumn frosts, West Indian hurricanes,
northeasters, Pacific Coast lows, and other weather types.[27] In contrast to the jubilant Pentecostalism of the Bergen apostles, Henry's
preface preached a dreary meteorological Calvinism:

> The book will be a disappointment to those, if there be such, who
> have formed the expectation that it will solve the difficulties of the
> forecasting problem. The consensus of opinion seems to be that the
> only road to successful forecasting lies in the patient and consistent
> study of the daily weather maps. Wherein the book will be helpful,
> however, is in the fact that it gives the experience of those who have
> gone before, and it is in this sense that it will find its most useful ap-
> plication.[28]

Convinced that weather forecasting was not likely to get much better, at least not soon, Henry and Marvin were not at all receptive to zealous prophets with preposterous proposals.

Despite different approaches to weather forecasting, the Weather Bureau and the Bergen forecasters shared a profound interest in the upper atmosphere. While the Norwegians pursued a theory-based interpretation of data—surface fronts inferred from ground observations and cloud types are, after all, only an *indirect* representation of the upper atmosphere—the Americans had concentrated on measurement-based exploration of the upper atmosphere with kites and balloons. Through an improved understanding of vertical trends in temperature, pressure, and winds they hoped to increase the reliability and reach of their forecasts. As Marvin's predecessor Willis Moore observed in 1899, a breakthrough was inevitable if not imminent:

> Having reached the highest degree of accuracy possible with our present instrumental readings, it becomes necessary to invade new realms if we desire to improve the character of the forecast and to make it of greater utility. I have long realized this, and several years ago determined systematically to attack the problem of upper-air exploration with the hope of being able ultimately to construct a daily synoptic chart from simultaneous readings taken in free air at an altitude of not less than 1 mile above the surface of the earth.[29]

Moore's pursuit of an upper-air weather map was the responsibility of Frank Bigelow, the Weather Bureau's storm-track and weather-types maven. Bigelow was a versatile fellow, competent in thermodynamics and fluid mechanics but convinced that the painstaking analysis of relevant observations held the key to improved forecasts. In the late 1890s he used cloud data from 140 telegraphic stations to construct three-level wind maps describing for various weather types the movement of air at the surface as well as among both "lower" and "upper" clouds. As the arrows in figure 4.7 illustrate for a typical "New England Winter High," the characteristic clockwise flow of an anticyclone is evident in the lower clouds, at an altitude of about 3,000 feet, but not even vaguely apparent among the upper clouds, at an altitude of roughly two miles. To describe comparable patterns at different levels, Bigelow divided the country into 96 squares, aligned on a grid, and estimated average direction of movement within each square for 19 surface weather maps, 35 lower cloud maps, and 60 upper cloud maps.[30] Like the example in figure 4.7, his other three-level maps

New England High (3), Winter.

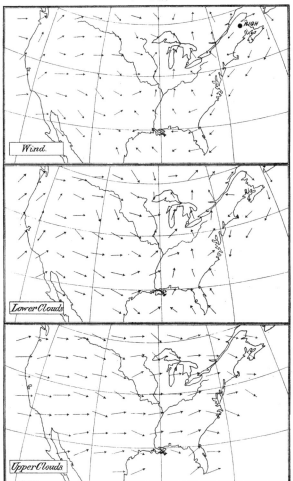

Fig. 4.7. Movement of upper and lower clouds and surface winds for a typical high-pressure cell centered over New England in winter. From *Report of the Chief of the Weather Bureau, 1898–99* (Washington, D.C., 1900), 2: n.p., chart 24.

demonstrated a greatly weakened circular flow in the upper atmosphere, where dominant west-to-east air currents were correlated with the generally eastward movement of cyclones and anticyclones.

In 1903, Bigelow carried Moore's wish a step closer to fruition with daily upper-air pressure maps. Mimicking the procedure for standardizing observed barometric pressure to sea level, Bigelow "re-

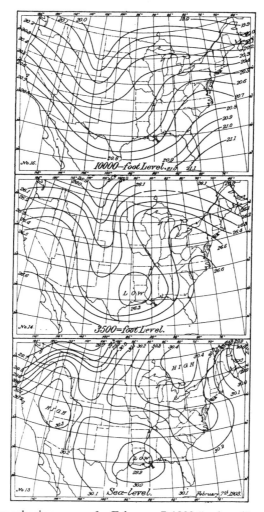

Fig. 4.8. Atmospheric pressure for February 7, 1903, "reduced" to 10,000 feet, 3,500 feet, and sea level. From Frank H. Bigelow, "IV. The Mechanism of Countercurrents of Different Temperatures in Cyclones and Anticyclones," *Monthly Weather Review* 31 (1903): 74.

duced" pressure readings to altitudes of 3,500 and 10,000 feet.[31] As the three-level maps in figure 4.8 reveal for February 7, 1903, cyclones and anticyclones that are readily apparent in surface isobars become progressively less prominent with increased altitude—a three-dimensional phenomenon he summarized graphically in figure 4.9.

To explain the formation of cyclones, Bigelow hypothesized the in-

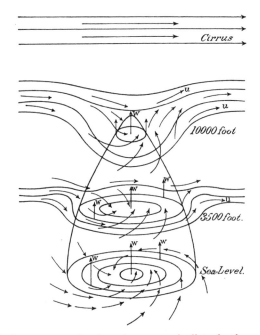

Fig. 4.9. Cyclone penetrating into the upper air disturbs the easterly high-altitude flow. From Frank H. Bigelow, "IV. The Mechanism of Countercurrents of Different Temperatures in Cyclones and Anticyclones," *Monthly Weather Review* 31 (1903): 74.

teraction of three independent air currents: strong high-level winds from the west, an "underflowing" (lower-level) polar current from the northwest, and an opposing low-level flow of warm air from the south.[32] In his schema, an upward thermal current creates a vortex as low pressure at the center draws in additional cold and warm air. Obviously aware of the importance of air masses, Bigelow made no effort to delineate their boundaries, or fronts. But his daily series of three-level maps demonstrated that upper-air pressure maps could help predict the direction and speed of storm centers.[33]

An important complement to Bigelow's three-level maps was the direct measurement of the upper atmosphere using kites.[34] The project began in 1893, under the direction of Charles Marvin. A mechanical engineering graduate who joined the Signal Office in 1884, the future bureau chief had advanced quickly to the rank of professor. Marvin's initial task was to develop kites and related apparatus for daily use. His experiments led to the Marvin-Hargrave Kite, an enormous box kite with 68 square feet of muslin lifting surface on a light

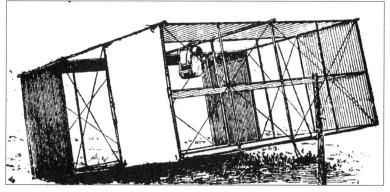

Fig. 4.10. The Marvin-Hargrave Kite with meteorograph mounted on frame. From Cleveland Abbe, "Meteorology," *Encyclopaedia Britannica*, 11th ed. (1911), 18: 279.

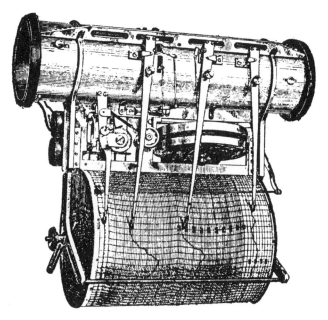

Fig. 4.11. The Marvin Kite Meteorograph. From Cleveland Abbe, "Meteorology," *Encyclopaedia Britannica*, 11th ed. (1911), 18: 280.

wooden frame (figure 4.10).[35] The kite's payload was the two-pound, aluminum-frame Marvin Kite Meteorograph (figure 4.11), which recorded a continuous profile of temperature, pressure, humidity, and wind speed. Another invention, the Marvin Kite Reel, doled out the thin piano-wire line and recorded the length and direction of the line as well as the pull of the kite. Powered manually or with a small gasoline engine, the reel helped the operator control the kite's ascent and descent, including 5 to 10 minute pauses to let the instruments adjust to selected altitudes. Kites could reach an altitude of two miles or more—if the wind was blowing.

Kite sampling was expensive and complex: different wind conditions required different size kites, flown from separate, specially equipped kite stations located well away from trees, power lines, and populated areas. Between April and November 1898, a network of 17 temporary kite stations yielded useful scientific data about vertical variations in temperature and humidity.[36] But because the untimely measurements could not justify the high operating costs—data were not available until the kite was reeled in—kites never gained a role in routine weather forecasting. By the 1930s, the airplane and the sounding balloon had made the kite obsolete even for scientific studies.

A boon to scientific meteorology, the airplane brought the Weather Bureau a new responsibility: reporting and forecasting the weather above, as well as at, the surface. The bureau's approach reflected the evolution of commercial aviation, for which the Department of Commerce developed a network of airways (figure 4.12), with airports and emergency landing fields spaced every 30 to 60 miles along railwaylike routes through major cities.[37] Because these new duties strained the bureau's limited resources, few airports had weather stations; and only selected airways, represented by the bold lines in figure 4.12, had teletype service for frequent, sometimes hourly, weather reports.

To monitor the upper air, the Weather Bureau maintained several dozen more or less uniformly spaced pilot-balloon stations (figure 4.13) at which observers released small hydrogen-filled balloons several times a day.[38] Because the balloon ascended at a known rate, the observer could estimate the height of the cloud ceiling with a watch. In addition, a precise surveyor's telescope called a theodolite allowed exact measurements of the balloon's direction and angle above the horizon, from which the observer could calculate wind speed. "Winds aloft charts" plotted for various altitudes from pilot-balloon data helped aviators plan routes and estimate flying time.[39] Radio tracking replaced the theodolite in the 1940s and 1950s, when weather services added an instrument package with a short-wave transmitter. Called *radiosondes*, these new sounding balloons provided timely upper-air data at low cost in clear or cloudy weather.[40]

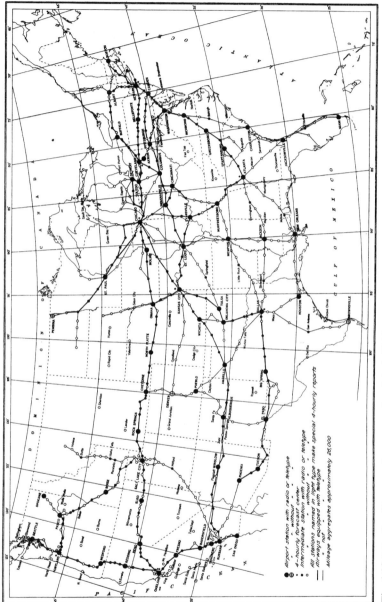

Fig. 4.12. Weather Bureau airways service, July 1, 1933. From Willis Ray Gregg, "History of the Application of Meteorology to Aeronautics with Special Reference to the United States," *Monthly Weather Review* 61 (1933): chart 2, preceding p. 169.

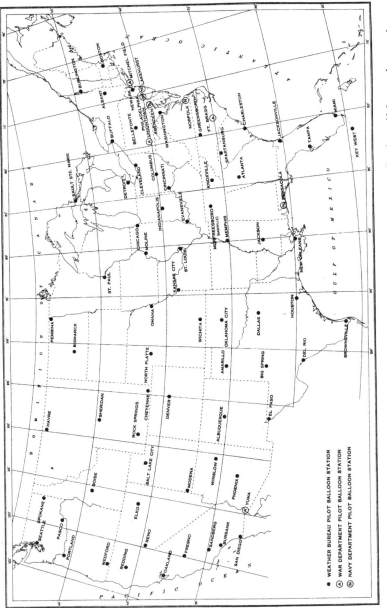

Fig. 4.13. Pilot-balloon stations, July 1, 1933. From Willis Ray Gregg, "History of the Application of Meteorology to Aeronautics with Special Reference to the United States," *Monthly Weather Review* 61 (1933): chart 1, facing p. 168.

Airlines appreciated the effort but demanded more—more airport stations, more frequent forecasts, and more reliable upper-air measurements.[41] They also wanted explicit depiction of air masses and fronts, which pilots and company meteorologists considered real and relevant. When Marvin resisted, they complained to President Roosevelt, who in July 1933 asked his Science Advisory Board, a panel of distinguished scientists, to investigate weather service operations.[42]

The board promptly formed a special committee, which issued a preliminary report in mid-November.[43] A 13-paragraph description of the Weather Bureau's crucial role in agriculture, commerce, and transportation carried a pointed endorsement of Bergen principles:

> During the last decade there has been very rapid progress in Europe in the development and general use of air-mass analysis methods. These require a knowledge of temperatures, humidities and pressures aloft as well as on the surface, but thus far no systematic attempt has been made to obtain at a given time upper-air measurements of these aerological conditions at a considerable number of stations scattered systematically throughout the country so as to make possible the drawing of a daily upper air map of the whole country similar to the surface maps now provided by the Weather Bureau.[44]

In addition to an enhanced aerological network and fuller army and navy cooperation with the Weather Bureau, the board recommended increasing the number of daily weather maps from two to four, moving local weather instruments out to the airport, and disseminating aviation forecasts by teletype or radio. Eager to have its recommendations implemented immediately, the board called for replacing Marvin as chief but allowing him to remain through August 1934 as associate bureau chief, thereby completing 50 years of service to the Signal Office and Weather Bureau.[45]

The new chief, recommended by the board, was Willis Ray Gregg (1880–1938). Head of the bureau's aerological division and author of a prominent textbook on aeronautical meteorology, Gregg appreciated the Bjerknes cyclone model and understood the need for more detailed upper-air data.[46] With the aid of the military, he quickly established a network of 20 stations with daily flights, at roughly the same time, to an altitude of three miles.[47] He also supplemented the 8 A.M. and 8 P.M. maps with 2 A.M. and 2 P.M. weather maps at selected airports, provided "sectional" maps tailored to des-

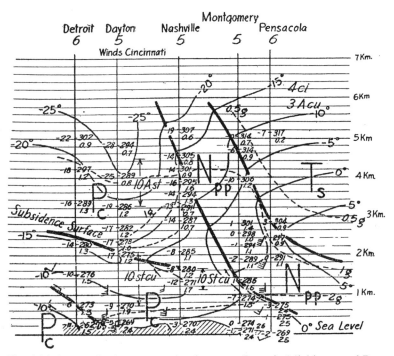

Fig. 4.14. Atmospheric cross-section between Detroit, Michigan; and Pensacola, Florida, for December 7, 1934. Heavy lines represent air-mass boundaries, or fronts aloft. From Hurd C. Willett, "Routine Daily Preparation and Use of Atmospheric Cross Sections," *Monthly Weather Review* 63 (1935): 5.

ignated airways every four hours, and experimented with daily weather maps showing fronts and "forecast maps" describing the next morning's weather.[48]

Despite improved airways reports and richer upper-air observations, Gregg's efforts to promote air-mass analysis encountered the resistance of an inbred forecasting staff skeptical if not openly hostile toward Bergen methods. Although he hired three young air-mass experts to construct daily synoptic frontal charts and brief the forecasting staff, the air-mass group was located well away from "the practitioners," who always issued their morning outlook before meeting with "the theorists."[49]

By contrast, the navy and the army air corps appreciated Bergen methods.[50] In addition, a Department of Agriculture grant supported a daily air-mass analysis at the Massachusetts Institute of Technology, where Carl-Gustav Rossby (1898–1957) established a meteorology department in 1928, at the request of the navy. A student of Bjerknes, Swedish meteorologist Rossby had worked at the Weather Bureau for 18 months, starting in 1926, as a Scandinavian-American Foundation fellow. Using the richer American upper-air data, he moved well beyond the Norwegian cyclone model with cross-sectional diagrams like figure 4.14, which describes a three-dimensional atmosphere far too complex for a conventional weather map.[51] Because upper-air observations were limited and profiles took time to construct, MIT researchers routinely constructed cross sections for a single east-west group of roughly aligned stations and for two north-south groups. The air-mass project moved to Washington in 1941, after Bergen convert Francis Reichelderfer (1895–1983), the navy's aerology expert, succeeded Gregg as Weather Bureau chief and appointed his old friend Rossby assistant chief for research and education. That year the bureau officially acknowledged the value of air-mass analysis by adopting a "more explicit and informative" general weather map—with warm fronts and cold fronts as standard features, portrayed by Tor Bergeron's lines of triangles and semicircles.[52]

Vilhelm Bjerknes lived until 1951—long enough to see his conceptualization of the atmosphere dominate weather forecasting. Never bashful about his work and its significance, Bjerknes summed up his contribution in two sentences:

> During 50 years meteorologists all over the world looked at weather maps without discovering their most important features. I only gave the right kind of maps to the right young men, and soon they discovered the wrinkles in the face of the Weather.[53]

However self-assured, Bjerknes's modest acknowledgment of his students' role was not the last word on fronts. Four decades after his death, meteorologists are questioning the need for fronts on synoptic weather maps. While no one challenged the reality of air masses or the polar front, atmospheric scientists were well aware of significant departures from the classic Bergen model.[54] Cyclones can form well away from the polar front, for example, and "cold fronts aloft"—missing from conventional weather maps because they never reach the ground—can trigger fierce storms. And fronts are often ambiguous, as demonstrated at the National Meteorological Center, in Camp Springs, Maryland, at a workshop on air-mass analysis, where different experts plotted markedly different frontal alignments from the same data.[55] Fronts can also confuse computers, as apparent in the zigzag fronts plotted (plate 19) by cartographic software not conditioned by theory to render air-mass boundaries as smooth, gently curved lines. To some meteorologists, these inconsistencies are a clear sign that instead of imposing fronts on the map, forecasters should let the data speak for themselves.

Looking Ahead

Aviators' needs are significant but short-sighted. With lives and livelihoods dependent on surface and upper-air conditions over the next three to six hours, pilots have little in common with farmers and contractors, who must weigh the risk of planting or painting today against the cost of waiting until tomorrow or next week. Meteorologists, too, think in days, not hours, partly because public assessments of their professional worth rest largely on how well—and how far ahead—weather forecasters can map the future. No wonder then that university researchers and Washington's weather bureaucrats have taken up the technological-scientific challenge of an accurate 8-to-14-day outlook.

Maps of future weather are a fundamental part of forecasting: although microchips have replaced paper as the cartographic medium, predictions of local conditions require the forward projection in time of the atmosphere's current geography. As examined in chapter 2, this strategy evolved in the nineteenth century as a two-stage process, in which the forecaster first projects the low-pressure center's likely movement along its assumed storm track and then assesses the effects of the cyclone's size, strength, and position on local and regional winds, temperature, and precipitation. These assessments were often guided by an algorithm, or set of rules, for matching projected pressure and temperature conditions to typical situations with specific consequences.

Some algorithms were verbal lists of conditional, *if-then* statements—*if*, for instance, heavy rain begins after the storm moves inland, *then* the storm is likely to continue—which could have been

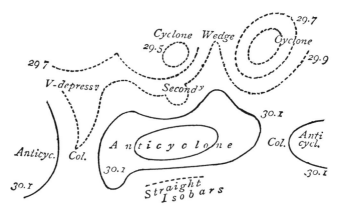

Fig. 5.1. Seven fundamental shapes of isolines. From Ralph Abercromby, *Weather: A Popular Exposition on the Nature of Weather Changes from Day to Day* (New York: D. Appleton, 1887), 25.

(but rarely were) described in a flow chart.[1] Others were graphic models similar to the seven fundamental shapes of isobars (figure 5.1) posited around 1883 by Ralph Abercromby (1842–1897). In addition to comparatively *straight isobars*, the Scottish meteorologist recognized the unique characteristics of the circular or oval isobars surrounding *cyclones* and *anticyclones*, the converging isobars of *V-shaped depressions* (troughs of low pressure) and *wedges* (of high pressure), the dimpled isobars marking *secondary* cyclones, and the *cols* (saddles of low pressure) found between anticyclones.[2] Each basic shape referred in turn to a more geographically refined model like figure 5.2, which describes varying weather conditions in the vicinity of a cyclone. Acclaimed by British forecasters in the late nineteenth century, Abercromby's ideas proved a useful supplement to the U.S. Weather Bureau's storm-track strategy.[3]

More useful still was the pressure-change chart: a plot showing areas with a significant increase or decrease over a 12- or 24-hour period.[4] Highlighted in Alfred Henry's 1916 manual, *Weather Forecasting in the United States*, pressure-change charts have their own terminology, with *allobar* describing an area where the barometric pressure has changed by 0.1 inch or more in 12 hours, and the Greek prefixes *an-* (upward) and *kat-* (downward) indicating the direction of change. Allobars are broad areas, which typically migrate from west to east, as illustrated in figure 5.3, where solid lines portray the leading edge and dashed lines identify the rear of a katallobar that crossed the continent in mid-November 1911. Dates linking front and

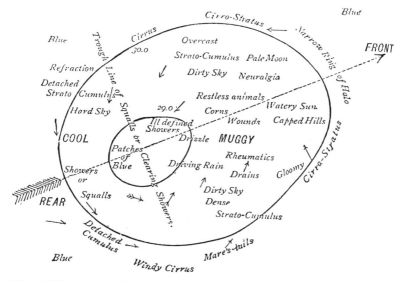

Fig. 5.2. Weather conditions in various parts of a cyclone. From Ralph Abercromby, *Weather: A Popular Exposition on the Nature of Weather Changes from Day to Day* (New York: D. Appleton, 1887), 28.

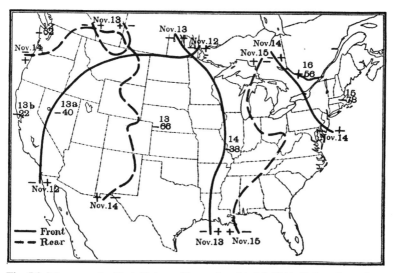

Fig. 5.3. Movement of katallobars, November 11–15, 1911. Pairs of numbers represent the center of the katallobar at 8 A.M. on various dates; for example, 14 over –38 indicates that on November 14 the greatest pressure drop (0.38 inches) occurred in southern Illinois. From Alfred J. Henry and others, *Weather Forecasting in the United States*, Weather Bureau publication no. 583 (Washington, D.C., 1916), fig. 22, facing p. 83.

rear boundaries indicate that on November 14, at 8 A.M., eastern standard time, a katallobar extended from the Pacific Northwest and eastern Rocky Mountains to central New York and eastern Pennsylvania, a distance well over a thousand miles. This region of declining pressure heralded a low-pressure system, which was centered over eastern Nebraska on the morning of the 14th and about 300 miles north of Toronto a day later. Of greater import, the leading edge evident on the 12th had forecast by 24 hours the pressure center's initial development over northern Wyoming.

Forecasters considered the position and extent of allobars important because katallobars usually precede cyclones while anallobars run ahead of anticyclones. Moreover, allobars have centers—points of greatest increase and decrease—which can be plotted and tracked, just like centers of low and high pressure. But as plate 2 demonstrates, the trace of an allobar center only vaguely approximates the path of a storm or anticyclone. More revealing is the katallobar that assumes the shape of a circle or ellipse, thereby signifying the increased geographic definition of its associated storm.[5] Although terminology has changed—meteorologists now call the lines isallobars—maps of pressure-change contours remain an important forecasting tool.

In the 1930s, meteorologists applying Bergen methods based their forecasts on independent projections of two weather maps: one for barometric pressure and winds at 10,000 feet and the other for sea-level pressure as well as surface temperature, fronts, winds, and precipitation.[6] After checking the two prognostications for consistency, the forecaster interpreted the effects on local weather of terrain and nearby bodies of water. Although the process included a 24-hour pressure-change chart as well as cross-section diagrams, maps of average pressure, and other climatic data, the short-period forecast focused on the displacement of fronts as well as their development and decay.[7]

The complementarity of surface and upper-air weather maps is apparent in their differences and similarities.[8] Surface weather charts are comparatively complex, as illustrated in figure 5.4 by a simplified version of the morning weather map for March 29, 1939. Redrafted as a compact, one-page illustration for a short chapter on weather forecasting in *Climate and Man*, the 1941 Yearbook of Agriculture, the map shows a family of three cyclones stretching from Mexico to the north Atlantic, off Labrador—a classic early spring portrait of the

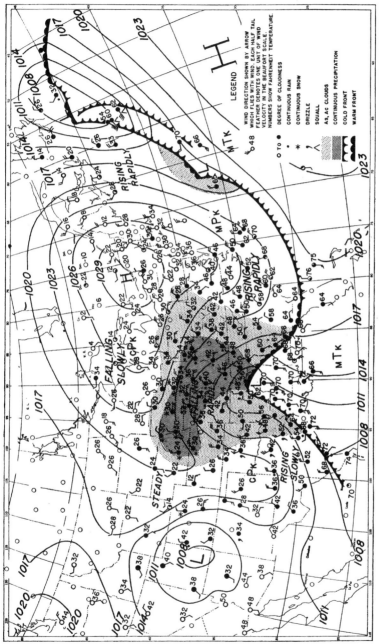

Fig. 5.4. Surface weather map for 7:30 A.M., March 29, 1939. Barometric pressure is in millibars. From C. L. Mitchell and H. Wexler, "How the Daily Forecast Is Made," in *Climate and Man*, Yearbook of the Department of Agriculture, 1941 (Washington, D.C., 1941), 587.

polar front.[9] Additional features of note include a low-pressure center over Utah and regions of high-pressure centered over the mid-Atlantic Ocean, southern Ontario, and (as inferred from isobars at the upper left) the north Pacific Ocean, off the coast of Washington. By contrast, the corresponding 10,000-foot chart in figure 5.5 depicts a far simpler pattern two miles above sea level, where the air is less humid, less dense, and less influenced by terrain and temperature differences at the surface. The upper-air isobars are not only more gently curved than their surface counterparts, but the headless arrows representing wind direction indicate air currents moving parallel to the isobars—*geostrophic* (earth turned) winds, for which the Coriolis force balances the pressure-gradient force.[10] Except for the trough of low pressure reaching through the Dakotas into Kansas and a weaker wedge of high pressure extending into Minnesota from the southeast, the upper-air currents steering the more complex weather down below generally mimic the west-to-east direction of the prevailing mid-latitude surface winds, aptly named the "westerlies."

Compared to surface-weather maps, upper-air charts are simple, stable, and easier to project forward in time. Because of the upper air's impact on surface conditions, forecasters scrutinized their 10,000-foot data for signs of change. In the late 1930s, careful study of upper-air charts led Carl-Gustav Rossby to observe that the alternating troughs and wedges on the upper-air chart could be represented mathematically as waves, governed by the conservation of vorticity (rotational energy) along the boundary between polar and tropical air masses.[11] *Rossby waves*, as they are now called, are also apparent in maps of the jet streams—strong, narrow, and meandering high-altitude air currents discovered by bomber pilots late in World War II.[12] A wave more than a thousand miles across might settle over the United States for several days, perhaps to bring abnormally high temperatures to the eastern half of the nation (where a wedge, or ridge, of high pressure might promote a northward flow of warm air) and comparatively cool weather to the West (where a trough, or valley, of low pressure would carry polar air southward). This paired ridge and valley are part of a hemispherical system of waves, which a week later might shift half a wavelength eastward, to deliver cooler temperatures to the East and warmer weather to the West. Rossby's mathematical analysis provided not only an explanation of these sinuous "displacements"—eastward or westward—of upper-level isobars but also a method whereby forecasters could calculate the wave's velocity and projected position.[13] Hydrodynamics and Bergen theory also provided a calculus for the displac-

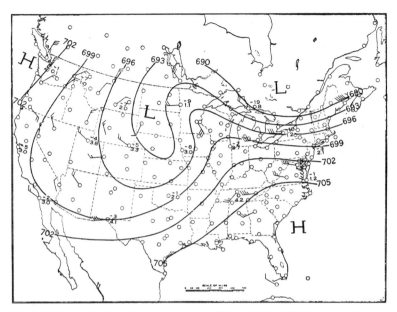

Fig. 5.5. Upper-air data for the morning of March 29, 1939. Barometric pressure is in millibars. From C. L. Mitchell and H. Wexler, "How the Daily Forecast Is Made," in *Climate and Man*, Yearbook of the Department of Agriculture, 1941 (Washington, D.C., 1941), 590.

ing pressure systems, fronts, and air masses on forecast maps of surface weather.[14]

Despite the incorporation of physical principles into complex mathematical equations, weather forecasting in the 1940s and 1950s was very much a manual-graphic process, in which clerks entered observations onto charts by hand and forecasters plotted displaced positions of fronts and zones of precipitation. Hand-eye coordination thus played a key role in drawing isolines, delineating fronts, and estimating latitude and longitude for features to be projected mathematically. Despite an improved understanding of atmospheric processes, numerous opportunities for imprecision limited the certainty with which weather forecasters could peer ahead 24 to 36 hours.

Electronic computing's slavish accuracy has vastly improved the forecaster's vision, but learning how to calculate the weather was a tough, often tedious struggle. Impediments included inadequate data, slow computers, and money—lots of it—because the hardware required for reliable long- and short-range meteorological forecasting is complex and expensive. And nature's own complexity obstinately resists efforts to model its behavior with electronic systems, even large ones, too small to simulate the myriad interactions of atmospheric constituents.

As some historians tell it, computational weather forecasting faced another obstacle: the embarrassing failure of Lewis Fry Richardson (1881–1953), who in 1922 presented a numerical weather forecast that failed badly.[15] That indictment seems harsh, though, for Richardson, an acknowledged mathematical genius, was widely accorded the respect and mild curiosity of an eccentric visionary. The subtle humor in his book, *Weather Prediction by Numerical Process*, reflects awareness that his innovation was impracticable, at least in the near future. But in the late 1940s, when experimental computers suggested that their descendants might one day be up to the task, prominent atmospheric researchers were well aware that the British scientist had been onto something.[16]

Richardson's approach was to impose a spherical grid of evenly spaced meridians and parallels, as in figure 5.6, and to manipulate pressure, temperature, moisture, density, wind velocity, and other atmospheric properties for the resulting cells. Mathematicians call this divide-and-conquer strategy a *finite-difference* technique because the grid treats a continuously variable phenomenon like barometric pressure as occurring only at a small number of points (the cell centers) separated by a specified (hence, finite) difference.[17] Although the map focuses attention on the cells, measurements and difference calculations refer to the cell centers.[18] Because the number of calculations is roughly proportional to the square of the number of cells, Richardson avoided an impossibly massive computational effort by positioning the parallels 200 km (120 miles) apart and separating the meridians by 2°48'45", the result of dividing the full circle (360°) into 128 equal parts.[19] Cells were approximately 120 miles wide at the southern tip of England (about 50°N) but progressively narrower farther north because of converging meridians. In cartographic parlance, the *digital map* described in figure 5.6 consists of *raster data*, so called because the parallel arrangement of its rows, although curved, has the scan-line structure of a television screen.

Richardson's model recognized the atmosphere's third dimension with five layers, with boundaries at altitudes of 1.2, 2.5, 4.3, and 7.08 miles (2.0, 4.2, 7.2, and, 11.8 km), where average barometric pressure is approximately 800, 600, 400, and 200 millibars, respectively. Each quadrilateral on the map was thus a stack of five three-

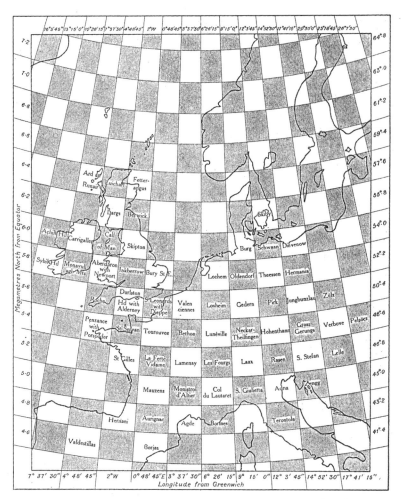

Fig. 5.6. Richardson minimized computational effort by treating the atmosphere as a coarse spherical grid. From Lewis F. Richardson, *Weather Prediction by Numerical Process* (Cambridge: Cambridge University Press, 1922), frontispiece.

dimensional cells, or prisms, containing roughly the same amount of air by weight. In addition, the model accommodated an input of solar radiation at the top of each stack (and its penetration and absorption at various layers, depending on cloud cover) as well as evaporation at the surface from open water or moist soil. In addition to the horizontal steps of its cartographic checkerboard and the vertical steps of its five layers, the algorithm moved forward in time in six-hour increments, with each new six-hour forecast serving as the starting point for another round of computation.

Calculations began with an initial map of atmospheric conditions, observed and recorded for specified places near the center of each cell. (Richardson reduced the computational demand even more by ignoring the nameless peripheral squares in figure 5.6, and calculating pressure only for the gray ones and wind velocity only for the white ones.) Cells centered over water were initialized from a nearby coastal weather station or by averaging, as in the case of St. Leonards and Dieppe, located on opposite sides of the English Channel. Systems of mathematical equations represented the adsorption of solar radiation, the settling of heavy air and rising of light air, the creation of turbulence in the lower atmosphere and the pull of gravity, the evaporation and condensation of moisture, and the effect on wind of pressure-gradient and Coriolis forces. Richardson's equations also obeyed the laws of hydrodynamics and thermodynamics as well as Newtonian rules for conserving mass, energy, and momentum. Wind direction had two horizontal components, so that each prism could transfer air to a neighbor immediately north or south and to another neighbor directly east or west. Vertical transfers were also recognized, as when air rose or subsided. Recall that each two-dimensional cell on the map represented a stack of five three-dimensional prisms, each with six sides, so that when an exchange occurred across one of these six boundaries, what one prism lost an adjacent prism gained. Conditions within a prism were thus linked to conditions at its neighbors, which were linked in turn to—well, by now you have the picture of a very complex process, and in the precomputer era, a very labor-intensive operation as well. To get the process started, the book included a set of 23 computing forms!

Richardson envisioned his algorithm as an enormous "forecast factory" inhabited by thousands of "human computers and computresses" armed with 10-inch slide rules and 5-digit tables of logarithms—with each cell requiring 32 arithmetically agile clerks just to keep abreast of the evolving weather. Tongue in cheek, he noted the need for

> a large hall like a theatre, except that the circles and galleries go right round through the space usually occupied by the stage. The walls of this chamber are painted to form a map of the globe. The ceiling represents the north polar regions, England is in the gallery, the tropics in the upper circle, Australia on the dress circle and the antarctic in the

pit. A myriad of computers are at work upon the weather of the part of the map where each sits, but each computer attends only to one equation or part of an equation. The work of each region is coordinated by an official of higher rank. Numerous little "night signs" display the instantaneous values so that neighboring computers can read them. Each number is thus displayed in three adjacent zones so as to maintain communication to the North and South on the map. From the floor of the pit a tall pillar rises to half the height of the hall. It carries a large pulpit on its top. In this sits the man in charge of the whole theatre; he is surrounded by several assistants and messengers. One of his duties is to maintain a uniform speed of progress in all parts of the globe. In this respect he is like the conductor of an orchestra in which the instruments are slide-rules and calculating machines. But instead of waving a baton he turns a beam of rosy light upon any region that is running ahead of the rest, and a beam of blue light upon those who are behindhand.

Four senior clerks in the central pulpit are collecting the future weather as fast as it is being computed, and despatching it by pneumatic carrier to a quiet room. There it will be coded and telephoned to the radio transmitting station.

Messengers carry piles of used computing forms down to a storehouse in the cellar.

In a neighbouring building there is a research department, where they invent improvements. But there is much experimenting on a small scale before any change is made in the complex routine of the computing theatre. In a basement an enthusiast is observing eddies in the liquid lining of a huge spinning bowl, but so far the arithmetic proves the better way. In another building are all the usual financial, correspondence and administrative offices. Outside are playing fields, houses, mountains and lakes, for it was thought that those who compute the weather should breathe of it freely.[20]

In this context of self-deprecating brilliance, Richardson's failed forecast was anticlimactic. To make his model manageable, he had taken a few short cuts, including the much-reduced grid in figure 5.7, for which he calculated only two parameters: momentum (velocity) for the cells labeled M and pressure for the cells labeled P. The 25 cells, each covering 3 degrees of longitude and stretching 120 miles from north to south, were slightly wider than the cells in his ideal model. Using data for May 20, 1910, he ran the model forward just one six-hour step, and in several months of calculating worked out just two predictions: wind speed for the cell at the center and pressure for the

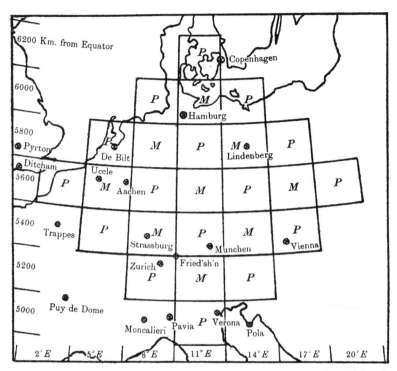

Fig. 5.7. Reduced grid for Richardson's trial forecast. From Lewis F. Richardson, *Weather Prediction by Numerical Process* (Cambridge: Cambridge University Press, 1922), 184.

adjoining cell directly south. His results were not only incorrect but preposterous: fierce winds and an unheard-of rise in surface pressure of 145 mb—more than a hundred times the observed change. Even so, these outrageous errors did not deter a reviewer for the *Monthly Weather Review* from praising Richardson's "remarkable" book as an "admirable ... first attempt [that] indicates a line of attack ... and invites further study with a view to improvement and extension."[21]

Although Richardson blamed these errors on faulty wind data, his flopped forecast largely reflects gaps in what was known then about atmospheric processes. According to University of Chicago meteorologist George Platzman, compensating for the "convergence of mass in individual layers" by applying a correction factor of 1/100 would have reduced the six-hour pressure increase to a more realistic 1.45 mb.[22] Better yet, had Richardson been aware that the atmosphere is always at very close to perfect equilibrium, with the Coriolis effect balancing the pressure-gradient in the upper air, he could have included the "quasi-geostrophic balancing equation" in his formulas. And had he known of simplifying assumptions validated in the 1950s, when electronic computers encouraged atmospheric scientists to experiment with numerical forecasting, the British model-builder could have streamlined his calculations and attempted a more realistic simulation, covering a larger area or a longer time.[23]

Richardson was lucky he quit when he did: had he projected weather conditions more than six hours into the future, his model no doubt would have "blown up"—mathspeak for yielding unreliable, absurdly large (or small) results—as a consequence of what mathematician Richard Courant called *computational instability*. In 1928, Courant and his colleagues reported that numerical solutions of finite-difference simulations of wave phenomena (such as weather) become unstable when too long a time step allows a wave (such as a front, cyclone, or other atmospheric disturbance) to travel from the center of one grid cell to the next before the start of the next round.[24] When a cell is thus unaware of conditions at neighboring cells, as the equations require, the model no longer reflects reality, and its calculations become meaningless. In Richardson's case, winds stronger than 20 miles per hour could cross a cell 120 miles wide in less than six hours, and threaten stability. Because much higher velocities are common in the upper atmosphere, where winds frequently exceed 100 mph, he unknowingly avoided further embarrassment by not projecting the weather beyond a single six-hour round of calculations. Later numerical models addressed the "Courant condition" with a shorter time interval or larger cells.

Richardson's greatest need was an electronic computer, with which he might have discovered and dealt with computational instability as well as experimented with mathematical short-cuts and filters for removing sound waves and other irrelevant phenomena. Whether he would have refined and perfected his model is iffy history, of course, but trial-and-error experimentation accounted for much of the progress in numerical meteorology in the 1950s and 1960s, when digital computing allowed computational models to not only keep pace with the weather but project conditions hours, even days, ahead in time.[25]

Perhaps the strongest argument that Richardson's failure did not retard progress was the early attention accorded weather forecasting by John von Neumann (1903–1957), the brilliant Hungarian-born mathematician who in the late 1940s presented the first detailed description of an electronically stored computer program.[26] As director of the Electronic Computer Project at Princeton University's Institute for Advanced Study, von Neumann hoped to build a digital computer more powerful than the fabled ENIAC (Electronic Numerical Integrator and Computer), the monster vacuum-tube computer developed in the mid-1940s at the University of Pennsylvania. To secure adequate government funding, he needed a substantial scientific challenge with a significant public benefit. By early 1946, Rossby, Reichelderfer, and electronics innovator Vladimir Zworykin had convinced him that weather forecasting was the ideal problem—enigmatic, tractable, and relevant to the military.[27] With the flaws in Richardson's work understood and its limitations soon to be rendered obsolete by the institute's new computer, numerical weather predication was clearly within von Neumann's reach.

To develop the software, von Neumann set up the Weather Project as a separate unit and hired Jule Charney (1917–1981) as its director. A meteorologist trained in mathematics, Charney devised a simplified model that not only incorporated an approximate balance between Coriolis and pressure-gradient forces but was less susceptible to computational instability.[28] In 1949, his staff tested their single-layer model on the ENIAC, which had been moved to Aberdeen Proving Ground, in Maryland. According to a paper published the following year by Charney, von Neumann, and visiting Norwegian meteorologist Ragnar Fjörtoft, the preliminary model had succeeded—sort of—where Richardson had failed: eight rounds of calculations carried out for a time interval of 3 hours and a 15 x 18 grid with cells 736 km (457 miles) on a side yielded plausible, if not wholly accurate, 24-hour forecasts.[29] Although computation consumed 24 hours, mostly for handling thousands of punched cards, the best of four trial projections, shown in figure 5.8, forecast a pressure surface (left) generally similar to actual conditions (right). With an estimated computation time of only a half hour on the institute's new machine, Charney and his colleagues had "reason to hope that Richardson's dream of advancing the computation faster than the weather may soon be realized."

Richardson's dream was quickly fulfilled. Freed from intermediate punch-card storage, the institute computer ran the one-layer model in a mere five minutes, not including the time to compile and enter the data and prepare the initial analysis. To better represent vertical movement and the development of cyclones, the Princeton team devised six new models, some with two layers and others with three, and successfully predicted (retrospectively, of course) the sud-

Forecast surface   Observed surface

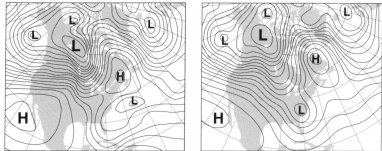

Fig. 5.8. Comparison of the observed 500-mb pressure surface (right) for February 1, 1949, with the 24-hour computer projection (left) of the previous day's map. Lines describe the height above sea level of the 500-mb pressure surface, where barometric pressure is a constant 500 millibars. An undulating three-dimensional surface with an elevation of roughly 18,000 feet, the 500-mb surface divides the atmosphere into two parts roughly equal in mass. (The lower half of the atmospheric mass extends downward to the ground, where air pressure is approximately 1,000 mb; the upper half extends skyward well over a hundred miles to the outer limit of the ionosphere, where pressure is 0 mb.) Represented cartographically by height contours, rather than isobars, the 500-mb surface is comparatively high over areas with high surface pressure—where the layer down to the ground is relatively thick, and hence heavier—and low over areas of lower surface pressure. Compiled from J. G. Charney, R. Fjörtoft, and J. von Neumann, "Numerical Integration of the Barotropic Vorticity Equation," *Tellus* 2 (1950): selected elements from fig. 4, p. 248.

den and severe 1950 Thanksgiving Day storm. Impressed with these results, the U.S. Weather Bureau and the navy and air force weather services formed the Joint Numerical Weather Prediction Unit in 1954 and bought an IBM 701 computer, similar in power to von Neumann's computer. Experimentation with the three-layer Princeton model was disappointing—according to Frederick Shuman, former director of the Weather Bureau's National Meteorology Center (NMC), it "provided little or no useful information to the forecaster"—but further refinement led in 1958 to an operational model that proved a timely and generally reliable forecasting tool, even though it treated the atmosphere as a single layer.[30]

The next three decades witnessed ever more powerful computers, more complex and representative atmospheric models, and impressive improvements in numerical forecasting.[31] This progress is well documented by the *skill score*, an index that the NMC devised in the

Skill score (percent) for 36-hour 500 mb forecast map, 1955–1992

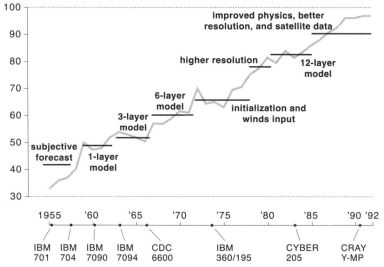

Fig. 5.9. Improvement in reliability of the 36-hour 500-mb prognostic chart, as measured by the National Meteorological Center's skill score, reflects advances in modeling and computing technology. Compiled from Ronald D. McPherson, "The National Centers for Environmental Prediction: Operational Climate, Ocean, and Weather Prediction for the Twenty-first Century," *Bulletin of the Meteorological Society of America* 75 (1994): 364; and Frederick G. Shuman, "History of Numerical Weather Prediction at the National Meteorological Center," *Weather and Forecasting* 4 (1989): 288.

early 1950s to evaluate the performance of forecast personnel.[32] Readily adapted to numerical forecasts, the skill score compares a forecaster's—or a computer model's—prognostic chart with the observed weather chart it was intended to predict. To address significant elements of the forecast—strong winds and large-scale cyclonic systems, in particular—the index focuses on pressure differences between points 270 to 525 miles apart within the forecast region and considers the difficulty posed by meteorological conditions as well as the forecaster's ability. As figure 5.9 describes for the 36-hour 500-mb pressure map of North America (the National Meteorological Center's first automated product), between 1955 and 1992 the skill level rose from less than 35 to over 95 percent. The sea-level prognostic chart has registered a comparable improvement, and three-day forecasts are now as accurate as 36-hour forecasts were just a few decades ago.[33]

Labels along the trend line and time axis in figure 5.9 underscore the computer's role as both a prerequisite and a stimulus. The larger memories and faster processors essential to an accurate representation of the atmosphere's vertical structure also supported a shorter time step, which in turn promoted the more reliable simulation of interactions among diverse physical processes. In the geographic realm, advances in electronic computing fostered both the finer grid needed for an accurate treatment of fronts and the broader, hemispheric grid basic to medium-range forecasting. In addition, high-speed computing accommodated rapid integration of data from ground stations, radiosondes, satellites, and other sources as well as a mathematical "initialization" stage to deal with inconsistencies in the data. That computers kept pace with much of what research meteorologists thought of adding made numerical modeling a focus for varied efforts in data gathering and experimentation, and led indirectly to richer data and a fuller understanding of atmospheric physics.[34]

The most valuable of these enhancements is the *global model*, which extends the forecast period out to five days or more by avoiding errors likely to creep inward from the edges of a less extensive grid.[35] For example, a model covering only North America and an adjoining strip of the Pacific Ocean might yield generally reliable 24-hour prognostic charts, but its 48- or 72-hour maps would not adequately reflect eastward-moving cyclones and anticyclones situated west of model's domain at the beginning of the forecast period. Because a high-resolution global forecast model can tax the most powerful of supercomputers, atmospheric scientists developed the composite, *nested-grid model*, which links a comparatively coarse hemispheric model to a fine-mesh regional model with cell centers 30 miles (50 km) or less apart. As described in figure 5.10, nesting not only captures smaller, "mesoscale" features missed by the hemispherical grid but accounts for large weather systems originating outside the finer "forecast grid."[36] Another promising nested-grid approach is the *floating-grid model* that forecasts the development and path of a northeaster or tropical hurricane by following the storm with a relatively fine but less extensive (and less computationally demanding) moving mesh.[37]

Advances in numerical modeling have given the weather forecaster a diverse array of tools, ranging from distinctly different models for short-range (0 to 3 days) and medium-range (1 to 10 days) prediction to long-range models for assessing climatic change and customized models for projecting storm tracks, seasonal snowfall, and growing-season precipitation.[38] Short- and medium-range models that project present conditions forward in time in steps of five

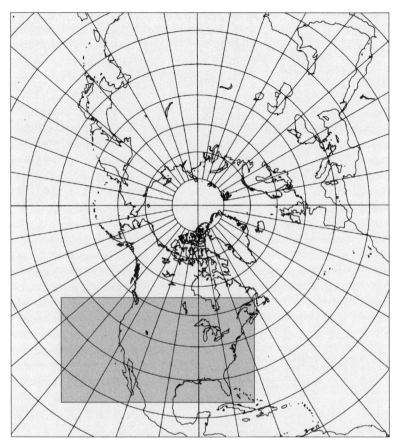

Fig. 5.10. A nested-grid model for North America would supplement a comparatively fine grid providing a more detailed treatment for the forecast area with a coarse hemispherical grid intended to dampen boundary errors certain to creep inward from outside the region and undermine forecasts looking ahead more than 24 or 48 hours.

minutes or less typically report predictions at one- to three-hour intervals. A meteorologist can not only check the forecast map for 4 P.M. tomorrow afternoon, for example, but gain insight into the model's behavior by viewing an animated sequence of hourly maps. In addition, visualization software and revved-up personal computers encourage forecasters to inspect prognostic charts for upper-air conditions at specific levels, examine maps of precipitable moisture and other specific factors, view cross sections along flight lines, and compare results from (and thus better understand) models based on dif-

ferent assumptions, different data, and different domains. Largely because of the computer and electronic graphics, weather forecasting
has evolved from a technology of making and interpreting four or
five maps of the recent past to the art of evaluating dozens of maps
and cross sections of the past, present, and near future.

Despite increased mathematical refinement and high-performance
computing, forecast models have a limited reach in both detail and
time. A comparatively precise regional model's 30-mile mesh is, after
all, too coarse to show, much less predict, individual thunderstorms,
tornadoes, and other local phenomena with strong vertical movement. And while models with a global domain can look 10 or more
days into the future, the typical medium-range forecast map is a relatively vague small-scale portrait showing where temperature or precipitation is likely to be above or below average over the next one to
four days. Among the many factors limiting the temporal reach of
forecast models are complex interactions among atmospheric variables; the model's inability to account for the effects of local terrain,
thunderstorms, and other smaller features; and imprecise data with
numerous tiny errors that grow into huge discrepancies when "prop-
agated" over time and space by the "nonlinear governing equations"
describing atmospheric processes.[39]

In the early 1960s, MIT meteorologist Edward Lorenz dampened
the optimism of fellow modelers with a critical look at how rapidly
errors compound. Although his experiments with a relatively unso-
phisticated forecast model suggested that errors doubled in roughly
four days—even more rapidly for more complex storms—a
markedly more sophisticated model decreased the error-propaga-
tion rate by only a day. "If it really requires as long as five days for
typical errors to double," Lorenz wrote, "moderately good forecasts
as much as two weeks in advance might some day become a reality.
If it requires no longer than five days for errors to double, accurate
detailed forecasts for a particular day a month or more in advance
belong in the realm of science fiction."[40] Exploration of the unpredictability of complex phenomena like the weather led to the development of *chaos theory*, a new branch of mathematical physics
concerned with the nondeterministic behavior of natural systems.[41]
In addition to stimulating the discovery of chaos, Lorenz contributed the "Butterfly Effect," a vivid, widely cited image of unpredictability concisely described in the subtitle of his 1979 address to
the American Association for the Advancement of Science: "Does

the Flap of a Butterfly's Wings in Brazil Set Off a Tornado in Texas?"[42]

Lorenz's butterfly is flapping its wings a lot these days at the Environmental Modeling Center (EMC), one of nine units of the National Centers for Environmental Prediction, formed in 1995, when the National Oceanic and Atmospheric Administration (NOAA) reorganized and renamed the National Meteorological Center.[43] EMC researchers have, in effect, embedded the Butterfly Effect within their numerical predictions through a new approach called ensemble forecasting.[44] An ensemble is a set of forecasts, each based on the same initial conditions but injected with a small but somewhat different computer-generated error, or "perturbation." How widely the resulting maps vary allows an assessment of the stability of a composite or dominant forecast. If the forecasts are generally similar, the forecaster can be relatively confident of their reliability. If not, the ensemble at least warns of chaotic conditions requiring extra careful monitoring. Recent experiments have demonstrated skill over a forecast interval as long as 14 days. Although some weather situations do not converge to a single, presumably trustworthy prediction, EMC director Steve Lord is convinced "we're knocking at the door" of Lorenz's two-week forecast.[45]

Downwind Dangers

Weather forecasters are not the only atmospheric cartographers who covet a crystal ball: air quality and emergency management officials responsible for regulating airborne hazards also rely heavily on cartomathematical models. But instead of simulating upper-air circulation and transfers of energy, they treat the atmosphere as a purveyor of poisons. Largely unconcerned with next week's or tomorrow's weather, hazards experts deal in forecasts covering decades, not days, for worst-case events everyone hopes never happen. Although we occasionally see their hand in maps warning of respiratory hazards, to view their most important work one must read environmental impact statements and attend pubic hearings.

In departing temporarily (and somewhat abruptly) from an examination of weather charts, this chapter explores the use of computer models to extrapolate broad trends from limited data and pose as well as answer iffy questions about toxic chemicals stored in tanks and transported by truck or train. Grounded in atmospheric physics, air-quality maps and toxic-plume models afford rational assessments of risk and vulnerability as well as visually forceful arguments for or against environmental regulation. Because they affect locational decisions, tax rates, and utility bills, air-quality maps are inherently more controversial than conventional maps of the atmosphere. However peripheral to the meteorologist's concept of weather and climate, air-quality charts are an essential element of any socially relevant discussion of atmospheric cartography.[1]

The chapter focuses on point sources like factories, power plants, and municipal incinerators, for which air-quality models generate large-scale maps of small areas. By contrast, non-point-source models representing vehicular traffic or the combined effects of numerous factories or power plants yield less geographically detailed portraits encompassing a metropolitan area, region, or continent.[2] Although mixing and dilution are key elements in modeling contamination from both single and multiple sources, a map focused on one location can either present the model plume for a particular hypothetical accident under specific meteorological conditions or describe composite, average vulnerability for representative variations in local weather.

Making plume maps for hypothetical accidents is an effective way to appreciate the hazardousness of a local farm store or railway line. So says the Citizens' Environmental Coalition, an Upstate New York group that published an informative booklet titled *How to Create a Toxic Plume Map*.[3] Straightforward instructions tell readers how they can construct a circular vulnerability zone around a potentially dangerous facility by overlaying on a large-scale topographic or street map the standard plume developed for chlorine by the National Transportation Safety Board using the U.S. Coast Guard's Hazards Assessment Computer System (HACS). Shown in figure 6.1 at half its original size, the graph describes the area engulfed by toxic concentrations of chlorine from a punctured railroad tank car. As the diagram's times and distances indicate, the ever widening plume moves rapidly, covering a third of a mile in a minute and a half, and carrying the deadly gas slightly more than 2 miles in less than 11 minutes. The dashed line beyond 10,700 feet signifies the difficulty of predicting the effects of turbulence and terrain on the plume's shape and extent over long distances—in this scenario, toxic concentrations of chlorine affect areas nearly 23 miles away—as well as the additional time for warning people in the path of a somewhat diluted toxic cloud. In describing the impact of a very bad accident or terrorist attack, the model provides, as the coalition's booklet notes, "a very useful teaching device."[4]

The graph adapts readily to local hazards. Plotted at 1:24,000, the standard scale of large-scale base maps, the model plume "can be photocopied onto a transparency sheet in a standard copy machine [and then] laid over a standard 7.5-minute community topographic (quadrant) map from the U.S. Geological Survey."[5] The reader can then "place the release point over the source of the chemical, and rotate the model to show a full vulnerability zone." Because winds vary,

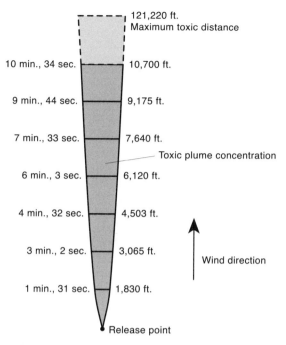

Fig. 6.1. Model toxic plume for chlorine describes times and distances from release point for concentrations of 100 parts per million or higher. Redrawn at half the original scale from diagram in Meaghan Boice, *How to Create a Toxic Plume Map*, CEC Fact Sheet no. 2 (Albany, N.Y.: Citizens' Environmental Coalition, 1992), 22.

the coalition recommends using the distances in the model plume as radii for circles describing the threat for all possible wind directions around the industrial plant, railway siding, grade crossing, or other plausible accident site. Concentric circles show relative risk and identify areas where residents have little or no time to close windows or evacuate.

As the coalition's booklet readily concedes, circular vulnerability zones are entirely hypothetical. The hypothesis for the model plume in figure 6.1 is a 12-inch-diameter puncture in a tank car containing 17,100 gallons of chlorine—a typical load in transit or parked at a factory. Meteorological conditions contributing to this worse-case scenario are stable air, a temperature of 36°F, and a wind speed of 11.5 mph. A stable atmosphere, in which warm air overlies cooler air, keeps the heavier-than-air chlorine near the ground, and strong, steady (but not highly turbulent) winds assure a long, thin plume and a large vulnerability zone.

Different chemicals and weather conditions generate surface plumes, or *footprints*, varying in shape and length. A relatively dense and highly toxic gas like chlorine—its use in World War I made trench warfare a deadly game of waiting and watching wind direction—travels far under a steady wind and has a wide vulnerability zone, whereas an equally deadly but buoyant gas like ammonia more readily rises and mixes with the surrounding air, and thus endangers a smaller neighborhood. The amount of material available for release is also highly relevant, as are the type of container and its location.

Local emergency management officials now have detailed inventories of dangerous chemicals stored nearby. After more than 2,200 people perished in 1984 following the accidental release of 40 tons of methyl isocyanate from a Union Carbide fertilizer plant in Bhopal, India, Congress passed the Emergency Planning and Community Right-to-Know Act (EPCRA), called SARA Title III for short.[6] (The law is Title III of the Superfund Amendments and Reauthorization Act, or SARA.) EPCRA not only requires users of hazardous chemicals to file yearly reports with states and the EPA but makes the information available to ordinary citizens and community groups. *How to Create a Toxic Plume Map* describes how Title III data and simple graphic plumes can help nearby residents assess their vulnerability.

Instead of constructing plume maps by hand, emergency response officials in state and local government rely on user-friendly, menu-driven software to link SARA Title III information and meteorological data with toxicity databases, electronic maps, and mathematical models.[7] Two types of models are required: a source model that describes the substance, its storage vessel, and the circumstances of its release, and a dispersion model that estimates concentration at points downwind for specific times since the release began.[8] The source model reckons the state (liquid or gas), concentration, and release rate of the escaping fluid from the size and elevation of the tank, the amount in storage, the size and position of the opening, and temperatures inside and outside the tank—whether a gas escapes slowly or rapidly is important, as is the likely rate at which a refrigerated liquid is likely to boil to a vapor. (Source models can also describe leaks from a pipeline or processing plant as well as fumes or combustion products from volatile liquids.) The dispersion model then simulates dilution and transport by the atmosphere,

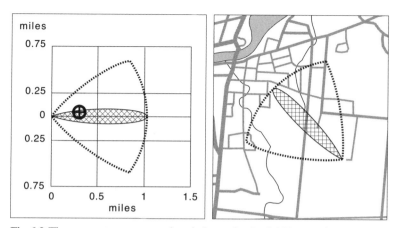

Fig. 6.2. The computer-generated toxic footprint for 2,000 pounds of chlorine escaping in one minute in a 4-mph wind under stable nighttime conditions (left) can be superimposed on a hazardous site (right) to identify the area at risk. The cursor (left) lets an operator assess toxicity at selected points. Model plume courtesy New York State Emergency Management Office.

under specific assumptions about wind speed, air temperature, and other meteorological factors.

A few simple graphics illustrate the process. The user describes a hypothetical release, and the computer plots a plume like the cigar-shaped footprint in figure 6.2 (left). In this example, 2,000 pounds of chlorine escaping in one minute from a source 10 feet above the ground into stable air moving at 4 mph over level terrain in open country yields a narrow plume slightly longer than a mile.[9] This footprint represents concentrations of 30 ppm (parts per million) or more, which the EPA considers "immediately dangerous to life and health."[10] A wider vulnerability zone surrounding the footprint allows for a 30° wind shift. Superimposing the model plume on a map (figure 6.2, right) dramatizes the risk to a hazardous facility's neighbors, especially schools, hospitals, nursing homes, and other "special needs" populations requiring prompt notification and perhaps evacuation.[11]

In addition to plotting plume footprints, interactive emergency response software can estimate outdoor and indoor concentrations at specific sites indicated by the user. For example, the outdoor concentration at the cursor position in figure 6.2, representing a point 528 yards downwind and 88 yards above the plume's centerline, is 81.8 ppm. But inside a single-storied building with 0.44 air exchanges per hour the concentration here is only 0.775 ppm—well below the dan-

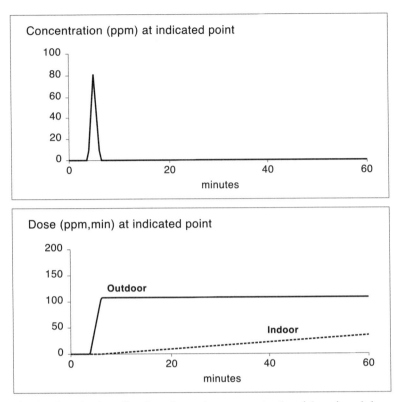

Fig. 6.3. Graphs describe time-dependent concentration (above) and dose (below) at the point identified by the cursor in fig. 6.2. Interactive plume-modeling software helps emergency management officials assess risk at selected locations within the vulnerability zone. Courtesy New York State Emergency Management Office.

ger level. In many cases, sheltering inside a well-insulated structure with the windows closed is safer than trying to evacuate.

The effectiveness of sheltering depends, of course, on duration and release rate as well as distance from the source. Graphs of time-dependent measures such as concentration and dose are especially useful in developing flexible emergency response plans that include both sheltering and evacuation. For the point at the cursor in figure 6.2, the upper chart in figure 6.3 shows that the toxic cloud from a one-minute release of chlorine travels 530 yards in less than five minutes: outdoor concentration rises rapidly to the 81.8 ppm maximum and then drops quickly to a safe level. The lower graph, which describes a sustained release at the same rate, suggests that sheltering can give potential victims three-quarters of an hour of breathing room—long enough for officials to shut off the source or arrange an evacuation. In a chemical or radiological emergency, a competent air-dispersion model can help officials minimize the dose from a passing plume.[12]

The thought of being caught outdoors or at home asleep with the windows wide open should be daunting for nearby residents. Even so, those living near a hazardous facility might welcome mitigation measures readily identified through experimentation with the software. For example, reducing the maximum amount of chlorine stored at a municipal water treatment plant from 800 to 500 pounds can cut the radius of the zone of vulnerability from 10 miles to 1 mile.[13] An EPA analysis of a hypothetical community showed how this 38 percent reduction in tank size decreased the population within the vulnerable zone from 125,000 to 1,250. Other uses of dispersion modeling as an exploratory tool have demonstrated the wisdom of using less hazardous materials in the industrial process, installing a more reliable monitoring network and warning system, or relocating the plant to a less densely populated area.

Vulnerability modeling is not foolproof. Emergency managers need to experiment with different models and understand their simplifying assumptions, which usually ignore the influence of terrain on toxic clouds of heavy gasses like chlorine and phosgene. They should also be wary of what the models might not tell them.[14] Especially disconcerting was the 1993 revelation that EPA data distributed to local emergency planning councils overlooked possible reactions with the atmosphere.[15] For example, the 0.6-mile vulnerability radius for fluorine escaping at 10 pounds per minute in stable air with a 3.4 mph wind expands to 9 miles when the gas reacts with water vapor in the air to produce hydrogen fluoride. And boron trisulfide, a compound not on the list of extremely hazardous substances, reacts with atmospheric moisture to produce hydrogen sulfide, which under similar conditions has a toxic reach of half a mile.

Air-dispersion models play an equally important role in assessing the environmental impact of incinerators and other smokestack facilities. Because continued exposure to carcinogens and other hazardous substances can have cumulative effects as detrimental as a single exposure to a massive accidental toxic release, local planning commissions and state environmental protection agencies require every environmental impact statement to include an air-quality analysis. Key concerns are the amounts of toxic substances emitted, their dispersion throughout the region, and the likely effects on humans and other species.[16] Because terrain and local winds can yield radical geographic differences in exposure, air-dispersion models must address these effects directly.

Although a proposed plant's immediate neighbors are typically its greatest opponents, air pollution and visual pollution have very different geographies. However unsightly and detrimental to property values, a tall smokestack almost always pushes the more severe health impacts well beyond its immediate vicinity. In general, the higher off the ground the point of release the fuller the mixing of emissions and the lower the concentration of airborne toxins everywhere, nearby and far away. But because of predictable variations in local weather, on some days still air will trap emissions near the plant, and on others the plume will not rise. Moreover, a hillside several miles downwind can intercept pollutants from a plant far enough away for residents to ignore its visual impact.

To assess the effects of complex terrain and whimsical weather, planners and environmental engineers use hourly data on wind, temperature, and humidity to simulate the proposed facility's operation throughout a more or less typical year. The simulation will treat the region as a huge grid of half-mile-square cells extending ten miles or more outward from the plant—the grid's resolution and extent depend upon terrain as well as the height of the smokestack—and use one or more air-dispersion models to estimate fallout at each square for every hour of every day throughout the period.[17] In addition to compiling cumulative totals, which are used to make maps of average annual deposition, the simulation will keep track of highest and second-highest concentrations for 1-, 3-, 8-, and 24-hour periods. Several averaging periods are necessary because some toxins can be immediately deadly whereas others typically have a cumulative effect. For instance, the average concentration of carbon monoxide may not exceed 10,000 micrograms per cubic meter ($\mu g/m^3$) in 8 hours or 40,000 $\mu g/m^3$ in 1 hour, whereas sulfur dioxide has three danger levels: 80, 365, and 1,300 $\mu g/m^3$ for periods of one year, 24 hours, and 3 hours, respectively. The second-highest concentration is calculated because air-quality regulations permit one "exceedance" a year, but not two. Hence a city in which CO exceeds an hourly average of 40,000 $\mu g/m^3$ twice in a single year becomes a "nonattainment area," and therefore subject to special restrictions on development, traffic, and motor-vehicle fuels. Where air quality is already fragile, a relatively sophisticated simulation must assess the combined effect of existing pollution and the new facility's likely contribution.

Incinerators are especially troublesome because of uncertainty about the nature and amount of combustion products.[18] How effectively can the local solid-waste-disposal agency divert construction debris, batteries, and other objectionable materials from its burn plant's waste stream? What might happen if the waste received falls below the amount essential for efficient operation? (Recycling and waste reduction can undermine the economics of an incinerator costing over a hundred million dollars.) Will income from the sale of electricity encourage the agency to import additional municipal waste, or even industrial waste, from neighboring counties or a distant metropolis? (Modern incinerators are largely waste-to-energy plants, which appeal to both fiscal conservatives and energy conscious environmentalists.)

To address these doubts, environmental engineers generate dispersion maps based on "unitized concentrations," which can be applied to a variety of smokestack toxins. Figure 6.4, the annual-average map from the air-quality study for a waste-to-energy plant recently built several miles from where I live, illustrates this approach: isolines represent dispersed concentrations for a hypothetical substance leaving the stack at a rate of 1.15 grams per cubic meter per second—a 15 percent safety factor inflates the unit concentration of 1.0 g/m^3/sec.[19] East of the plant site (portrayed by the large dot southeast of Syracuse, just beyond the city border) the isoline labeled 0.04 represents a dilution of the 1.0 g/m^3/sec unitized emission to a mere 0.04 µg/m^3. (One microgram (µg) is one millionth of a gram.) Simple multiplication adapts this isoline to stack-emission rates for particular pollutants. If, for instance, the plant were likely to emit generic guck at 100 g/m^3/sec, the estimated concentration at places along the 0.04 isoline would be 4 µg/m^3 (100 × 0.04), whereas if the emission rate for carcinogenic crud were only 0.2 g/m^3/sec, the local concentration would be 0.008 µg/m^3 (0.2 x 0.04). Using the highest unitized concentration (0.097 a little over four miles east-southeast of the plant) and maximum plausible emission rates for various substances, the county's environmental consultants estimated that nitrogen dioxide from the new facility would at most reach 1.98 percent of the EPA's yearly maximum, with even lower levels for other pollutants with an annual-average nonattainment threshold.

Dispersion contour maps provide an insightful comparison of the four short-term time intervals. Whereas the highest annual-average concentrations occur about four miles east of the incinerator—light

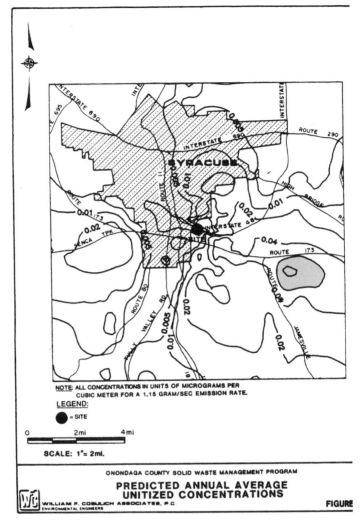

NOTE: ALL CONCENTRATIONS IN UNITS OF MICROGRAMS PER
      CUBIC METER FOR A 1.15 GRAM/SEC EMISSION RATE.

LEGEND:

● = SITE

0   2mi   4mi

SCALE: 1"= 2mi.

ONONDAGA COUNTY SOLID WASTE MANAGEMENT PROGRAM

**PREDICTED ANNUAL AVERAGE
UNITIZED CONCENTRATIONS**

WILLIAM F. COSULICH ASSOCIATES, P.C.
ENVIRONMENTAL ENGINEERS

FIGURE

Fig. 6.4. Unitized annual average concentration estimated for a municipal incinerator just south of Syracuse, New York. From County of Onondaga, New York, Solid Waste Management Program, *Waste-to-Energy Facility: Draft Supplemental Environmental Impact Statement* (June 1988), 5.12.

gray shading added to figure 6.4 highlights areas with unitized concentrations greater than 0.06 μg/m³—the plant's worst short-term effects are closer and more nearly to the south. Figure 6.5, on which a similar gray shading points out each map's worst concentrations, reveals a shift from south to east as the time period for the second-high-

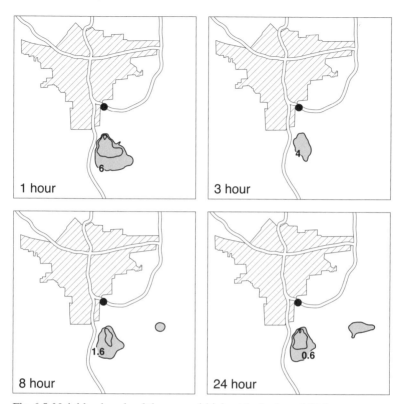

Fig. 6.5. Neighborhoods of the second-highest 1-, 3-, 8- and 24-hour concentrations, as simulated for the municipal incinerator southeast of Syracuse, New York. Compiled from County of Onondaga, New York, Solid Waste Management Program, *Waste-to-Energy Facility: Draft Supplemental Environmental Impact Statement* (June 1988), 5.13–5.16.

est concentration increases from 1 to 24 hours. Hills two to three miles south of the site receive the greatest 1-hour and 3-hour doses, but the maps for intervals of 8 and 24 hours show the emergence of high concentrations farther away to the east-southeast. This new outlier of relatively high concentration reflects the west-to-east air movement more readily apparent in the annual-average map.

The effects of wind and terrain are unmistakable. Instruments installed 300 feet above the plant site revealed winds principally from the north and west. Elevations high enough to catch pollution from the plant occur in isolated patches to the south, east, and west. Over comparatively long periods, according to the maps, the worst concentrations will occur about four miles east of the plant, past the county prison in a hilly area of large lots with expensive homes, many occupied by affluent physicians, a few of whom have become vocal skeptics of the plant. For shorter intervals, the worst concentrations will affect the lower, slightly closer hills south of the plant. But in no case, the maps declare, would the incinerator pose a threat to health. Moreover, the isolines promise an inconsequential added impact in the more heavily populated, low-lying city and suburban neighborhoods north and west of the plant—areas threatened far more by emissions from cars, trucks, and nearby factories.

Can residents and local officials trust the maps and the planners' conclusions? Federal and state regulations suggest they can. After all, the air-quality analysis was conducted according to EPA guidelines under state supervision by air-quality scientists at the New York State Department of Environmental Conservation, and the estimated risk proved minimal, not borderline. In one sense at least, the caveat "garbage in, garbage out" does not apply: environmental engineers supplemented data from the weather station north of the city by erecting a 300-foot instrument tower at the site and measuring wind direction, wind speed, temperature, and atmospheric stability every hour for a full year. Using these locally specific data, they simulated deposition hour by hour for a 12.4-by-12.4-mile (20-by-20-km) area centered on the plant under the worst-case assumption of continuous, 24-hour operation at full capacity: 990 tons of trash a day, 361,350 tons a year. They also supplemented a coarse screening grid of "receptor (sample) points" 1 km (0.62 mi) apart with two finer grids: a mesh of points 0.25 km (0.15 mi) apart extending 2 km (1.24 mi) north and 3 km (1.86 mi) south of the stack, and a finer grid of receptors 0.1 km (330 feet) apart for an area within 1 km (0.62 mi) of the stack. To determine accurately the points of maximum concentration, they not only applied the intermediate 0.25-km grid to areas of high impact identified with the coarse grid but used the fine 0.1-km grid to refine further the high-impact areas identified by the intermediate grid.[20] And to anticipate questions about the incinerator's impact on local schools, nursing homes, parks and other sensitive areas, the engineers modeled fallout at 38 additional points. Moreover, they estimated elevations conservatively—taking the highest point at a site, not the lowest—to increase rather than decrease the predicted concentration.

To adequately reflect local terrain, the engineers employed two air-diffusion models. The Industrial Source Complex (ISC) model used the hourly meteorological data to account for atmospheric mixing, the buoyant dispersion of emissions, the settling and dry deposition of very fine particles, and a turbulent effect called stack-tip downwash, which pushes the lower part of the plume below the top of the stack. A preferred EPA technique for estimating short-term dispersion in both urban and rural settings with more or less level terrain, the ISC model does not reliably predict ground-level concentration at elevations greater than the release height.[21] To compensate, the simulation also applied the EPA's COMPLEX I model at all receptors higher than the top of the stack and recorded whichever hourly estimate was greater.[22] A standard EPA strategy for dealing with "complex terrain," the COMPLEX I model better reflects the likely interception of pollutants at elevated receptors in the path of the rising plume.

How reliable are these models' predictions? Highly reliable if safety is more a concern than accuracy, according to a report by the U.S. General Accounting Office.[23] All but three of the 48 models used by the EPA in regulating air quality had been evaluated by comparing model predictions with measured concentrations. A few of the models tended to underpredict pollution levels, but the two used to simulate the county incinerator performed satisfactorily. Specifically, the ISC model estimated concentrations within 10 percent of observed levels—well within the 15 percent margin of error the engineers had allowed—whereas the COMPLEX I model introduced an added safety factor of sorts by overestimating observed concentrations by more than 800 percent. Such gross overestimation suggests, in retrospect, that the air-quality simulation played a largely rhetorical role in providing graphic and numerical evidence of the plant's safety. Why take our word for it, officials could say—look at the maps and data.

But how can two models yield such radically diverse predictions? Part of the explanation lies in the fact that the ISC and COMPLEX I models are both Gaussian plume models, based on the symmetrical, bell-shaped curve devised by Karl Friedrich Gauss (1777–1855). Widely used in science and engineering, Gaussian curves conveniently describe sets of numbers for which values near the mean (average value) are comparatively common and values progressively farther from the mean are increasingly less frequent. In air-dispersion modeling, these numbers are emissions concentrations, which are greatest along the centerline of the plume and spread out laterally, and to a lesser extent vertically, with increasing distance downwind, as figure 6.6 describes for an ideal plume rising to altitude H from a stack of height h. Dispersion ellipses encompassing 99 percent of all guck

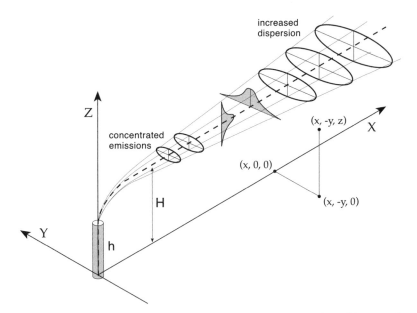

Fig. 6.6. Gaussian plume models generate families of dispersion ellipses and Gaussian curves for estimating emissions concentration at downwind locations.

grow progressively broader and thicker as density decreases with increasing distance downwind. As the axes of the ellipses suggest, pollution concentration is described by two bell curves, one vertical and the other horizontal. Using various physical constants and empirical coefficients, the model customizes these two curves at downwind distance x for a given wind speed and level of atmospheric stability, so that the computer can then estimate concentration at distance y outward from the center line and elevation z above the stack base. The diagram's smooth lines and symmetric curves suggest that a Gaussian plume model, however sophisticated its mathematics, is at best a pragmatic simplification of closely linked atmospheric processes interacting with conveniently ignored local conditions. Applied to simpler situations, with better behaved wind fields, the ISC model is inherently more reliable than the COMPLEX I model, which attempts to account for complex terrain with only a vague description of local relief: the elevations at receptor points. The result is bigger, broader ellipses, predicting higher, wildly overstated concentrations.

Not surprisingly, more precise data can produce more accurate results. For example, an air force study demonstrated that even a crude map of elevation, slope, and surface roughness helped a complex ter-

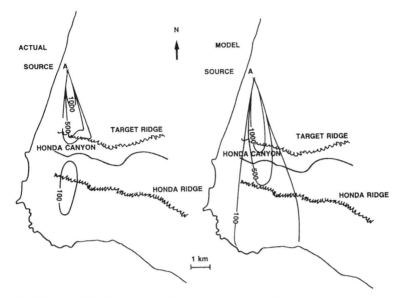

Fig. 6.7. Actual (left) and model (right) plumes describe the early evening release of a trace gas to a northerly wind under stable conditions near Vandenberg Air Force Base. From Bruce A. Kunkel and Yutaka Izumi, *WADOCT: An Atmospheric Dispersion Model for Complex Terrain*, Environmental Research Paper no. 1062 (Hanscom AFB, Mass.: U.S. Air Force Geophysics Laboratory, 1990), 34.

rain model produce a useful, generally reliable prediction of concentrations of inert tracer gasses released in hilly terrain near Vandenberg Air Force Base, in California.[24] In one series of tests, the model was provided a 56-by-61 array of terrain data, with elevations sampled at grid intersections 200 m (550 ft) apart. Figure 6.7, a representative comparison of actual and model plumes, shows that while the model overestimated the length and width of plume footprints, it predicted the rightward veering of the centerline. Even so, discrepancies between actual and model plumes hint strongly that nonlinear near-ground turbulence and wind shear will thwart precise predictions of downwind drift over rough terrain in the same way that the chaotic global atmosphere hobbles long-range weather forecasting.[25]

As weather maps have helped reveal the workings of air masses and global circulation, plume maps and air-quality models can clarify the consequences of public decisions about where and how we generate power, manufacture chemicals, and incinerate waste. In both cases a coherent series of cartographic images offers invaluable insights to process and possibilities. Weather, of course, acts as its own mapping engine, generating a continual stream of informative graphics. Regulatory decisions, by contrast, must rely on the eagerness of engineers, public officials, and concerned citizens to consider numerous designs and locations and thereby use air-dispersion modeling as an exploratory tool, not just as a means of conjuring up a few maps to appease clean-air statutes. However imperfect, air-dispersion modeling can be a highly effective tool if (and this is always a big *if*) the user understands how to use it and how well it performs. Atmospheric scientists need not only to develop more sophisticated models but also to improve their methods for both assessing and communicating a model's expected performance.[26] After all, a flawed forecast based on a poorly understood environmental-impact or emergency-response model can have consequences far more tragic than an inaccurate five-day weather outlook.

7

Looking Down

One of the most vivid memories of my adolescence is standing with neighbors in front of our house in Baltimore on a warm summer night in 1958, looking upward in hope of spotting Russia's 3,000-pound satellite Sputnik III. The previous year the Soviets had taken an early lead in the space race by placing Sputnik I in an elliptical path roughly 500 miles above the earth. But even though America saved face four months later by successfully launching the 18-pound Explorer I into a similar orbit, our Cold War rival clearly had the more powerful rockets. This memory survives because the moving speck of light meant more than a blow to national pride: the oceans that once made isolationism an option offered no protection against intercontinental missiles with atomic warheads. Without massive investment in military technology and science education, defense experts argued, the United States might lose much more than the race to the moon. Perhaps they were right, but it's plain today that the principal beneficiaries of the congressional largess that followed have been telecommunications, mapping, and meteorology.

While Russia held the lead in lofting large payloads, our newly established National Aeronautics and Space Administration (NASA) scored a significant technical triumph on April 1, 1960, by launching TIROS I, the world's first weather satellite. The first of 10 Television and Infra-Red Observation Satellites, the 263-pound TIROS I was still very much an experiment.[1] Placed in a nearly circular orbit 450 miles above the earth, the satellite carried two miniature black-and-white TV cameras designed for mapping cloud cover. While a wide-angle camera with a 104° field of view recorded scenes over a thousand miles wide, an otherwise identical system with a 12° lens snapped more detailed shots of a smaller area at the center. Unlike commercial television cameras, which capture talking heads and running bodies with 30 pictures a second, the slow-scan TIROS cameras grabbed a new image only once every 10 to 30 seconds, as directed from below. Within range of a ground station, the satellite transmitted its pictures immediately. Out of range, TIROS saved the images on a tape recorder, which replayed the scenes for broadcast on a later orbit. Traveling at 17,000 miles an hour, the satellite circled the earth every 99 minutes. During the 77 days its cameras and relay systems performed properly, TIROS I took 22,952 cloud snapshots, of which 19,389 were deemed useful. Relayed to the Goddard Space Flight Center near Washington, the images were recorded on 35mm film for visual analysis.

NASA engineers could tell their cameras when to shoot but had little control over where the lenses pointed. To avoid a wobble that would have made picture-taking unpredictable as well as troublesome, TIROS was a spin-stabilized satellite: a cylindrical hatbox 19 inches high and 42 inches in diameter spinning like a top 10 times a minute about an axis aimed at a fixed point in space. Powered with energy collected by solar panels that coated the cylinder, the satellite passed over the illuminated side of the globe with its cameras pointed downward—more or less—as described in figure 7.1.[2] Circling the more slowly rotating earth in an orbit inclined at 48° to the equator, TIROS I traced overlapping zigzag ground tracks within a belt between 48°N and 48°S. Commands from the ground turned off the cameras when the lenses pointed outward over the unilluminated hemisphere. But with cameras aligned to the satellite's spin axis, many shots were low-angle oblique views, dominated by a visible horizon and copious sky.

Because of their awkward perspective and modest resolution, TIROS snapshots were not easily related to the viewer's mental map. To make the pictures more readily meaningful, NASA added grid lines, coastlines, and political boundaries.[3] By integrating radio observations of the satellite's tracking beacon with the map coordinates of identifiable shoreline and terrain features, mathematicians reconstructed the relative geometries of earth, satellite, and image. The precise time at which a scene was recorded yielded a location and camera orientation, from which a computer program estimated the position and shape of meridians and parallels. Plotted by hand on an overlay, the spherical grid provided a geographic framework for transferring features by eye from a conventional map. However in-

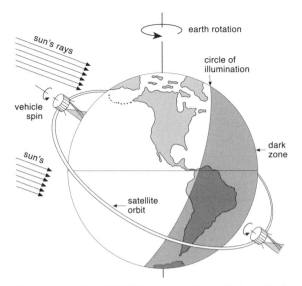

Fig. 7.1. The spin-stabilized TIROS I circled the earth in an inclined orbit. Although its cameras pointed generally downward when the satellite was over the illuminated side of the earth, some images consisted largely of the sky above the horizon. Redrawn from Goddard Space Flight Center, *Final Report on the TIROS I Meteorological Satellite System*, NASA Technical Report R-131 (Washington, D.C.: National Aeronautics and Space Administration, 1962), 20.

formative, postmission cartography was slow and uncertain, and until the space agency refined and automated the process later in the decade, cloud snapshots had little influence on weather forecasts.

Despite the lack of a ready grid, TIROS I demonstrated the enormous meteorological value of satellite observation. Cartographically enhanced images like figure 7.2 not only presented dramatic evidence of cyclonic circulation but raised absorbing questions about the vertical structure of storms. Recorded over the eastern Pacific between Hawaii and California, this snapshot revealed alternating bands of cloudiness (bright areas) and cloudlessness spiraling inward toward the center of the storm. Although the inner part of the spiral confirmed the classic cyclonic pattern described a half century earlier by Vilhelm Bjerknes, Weather Bureau researchers who compared satellite images with conventional weather charts noticed that several hundred miles out from the center of low pressure the bands were nearly perpendicular to wind direction, surface-level isobars, and contours of the 700-mb pressure surface.[4] Could this apparent anomaly reflect a secondary cold front that had passed through the

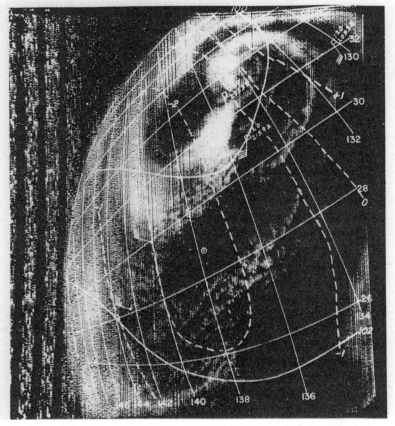

Fig. 7.2. TIROS I photo showing a spiral of clouds around a low-pressure center in the Pacific Ocean. Recorded at 0000 Greenwich Mean Time (7 P.M. eastern time) on April 6, 1960, the image was enhanced with grid lines, point symbols showing wind direction and speed, 200-foot contours (irregular solid lines labeled 100 at the top and 102 at the lower right) for the 700-mb pressure surface, and isolines (dashed lines with labels between +1 and -2) describing estimated vertical motion at 600 mb in cm/sec. From Jay S. Winston, "Satellite Pictures of a Cut-Off Cyclone over the Eastern Pacific," *Monthly Weather Review* 88 (1960): 314.

area 24 hours earlier, they wondered? Equally intriguing was the discrepancy between the marked spiral and the simplistic pattern of vertical motion estimated by a two-layer numerical forecast model. Might not a more accurate representation of vertical flow significantly bolster the predictive power of computer forecast models?

Officials were thrilled with the prospect of watching the weather from above. Hours after the launch, at a meeting of the American

Meteorological Society, NASA meteorological director Morris Tepper predicted that April 1, 1960, would "go down in meteorological history as one of its more important memorable days."[5] Equally enthusiastic was Weather Bureau chief Francis Reichelderfer, who averred that the "spectacular operation of TIROS I had opened a new era in weather surveillance."[6] And to Weather Bureau scientist Harry Wexler, who had raised the idea of operational weather satellites in a 1954 address to the British Interplanetary Society, TIROS meant going "from rags to riches overnight."[7]

Political leaders were impressed as well. Shown a sampling of cloud snapshots seven hours after the launch, President Eisenhower labeled the satellite a "marvelous development."[8] And Senator Lyndon Johnson called TIROS "the best space news we have had in a long time."[9] The news media also expressed wonderment and admiration. The day after the launch, the *New York Times* ran a pair of cloud photos and mused that for meteorologists TIROS "held some of the promise that the discovery of the telescope must have held for astronomers in the seventeenth century."[10] The following morning a *Times* editorial contrasted the satellite with previous NASA missions, which had focused on space exploration, and praised TIROS as "the forerunner of objects in space whose chief purpose is to add a new dimension to our abilities to do things on this earth we inhabit."[11]

An eye in space was only part of the promise of TIROS. Just as valuable were cameras that could extend human vision by operating in the dark and penetrating clouds. To understand these possibilities, we must recognize that the human eye is sensitive to only a very narrow range of wavelengths near the middle of the electromagnetic spectrum. As shown in figure 7.3, the spectrum runs from the ultrashort wavelengths of cosmic and gamma rays to the enormous wavelengths of radio and television. The unit of measurement here is the *micron*, or micrometer (μm): one millionth of a meter. Sandwiched between ultraviolet radiation, responsible for suntan and melanoma, and reflected infrared light, advantageous in mapping vegetation, is the visible band, consisting of wavelengths between roughly 0.4 and 0.7 μm. Visible light is useful in meteorology because dense clouds, which reflect visible light, appear bright to both the eye and a black-and-white TV camera, as in figure 7.2. But equally revealing images exist in other parts of the spectrum, particularly in the thermal infrared band, which is influenced by differences in moisture, heat loss, cloud height, and carbon dioxide.

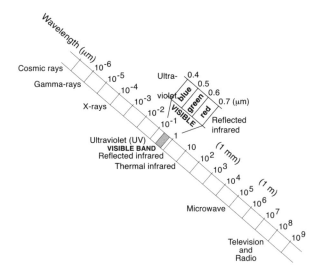

Fig. 7.3. The visible band is a small portion of the electromagnetic spectrum, which includes the thermal infrared band in which atmospheric gasses and the earth's surface reradiate solar energy.

Differences in intensity within a spectral band can be highly informative. Consider, for instance, the visible band, which color film and color television treat as separate red, green, and blue bands, described in figure 7.3. Healthy plants, which look green because chlorophyll in their leaves absorbs proportionately greater amounts of red and blue light, have a higher-than-average reflectance in the green band and thus would appear comparatively bright on film sensitive only to green light. Although our eyes don't notice, vegetation produces highly pronounced images in the near-infrared band, just beyond visible red, where green leaves are even more reflective. During World War II, military intelligence experts took advantage of this phenomenon to differentiate trees from camouflaged tanks and trucks. Although vehicles and structures painted green could fool both the naked eye and conventional color film, only real vegetation registered its diagnostic bright image on infrared film. This principle led to color infrared imagery, also called *false-color* imagery, which has greatly simplified land-cover mapping in agriculture and forestry with satellite pictures showing vegetation in bright red.

Farther along the spectrum, the thermal infrared band held similar promise in meteorology. It is here that atmospheric gasses and the earth's surface reradiate into space most of the incoming solar radiation. To an electronic infrared sensor called a *radiometer*, the atmosphere looks like a giant self-luminous light bulb, with higher emissions where the air is warmer or more dense. But within the thermal band individual gasses absorb and emit radiation at different preferred wavelengths. The result of this selective absorption and emission is a *spectral signature* that asserts the gas's relative abundance and warmth. Because each gas has a unique spectral signature, scientists design radiometers to measure energy in carefully chosen, comparatively narrow sub-bands called *channels*.

TIROS II, launched on November 23, 1960, carried a five-channel infrared radiometer in place of the narrow-angle TV camera. Each channel captured a different image of the same scene: an image that was bright where radiant energy within the channel was relatively intense and dark where energy was weak or absent. Channel 1 probed temperature as well as the top of the water vapor layer by measuring wavelengths between 5.9 and 6.7 μm, where water molecules absorb much if not all radiant energy from the earth's surface. Where clouds were present, a bright zone on the channel 1 image signified a thin vapor layer with a low top, whereas a dark region revealed a comparatively thick vapor layer extending well above 25,000 feet. But if the sky was clear, the infrared image described surface temperature, which was high in the bright areas and low in the dark. Although accurate interpretation depended upon weather stations or the wide-angle TV picture to indicate the presence or absence of clouds, channel 1 provided valuable information on the amount of moisture available for rain or snow.

Channel 2 also played a dual role. Sensitive to radiation between 8 and 12 μm, where the atmosphere is comparatively transparent, these images not only indicated temperature at the surface or the tops of clouds but also detected the presence of clouds at night. In contrast, channel 3, which responded to ultraviolet, visible, and infrared radiation in the 0.2–5 μm range, measured reflected sunshine. Meteorologists and climatologists use these measurements in estimating the earth's *albedo*, the proportion of incoming solar energy reflected directly back into space. Varying from place to place and hour to hour, the albedo is important in studying transfers of energy from the tropics, which receive more energy than they reflect or reradiate, to polar areas, which lose more energy than they receive. Energy imbalances underlie the development and movement of air masses and fronts, which affect winds and precipitation. Channel 4, 7–30+ μm, also contributed to energy balance calculations by measuring the total amount of long-wave energy reradiated to space—expelled heat picturesquely described as the earth's exhaust.[12] Channel

5, the narrowest channel, focused on 0.5–0.7 µm, the green and red parts of the visible band, and tested the scanner's potential as an alternative to the TV camera.

Although TIROS II's television camera and channel 5 radiometer were similar in spectral sensitivity, they differed substantially in both operation and image geometry. As implied by the word "camera," the TV system shared key characteristics with the photographic cameras we use to record birthdays and family vacations: a lens focuses the image on a flat plane, all parts of which are exposed simultaneously when we snap the shutter. Because TIROS was spinning in space, its camera avoided blurred pictures by opening the aperture for a mere 1.5 milliseconds. And because parachute recovery was impractical, NASA's weather eye recorded cloud snapshots not on film but on a "sticky" photoconductive material that held an electronic charge for the two seconds required to scan the image electronically, line by line, and transfer the scene to tape.[13] By contrast, the infrared radiometer did not capture large scenes as discrete frames but scanned the earth and atmosphere below continuously, measuring and recording radiation in sequence for a string of adjacent squares, or ground spots, arranged in scan lines perpendicular to the ground track, as portrayed schematically in the left part of figure 7.4.[14] The picture thus grew incrementally, scan line by scan line, as the satellite circled the earth. Because a scanning radiometer is analogous to a person advancing steadily along a path while sweeping back and forth with a broom, the length of a scan line is called its *sweep*.

To compensate for a spin axis aimed at a fixed point in space, rather than directly downward at all times, the TIROS II scanner looked outward 45° away from its axis in opposite directions, one tilted forward and the other backward. As described by the thick curved lines in the right part of figure 7.4, the resulting scan lines varied widely in sweep, shape, and coverage. When the spin axis pointed directly downward, toward the earth's center, the dual scans defined adjoining half circles. At this position the squarish ground spot was roughly 40 miles on a side directly below the satellite but progressively larger with increasing distance from the ground track. Farther along the orbit, the scan lines separated into distinctly different parabolas with ever broader sweeps, as first one and then the other lifted away from the surface toward space. And as the spin axis pointed farther away from the earth's center, the ground spots grew ever larger and less precise. Even so, the space agency considered its five-channel scanner a "medium-resolution" radiometer in contrast to TIROS II's other infrared imaging system, a nonscanning two-

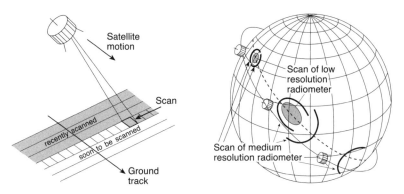

Fig. 7.4. Schematic representation (left) of the scan lines, sweep direction, and ground spots of the TIROS II medium-resolution infrared radiometer. The pictorial diagram (right) showing the satellite in three positions along its orbit describes the varying shape and extent of the infrared radiometer's scan lines (thick lines) and the low-resolution radiometer's ground spot (in gray). Right-hand portion adapted from W. R. Bandeen and others, "Infrared and Reflected Solar Radiation Measurements from the TIROS II Meteorological Satellite," *Journal of Geophysical Research* 66 (1961): fig. 2 on p. 3171.

channel "low-resolution" radiometer, which merely looked down along the spin axis to measure reflected energy for a single, much wider ground spot, shown in gray in figure 7.4. Like the sweep of the medium-resolution scanner, the huge ground spot of the low-resolution radiometer varied widely in size and shape.

To examine the reliability and potential of satellite radiometry, NASA scientists compared measurements from TIROS II's first few orbits with weather maps based on traditional data. Particularly gratifying was the visual correlation in figure 7.5, which superposed channel 2 data from orbits 3 and 4 onto a cartographic snapshot of cloud cover and fronts. The map shows surface weather for 1 P.M. eastern time, about an hour and a half after the satellite's third crossing of the continent but only a few minutes before its fourth overflight. For each orbit, the plot represents selected scan lines with three dots, one near the ground track and the others 10° to the side. Next to each dot an estimate of absolute "blackbody temperature" in degrees Kelvin describes heat loss from the surface.[15] As expected, heat loss is highest over clear areas, like the Great Lakes region in this scene, where infrared energy escaped the surface without interference from clouds. In contrast, low temperatures reveal the presence of clouds, which absorb energy in the 8–12 μm range and reradiate at other wavelengths in all directions, back to the surface as well as upward into space. Iron-

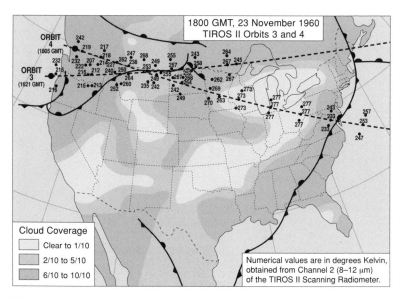

Fig. 7.5. Blackbody temperatures measured by channel 2 of the TIROS II medium-resolution radiometer reflected cloud cover and surface fronts. Redrawn from W. R. Bandeen and others, "Infrared and Reflected Solar Radiation Measurements from the TIROS II Meteorological Satellite," *Journal of Geophysical Research* 66 (1961): fig. 18 on p. 3182.

ically, surface temperature typically is higher in areas of lower blackbody temperature because the clouds act like blankets to trap heat. Because temperature contrasts between clear and cloudless areas were especially prominent at night, channel 2 radiometry gave forecasters a valuable new tool for tracking storms and air masses.

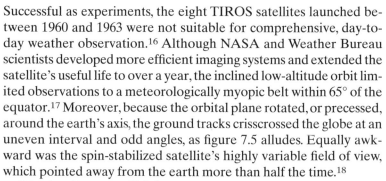

Successful as experiments, the eight TIROS satellites launched between 1960 and 1963 were not suitable for comprehensive, day-to-day weather observation.[16] Although NASA and Weather Bureau scientists developed more efficient imaging systems and extended the satellite's useful life to over a year, the inclined low-altitude orbit limited observations to a meteorologically myopic belt within 65° of the equator.[17] Moreover, because the orbital plane rotated, or precessed, around the earth's axis, the ground tracks crisscrossed the globe at an uneven interval and odd angles, as figure 7.5 alludes. Equally awkward was the spin-stabilized satellite's highly variable field of view, which pointed away from the earth more than half the time.[18]

Weather Bureau officials were eager to remove these impedi-

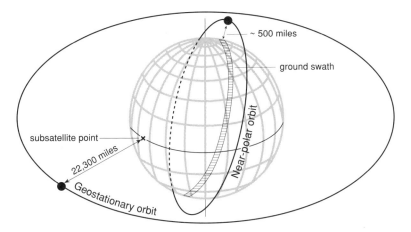

Fig. 7.6. Operational weather surveillance demands satellites in geostationary and near-polar orbits.

ments with the complementary orbits described in figure 7.6. Particularly crucial for operational surveillance is the *geostationary*, or *geosynchronous*, orbit, in which the satellite is "parked" 22,300 miles above a fixed "subsatellite point" on the equator. At this altitude, an object revolving about the earth's axis in exactly 24 hours perfectly balances the downward pull of gravity with the centrifugal force straining to fling it into space. Although more complex TV cameras and infrared scanners might not fully compensate for loss of detail, a fixed high-altitude platform over equatorial Ecuador, say, affords continuous coverage of North American weather. By contrast, a low-altitude satellite in a *near-polar* orbit addresses scientists' needs for high-resolution imagery, especially in the higher latitudes, where low-angle lines of sight severely distort the view from a geosynchronous platform.[19] Although the satellite's lower altitude produces a narrow field of view, the greater rotational velocity needed to offset gravity assures more frequent flyovers and allows complete worldwide coverage (except for the immediate neighborhood of the poles) as frequent as once a day.

NASA's Nimbus Technology program, named after the Latin word for cloud or rainstorm, combined a near-polar orbit with a control and stabilization system that not only pointed the satellite's cameras and scanners downward throughout the orbit but aimed its solar panels directly toward the sun.[20] Although Nimbus 1, launched in August 1964, became useless when its solar collectors locked in place after only a month, Nimbus 2, launched on May 15, 1966, transmitted useful images for two and a half years from a circular orbit 700 miles above the earth. Inclined 10° away from the poles, its orbital plane followed the sun slowly westward so that the satellite, which circled the earth every 108 minutes, passed overhead at about the same local time. In addition to keeping sunlight roughly comparable for successive ground swaths, this innovative *sun-synchronized* orbit allowed the advanced vidicon camera subsystem (AVCS) to provide complete, worldwide coverage every 24 hours with an array of three TV cameras: a center camera pointing downward and side cameras pointing outward for a combined field of view 170° wide. Because of surveillance by Nimbus and its successors, tropical storms—hurricanes and typhoons are notorious for forming undetected over the oceans—have not surprised sailors and coastal residents since the mid-1960s.[21]

Nimbus 2's various imaging systems offered improved resolution and wider dissemination. AVCS images, which resolved a ground spot 0.5 mile square directly below the camera, were markedly more detailed than TIROS I's wide-angle cloud snapshots, with only two-mile resolution (at best) along the ground track. Although the automatic picture transmission (APT) system on Nimbus 2 was similarly constrained by its two-mile resolution, APT images consisting of 800 scan lines covered a markedly larger area than TIROS wide-angle TV pictures, with only 500 scan lines. What's more, the APT system made its data more widely available to forecasters by broadcasting images live to 150 receiving stations worldwide. And whereas the TIROS II infrared scanner measured radiation for ground spots no narrower than 40 miles, Nimbus's high-resolution infrared radiometer, which monitored the comparatively transparent "atmospheric window" between 3.4 and 4.2 µm with 5-mile resolution, provided considerably more detailed images of cloud-top temperatures at night and reflected solar radiation during the day. While the medium-resolution infrared radiometer on Nimbus 2 offered no notable refinement in spatial resolution, NASA scientists had fine-tuned its five channels to obtain more revealing measurements. In particular, a "carbon dioxide channel," sensitive to strong absorption by CO_2 between 14 and 16 µm, collected valuable data for inferring atmospheric temperature in the upper troposphere, at altitudes of 10 to 20 miles.

Of the 24 American weather satellites placed in orbit in the 1960s, NASA's Applications Technology Satellite (ATS-1), launched on December 6, 1966, was the first to complement the synchronous weather map with timely cloud photos covering nearly a full hemisphere. Like most present-day geosynchronous satellites, its mission focused on telecommunications, not meteorology. But as a research satellite, ATS-1 carried an experimental camera that could transmit a cloud photo of a third of the earth every half hour—sufficiently frequent to monitor transitory thunderstorms and squall lines as well as track hurricanes.[22] Unlike the TIROS television system, which captured and subsequently scanned an instantaneous snapshot, the spin-scan cloud camera on ATS-1 divided the area between 55°N and 55°S into 2,000 horizontal scan lines and took advantage of the spin-stabilized satellite's rapid, 100 rpm rotation by transmitting one scan line per revolution as the camera tilted slowly from north to south to capture a complete picture in 20 minutes. By joining successive images into a short movie loop, meteorologists were able to study the motion of clouds in detail as well as validate cloud velocities measured with radar.[23] Another ATS-1 experiment demonstrated the practicality of using communications satellites to disseminate facsimiles of conventional weather maps and Nimbus snapshots.

While the effort to send men to the moon and back consumed much of the space agency's attention and budget, NASA eagerly pursued meteorological applications. ATS-3, launched in November 1967, continued the facsimile experiments and tested an improved spin-scan camera that transmitted full-disk, pole-to-pole color views of earth a year before the Apollo 7 astronauts took similar pictures during their 11 days in orbit.[24] And the experimental Synchronous Meteorological Satellite, SMS-1, launched in May 1974, carried a two-channel very high resolution radiometer (VHRR) that captured visible (0.55–0.75 μm) and infrared (10.5–12.5 μm) images with square ground spots as small as 3.5 and 7 miles on a side, respectively.[25] The first geosynchronous satellite to observe cloud cover at night, SMS-1 also functioned as a communications satellite, receiving and disseminating weather, hydrographic, and seismic data from thousands of buoys, balloons, and remote ground stations.[26]

However revolutionary, the ATS and SMS weather eyes were largely technological stepping stones to the Geostationary Operational Environmental Satellite (GOES) program, envisioned by NASA in 1959 as a series of long-lived "operational" spacecraft collecting data relevant to routine forecasting and managed by meteorologists, not aeronautical engineers. American satellite meteorology came of age in late 1975, when the space agency transferred control of GOES-1, launched on October 16, and two other satellites (SMS-1 and SMS-2) to the National Oceanic and Atmospheric Administration (NOAA), which oversees the National Weather Service. Between 1975 and 1987, NASA launched and transferred to NOAA

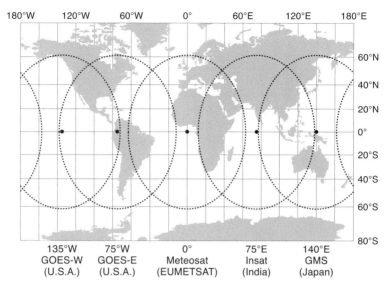

Fig. 7.7. Areas with useful cloud-cover surveillance by GOES-East, GOES-West, and the other three geosynchronous satellites of the World Weather Watch.

seven GOES satellites, which usually operated in pairs, with GOES-East positioned at 75°W to monitor Central America, South America, eastern North America, and the western Atlantic Ocean, and GOES-West located at 135°W to cover the central and eastern Pacific Ocean as well as the western half of the United States, including Alaska and Hawaii. As figure 7.7 describes, America's two GOES platforms complement strategically located geosynchronous weather satellites maintained by India, Japan, and EUMETSAT, a consortium of European weather services, to play an important role in the World Weather Watch, an international network organized by the World Meteorological Organization.[27]

What happens when a satellite dies, taking with it continuous surveillance of a substantial part of the globe? This happened in 1989, when the failure of GOES-6, parked at 135°W, left the western hemisphere covered, sort of, by a single geostationary satellite. Launched two years earlier, GOES-7 occupied the GOES-East position, too far east to monitor the West Coast weather previously viewed from GOES-6. No problem, said officials of the National Environmental Satellite, Data, and Information Service (NESDIS), NOAA's satellite division: merely move GOES-7 westward to an intermediate position at 98°W, for single-satellite coverage of both coasts as "GOES-Prime."[28] To provide fuller scrutiny of the birthplace of tropical Atlantic storms, they borrowed an older but still operative European satellite, Meteosat-3, and transferred it westward to 75°W.[29]

Repositioning an orbiting satellite is easier than it seems—or at least easier than getting the bird up there in the first place. Because there's no air friction 22,300 miles up, tiny jets called thrusters can give the satellite a slight but salient kick westward, eastward, downward, or upward—in whatever direction might be needed to correct an inaccurate orbit, move to a new parking spot, or retire a dying or hopelessly obsolete satellite to outer space.[30] As the spacecraft approaches the intended position, a thruster cancels its momentum with a well-directed squirt in the opposite direction—imagine yourself floating weightlessly in space, exploiting Newton's third law of motion ("for every action there is an equal and opposite reaction") with a can of hairspray. Weather satellites do much the same thing. Before turning a satellite over to NOAA, NASA typically parks it temporarily at 90°W, near the Gulf of Mexico, for testing. Once vetted, the satellite travels east or west as needed, to its assigned location.

Dying satellites were no joke for NOAA in the late 1980s. The Reagan administration had tried to sell the nation's civilian weather satellites in 1983, and although Congress had rejected efforts to privatize weather surveillance, budget cuts and mismanagement had delayed development of improved satellites and interim replacements.[31] Satellites have a useful life of five years or so, and because premature failure is a constant threat, spares are as important as planned replacements. Geosynchronous coverage was clearly in jeopardy after GOES-5 failed in mid-1984 and the rocket carrying its replacement had to be destroyed shortly after lift-off two years later.[32] Although GOES-7 went into orbit in February 1987, NOAA lacked a ready replacement for GOES-6, in operation since 1983. In July 1991, two years after the demise of GOES-6, the U.S. General Accounting Office (GAO) pondered the consequences of the inevitable if not imminent death of GOES-7, then in its fifth year.[33] If the old bird died or went blind, accurate forecasts would require renting at least one additional foreign-owned satellite.

Especially troublesome to the government auditors was the escalating cost of the next generation of geosynchronous satellites. Already more than three years behind schedule because of "design complexity, inadequate government management . . . and poor contractor performance," the first "GOES-Next" satellite would not be available until 1993, at the earliest.[34] Alarmed at the gridlock of two government agencies dealing with a large aerospace contractor recently purchased by another, now even larger aerospace contractor, the GAO suggested that "the Congress may wish to withhold fiscal year 1992 funds" for the GOES-Next program.[35] Despite or perhaps because of prodding by the GAO, NASA and the Weather Service muddled through to place GOES-8 and GOES-9 in orbit in April 1994 and May 1995, respectively.[36] GOES-7 cooperated by broadcasting images until January 1996, when NESDIS put it in storage at 94°W, ready for revival should GOES-8 or -9 fail.[37]

NOAA was proud of its new birds. Spiffy public relations brochures praised the duo launched in the mid-1990s as part of a five-satellite package designed to carry geostationary coverage "into the first decade of the 21st century."[38] In addition to scheduled replacements and an "on-orbit spare" (GOES-10, launched as GOES-K in April 1997, is ready to pinch-hit for GOES-8 or -9) the GOES I-M package introduced two substantially refined independent scanning instruments called the Imager and the Sounder as well as an improved stabilization system based on flywheels and miniature rockets rather than a rapidly spinning satellite.[39] For the first time a steady platform with a constant orientation to the ground was available for uninterrupted surveillance of tornado-producing thunderstorms, lake-effect snow, and other hazardous weather.

Simultaneously scanning one visible and four infrared channels, the GOES Imager combines flexible scene selection with improved spatial resolution. A computer-controlled telescope and movable mirrors focus the desired scene onto an array of detectors, which capture scenes ranging in size from the full-disk shot of the Earth in figure 7.8 to close-ups like the tornado-producing thunderstorm in figure 7.9. Under routine operation, each satellite scans a full-disk image once every three hours: America's contribution to an international, multisatellite worldwide map of cloud cover. A full-disk scan takes almost a half hour, but in the other two and a half hours the Imager can zoom in onto smaller areas, as needed, as well as capture "sector" views showing the edge of the earth, like the spectacular shot in plate 5 of a trio of tropical storms advancing on the East Coast.[40] Although detailed shots of small areas captured as frequently as one a minute can monitor severe storms or trouble spots, the most routinely useful images are the familiar views of the contiguous 48 United States, transmitted every 15 minutes to inform

Fig. 7.8. First GOES-9 full-disk visible image, scanned on June 12, 1995, starting at 1:45 P.M., eastern daylight time. Courtesy of the National Oceanic and Atmospheric Administration.

short-range forecasts as well as provide consecutive frames for the time-lapse animations known on TV as "satellite loops."[41]

Resolution varies by channel and size of scene. Visible-channel pictures can, at least in theory, reveal clouds and forest fires as small as 0.6 miles (1 km) across at the subsatellite point. By contrast, the water-vapor channel provides 5-mile (8-km) resolution, while the other three infrared bands offer 2.5-mile (4-km) resolution: not as detailed as visible GOES images, yet highly useful for short-range forecasting. The Imager can also detect minute differences in intensity in all five channels. In describing the visible channel's ability to discriminate cloud features only a half mile wide, GOES product manager Jamie Hawkins invoked the analogy of a beetle at a baseball game: "Stand at home plate in a major league ball park and scan the

Fig. 7.9. GOES-8 image taken at 4:45 P.M. on June 24, 1996, showing a massive thunderstorm near Washington, D.C. While the Imager was scanning the area a tornado touched down in Fairfax County, Virginia, southwest of Washington. A few minutes later a power outage caused by severe winds knocked out NASA's "ingester" receiving station at the Goddard Space Flight Center, in Greenbelt, Maryland. Courtesy of the National Oceanic and Atmospheric Administration.

center field fence. Can you make out a June bug crawling across the beer advertisement? If you've got the vision of Superman, you can. But now describe how dark or light the bug is—on a scale of 1 to 1024. The GOES-8 visible sensors can do it, representing a visual acuity forty times better than that of their NOAA satellite cousins in lower polar-orbit."[42] Too fine to be noticed on TV weathercasts, details this minute are priceless to the meteorologist with an interactive graphics workstation.

Designed to support numerical weather prediction, the Sounder is a 19-channel radiometer that enables computers to construct vertical profiles of temperature, moisture content, and the movement of moist air.[43] Although NESDIS estimates motion for only three atmospheric layers, its temperature and moisture profiles have a vertical resolution of about 2–3 miles (3–5 km)—more than adequate in the middle and upper atmosphere for the most demanding multilayer forecast models. There are trade-offs, though: all three types of profile have a spatial resolution of only 30 miles (50 km), and because 19-channel soundings require a longer "staring" than 5-channel images, the Sounder takes 45 minutes to scan the 48 states, whereas the Imager captures the same area in only 4 minutes. NESDIS provides several other Sounder products, including estimates of surface and cloud-top temperatures, total moisture, cloud height, and ozone distribution.[44] Because nonlayered estimates are less demanding than profiles, these images are allowed a 6-mile (10-km) resolution. And as the map of precipitable moisture in plate 6 illustrates, Sounder data displayed cartographically are contributing to the weather map's growing list of themes and formats.

Geostationary satellites have both altered and strengthened the map's role in meteorology. In the early 1980s, the weather map was like a graying screen actor: convincing in commercials yet no longer a star. Weather forecasts in newspapers and on TV required maps—lots of maps—but numerical models that manipulated fronts and pressure centers within an electronic black box threatened to reduce meteorological cartography from a process to a product. GOES helped reverse that trend. Although satellite data increased the reliability of predictive models, frequent and richly varied images from GOES and improved weather radar demanded high-interaction graphics workstations that focused largely on maps. More evolutionary than revolutionary, the rise of computer-based visualization was no doubt inevitable: electronic sensors and computers were overwhelming the forecaster with location-specific numbers, comprehensible only when organized spatially on maps.

Geostationary imagery also altered meteorological mapping's notions of time and scale. Within nearly a full hemisphere, GOES can scan where its handlers want at any time they want. Because satellite images are not snapshots—the Imager took 27 minutes to scan the full-earth view in figure 7.8—time refers, by convention, to the time at which the scan began. Time also figures importantly in trade-offs between continuous monitoring and size of scene, with NESDIS managers forced to choose between timely continuous coverage of small areas and routine scanning of larger ones. And because resolution is fixed, by channel, as the size of the square ground spot at the subsatellite point, display scale is not the principal indicator of accuracy. In other words, while zooming in at a workstation can make the image more temporally consistent, the zoom does not affect spatial precision. Moreover, as the visible grid of picture elements, or *pixels*, in plate 6 demonstrates, coarse close-ups are useful not only for their information content but also for this blocky built-in reminder of the data's precision.

Satellite imagery also expanded weather's role as art. In animations or short loops, or as single frames, spectacular God's-eye views of the atmosphere are eminently suited to the personae of storms named to connote gender and ethnicity. Electronic image processing, which can sharpen details and assign colors to nonvisible wavelengths, confers artistic license either to highlight noteworthy differences in the data or to exploit cultural associations of navy blue with water and green or brown with land. This power to colorize radiation intensities and estimated quantities like precipitable moisture can either help viewers use the map key to decode conditions at specific places, as in plate 6, or dramatically reify storms and their surroundings, as in plate 5. A gift of the NOAA public relations staff, this intriguing "color-enhanced" image of two hurricanes and a wannabe demonstrates vividly the secondary value of weather cartography as institutional propaganda.

Looking Around

*L*ike satellite surveillance, weather radar is a defense technology advantageously adapted to civilian needs. There are significant differences, though. Unlike space-borne TV cameras and multichannel radiometers, radar is an active sensor, which generates the energy it detects, in contrast to passive sensors, which rely on heat and reflected sunlight. Developed in the late 1930s as a secret weapon, radar exploits electromagnetic radiation with markedly longer wavelengths, in the microwave portion of the spectrum (figure 7.3).[1] Its meteorological application was discovered quite by accident during World War Two, when radar operators watching for enemy aircraft serendipitously identified splotches on their screens as rain clouds.[2] Weather radar has come a long way since the 1940s: a radically more informative version known as Doppler radar is the basis of the very-short-term storm-detection technology called "nowcasting" as well as the foundation of a sweeping reorganization of the National Weather Service.[3] In addition to examining the evolution and promise of radar maps, this chapter introduces the "modernized" real-time meteorological cartography that integrates radar with numerical models and satellite images.

What is radar? A contraction for "radio detection and ranging," radar is a technique for estimating distance from the observer to an object, perhaps an approaching airplane, by hurling out bursts of long-wave radiation and counting how long these pulses take to bounce back.[4] Because radio-frequency waves, like other forms of electromagnetic radiation, travel at the speed of light (186,000 miles per second), radar units must count rapidly, in microseconds. And their radiant energy must be able to bounce back without interrupting talk shows and police calls. Fortunately for airport operators and meteorologists, objects ranging in size from jumbo jets to water droplets reflect radiation in different parts of the microwave band, which encompasses wavelengths between roughly a meter and a centimeter (figure 7.2). Because an object interferes with radiation by scattering it in all directions, physicists prefer to say "backscatter" rather than "reflect," which implies a stronger, more direct return. Whether large or small, the radar "target" scatters back to the receiver very little of the original signal. But because large targets, like aircraft, will scatter longer wavelengths than much smaller targets, like raindrops or insects, radar engineers can select targets by controlling wavelength. For weather radar this means altering wavelength so that precipitation droplets or crystals will reflect radar beams.

A basic radar system consists of a moving antenna, an electronics unit, and a display. Rotating at a constant speed, the antenna scans the surrounding sky by transmitting a narrow stream of very short pulses, more than a hundred per second. Because the silences between pulses are sufficiently long for a pulse to bounce back from a target a hundred miles away, the antenna can receive as well as transmit. Aimed slightly upward at a small, constant angle, the pulses form an expanding cone, which reaches ever higher altitudes with increasing distance. As figure 8.1 describes for older radars, a circular cathode ray tube depicts the pulse beam with a moving *sweep line*, radiating from the center of the screen. Called a *plan-position indicator* (PPI), the screen image serves as a map on which numbered ticks on the circumference indicate azimuths, the point at the center represents the antenna, evenly spaced circular *range markers* show distances at an interval of 20 or 25 nautical miles, and luminous patches called *echoes* point out targets. Typically an echo remains on the screen for at least one return of the sweep line. Ephemeral reflections from birds and insects disappear quickly, whereas an intended target like a plane or rain cloud continually refreshes the echo by backscattering radiation from subsequent pulses beamed in the same direction. To help the operator locate the target, more advanced systems embellish the display with coastlines, boundaries, and other maplike reference features.

Although spurious echoes from nearby buildings, trees, and terrain can mar the display center with a luminous splotch of *ground clutter*, large echoes elsewhere on the radar scope provide valuable

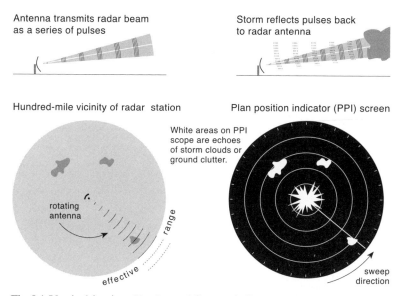

Fig. 8.1. Vertical (top) and horizontal (bottom) diagrams illustrating the principles of weather radar. The plan position indicator (PPI) scope parroted the sweep direction of the antenna, which was reversed from day to day to balance the wear on its bearings.

clues to the structure and severity of storms. Military meteorologists, puzzled in the early 1940s by "ghost echoes" too large to be aircraft flying in formation, learned to detect storms with radar as early as August 1943. Curious researchers, who began to photograph radar screens for comparison with weather maps, quickly discovered that different kinds of storms produce echoes distinctive in shape and brightness. Hurricanes, for instance, are easily identified by a clear "eye" more than 10 miles across at the storm's center, whereas squall lines appear as rows of rain clouds parallel to a cold front.[5] Severe thunderstorms, with updrafts carrying detectable moisture to heights as great as 50,000 feet, produce a unique "storm echo top," readily apparent on a *range-height-indicator* (RHI) screen, designed to display echoes obtained by locking the antenna on a particular azimuth and moving the beam up and down.[6] Moreover, fierce thunderstorms often move in a line revealed on a PPI display by a wavy band or an elongated blob. Equally significant is a bright V-shaped echo formed by hail or other dense, highly reflective precipitation moving in two converging lines.[7]

Most ominous are the hooked, fingerlike appendages associated with incipient tornadoes.[8] As shown in figure 8.2, a typical hook ex-

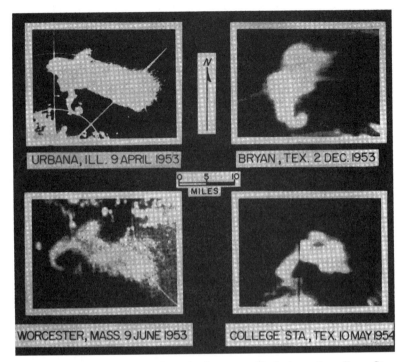

Fig. 8.2. Examples of hook echoes for known or likely tornadoes. From Stuart G. Bigler, "An Analysis of Tornado and Severe Thunderstorm Phenomena," Report for U.S. Air Force contract A.F. 19(604)-573 (1955).

tends one to six miles outward from a larger echo toward which it curls back in a counterclockwise, cyclonic direction like the number 6.[9] Produced by backscattering from precipitation aloft that is drawn into a whirling column of wind (precipitation-free air flowing into the storm forms the contrasting notch), a hooked configuration might last only a few minutes. Although not all tornadoes form hooks, and not all hooks reflect tornadoes, hooks typically signify well-structured tornado cyclones. The meteorologist who sees one on the radar scope recognizes a thunderstorm with extreme turbulence and knows twisters are likely. Not surprisingly, when the Weather Bureau began installing modified war surplus radars in the late 1940s, forecast offices in Oklahoma and other "tornado alley" states were among the first in line.

However helpful to data-hungry forecasters, converted military radars were not designed to monitor storms. After a decade of successful storm warnings and revealing experiments with wavelength and echo intensity, the Weather Bureau commissioned its own system in 1957. Called WSR-57—the acronym stands for Weather Surveillance Radar—the new design incorporated a powerful transmitter, two receivers, complementary PPI and RHI scopes, circuitry for identifying and suppressing ground clutter, and an automatic rapid-advance camera for capturing screen images.[10] Although its effective range varied depending upon the type of information sought, WSR-57 could discriminate tornado hooks and measure precipitation intensity out to 100 nautical miles.[11] At that distance, the slight upward tilt of its antenna lifted the center of the radar beam 6,500 feet above level ground so that the system could no longer reliably detect low-altitude disturbances such as tornadoes.

Radar units were expensive, and the Weather Bureau, which installed the first WSR-57 radar at the Miami Hurricane Center in 1959, could afford only a few new units each year. By 1971 (a year after the Department of Commerce changed the bureau's name to National Weather Service), 48 WSR-57 systems buttressed a network that included 37 modified war surplus radars, some filling in voids and others supporting storm detection at local forecast offices in tornado-prone areas.[12] Because radar has a limited range in rugged terrain, the weather service placed all but a few of its new, first-generation weather radars between the Rocky Mountains and the Atlantic Ocean, typically at nationally or regionally important airports.

To make the data more widely available, the weather service delivered screen images to forecast offices distant from a radar site through various techniques collectively dubbed "radar remoting."[13] Some locations were served by microwave links or telephone, which allowed immediate direct viewing, but without information on echo intensity and movement. Offices connected by facsimile transmission received hand-drawn, annotated pictures of selected radar images, usually from more than a single radar site. Because storms partly off the screen were difficult to analyze, even a forecast office with its own radar could benefit from viewing neighboring radars.

An operational system for integrating observations from different sites was decades away in 1955, when Stuart Bigler, a research meteorologist at Texas A&M University, constructed composite maps from partly overlapping radar images.[14] To explore their value as a forecasting tool, he juxtaposed a conventional surface weather chart with a concurrent radar map compiled for a network of six radar stations extending through Texas into Missouri. As shown in figure 8.3, the maps offer very different views of a squall line moving eastward ahead of a cold front. On the weather chart (left), a traditional barbed

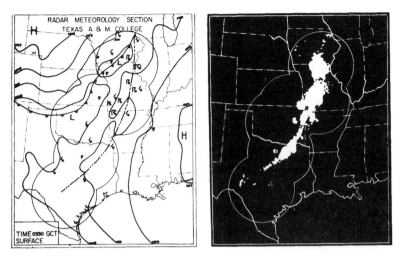

Fig. 8.3. Concurrent surface weather map (left) and composite PPI radar image (right) provide complementary views of a long squall line marked by thunderstorms. The perimeter of coalescing circles encloses the effective area covered by overlapping radars. On the conventional map a gently curved chain of long and short dashes portrays the squall line. From Stuart G. Bigler, "A Comparison of Synoptic Analysis and Composite Radar Photographs of a Cold Front and Squall Line," *Proceedings of the Weather Radar Conference*, Signal Corps Engineering Laboratories, Fort Monmouth, N.J. (September 12–15, 1955): fig. 5 on p. 118.

line delineates the front while bent arrows depict isolated thunderstorms surrounding the squall line. In contrast, the radar map (right) describes the full extent of a nearly continuous linear storm stretching almost a thousand miles from the Rio Grande to southwestern Illinois. Because of their complementarity with surface and upper-air weather charts, composite radar maps could help forecasters track disturbances and monitor thunderstorm formation. Four year later, Francis Reichelderfer, impressed with Bigler's research, lured him to Washington to direct the Weather Bureau's radar unit.[15]

Another cartographic advance was the isoecho map, on which contours representing equal echo strength delineate bands similar in rainfall intensity. Aware that heavy rain produced stronger reflections than light rain, meteorologists were eager as early as the late 1940s to map the rate as well as the extent of precipitation. To provide a meaningful map key, researchers aimed their radar beams at a low angle over a small network of highly precise, recording rain gauges and used the resulting data to calibrate an equation relating reflectivity to precipitation rate.[16] Validation of the radar-rainfall equation led to two-class white-and-black maps based on a variable intensity threshold, which allowed the operator to control the minimum rainfall intensity displayed on the PPI scope. With a comparatively high reflectivity threshold, for instance, the screen would focus on areas with very intense precipitation, whereas with a low threshold, the display would reveal the broader extent of precipitation.

Further enhancement of the electronics yielded radar maps based on two thresholds, which allowed the operator to highlight areas with light or moderate precipitation. WSR-57 radars provided even more advanced rainfall-intensity maps on which alternating light and dark bands defined several isoecho contours. Around 1970 experimental PPI displays with six discrete gray levels applied the brighter-is-more cartographic principle, which eliminated the need for frequent reference to the map key.[17] Electronic "integrators" that summed the rainfall estimate over space and time made weather radar a valuable tool for forecasting floods.[18] The weather service implemented these features in the WSR-74 system, an improved but hardly revolutionary generation of weather radar introduced in the mid-1970s.[19]

Weather radar's most important enhancement merited a new name: Doppler radar. To understand how Doppler radar works visit a busy high-speed rail line and stand a quarter mile or so up the right-of-way from an unprotected grade crossing, a bit beyond the sign with a large W (for whistle) that tells the engineer to sound the air horn. Listen carefully to the shrill warning, which appears distinctly different after the locomotive passes because of the Doppler shift, a wave phenomenon described in 1842 by Austrian physicist Christian Doppler (1803–1853).[20] As the train approaches, you'll notice a higher pitch—higher than when it's standing still or receding, and higher than the engineer hears—because the horn's forward motion compresses the sound waves, making the wavelength shorter and the waves more frequent. After the train passes, longer and less frequent waves produce a noticeably lower pitch. To observe Doppler's discovery, you need not risk life and limb—or a nasty fine for trespassing. As folks living near a railway know all too well, a marked change in pitch is noticeable a mile or more from the track if the engineer blasts the air horn with the train moving closer and hits the horn again, a few minutes later, with the train moving away.

A pair of diagrams illustrates weather radar's clever exploitation of the Doppler shift. The advancing and retreating rain clouds in figure 8.4 are analogous to a speeding locomotive with a strident air

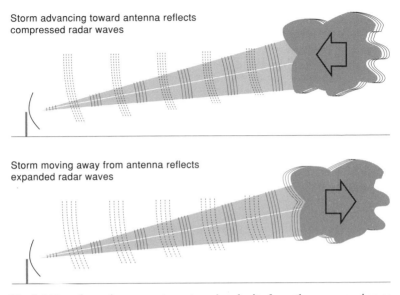

Storm advancing toward antenna reflects
compressed radar waves

Storm moving away from antenna reflects
expanded radar waves

Fig. 8.4. Doppler radar computes a storm's velocity from the compression or expansion of radar waves reflected by the precipitation-size droplets of an advancing or retreating storm.

horn. But instead of emitting sound waves, which are orders of magnitude slower than microwave radiation, these meteorological behemoths bounce pulsed microwaves back to the radar station, which then measures the shift in wavelength and computes the cloud's velocity. As with the train whistle, the approaching rain cloud (above) compresses the backscattered waves whereas the receding storm (below) spreads the waves farther apart. Necessarily exaggerated in this pedagogic portrait, the minute shifts produced by slow-moving rain clouds are a piece of cake for radar electronics.

But that's only half the story. Sensitive to microwave reflection from water vapor as well as particles of hail, snow, dust, and debris, Doppler radar also measures the speed and direction of winds within a storm.[21] Eager to apply this principle to tornado detection, the Weather Bureau obtained a surplus Doppler radar from the navy in 1956, installed it in a trailer with twin antennas on the roof, and towed the unit to Wichita Falls, Texas. Starting with the 1957 tornado season, researchers studied whatever squall lines or isolated thunderstorms nature threw their way.[22] Operating from Wichita, Kansas, on June 10, 1958, they noticed a suspicious hook echo on their conventional PPI scope, aimed the Doppler antennas toward the disturbance, and recorded a frequency shift that, according to subsequent calculations, revealed winds as strong as 206 miles per hour. Later evidence confirmed that they had detected a funnel cloud—an infant twister that quickly matured into a full-fledged tornado that struck El Dorado, Kansas, 30 minutes later, killing 15 people and destroying 150 buildings. However tragic, the incident underscored the promise of Doppler radar: an advanced radar warning system specially designed to calculate and display wind speed might have prevented most, if not all, of these fatalities. A half hour—or even ten minutes—is a long time in Tornado Alley, where forecasters understand the threat of violent winds and wary residentsy can seek shelter on a moment's notice.

Doppler radar's transition from research tool to operational network took three decades, two billion dollars, and a sweeping reorganization of the weather service. With obvious benefits to the air force and the Federal Aviation Administration (FAA) as well as the National Weather Service, the innovation faced institutional as well as technological hurdles. The first hint of bureaucratic progress occurred in 1979, when an interagency task force endorsed Doppler radar for operational forecasting.[23] Three years later, amid the Reagan administration's intense effort to cut costs and lower taxes, the government began the slow procurement process of setting performance standards, encouraging technologically savvy firms to contribute designs and submit bids, and selecting a contractor to build the Next Generation Radar, dubbed NEXRAD. Three firms submitted designs, two received support to build prototypes for the WSR-88D system—the D stands for Doppler—and in 1987, Unisys Corporation won a $359 million contract to build 165 units, with the first installation planned for 1990.

Although the adversary was Mother Nature, not the Evil Empire, NEXRAD was very much a defense system. It was complex and expensive; its key bidders, Raytheon and Unisys, were big names in defense electronics; and it soon experienced delays and cost overruns. By the end of 1987, the price had swelled to $418 million, and over the next two years various amendments to the contract added another $20 million. Equally troubling were the poor performance of a prototype and the financial instability of Unisys, which in 1990 reported a third-quarter deficit of $357 million.[24] Dismayed by software problems, the National Weather Service refused to accept the initial unit, scheduled for installation that October at the National Severe Storms Laboratory, in Norman, Oklahoma. August 1991 was the low point in a summer of discontent: Unisys begged for more money; NOAA griped about the firm's failure to meet contract specifications; NEXRAD had fallen two years behind schedule; the House Committee on Science, Space, and Technology demanded the bureaucratic equivalent of a human sacrifice; Commerce Department officials threatened to sack Unisys and turn the project over to Raytheon; and the National Weather Service limped along with an unreliable network of aging radars and a declining supply of vacuum tubes and other spare parts.[25]

Money and perseverance apparently triumphed. By 1993, WSR-88D was up and running at strategic locations in Tornado Alley and the Southeast, and National Weather Service officials were touting NEXRAD's revealing maps and timely warnings. "It's incredible," Dennis McCarthy, meteorologist-in-charge at Norman, Oklahoma, told a reporter for *Science*.[26] "We see things we always knew were there but couldn't see." No less important was when and where they saw it. In May 1992, for instance, forecasters at Norman spotted a nascent tornado 20 minutes before the storm rampaged through Kingston, Oklahoma, 90 miles away. "In past years," the death toll would have left emergency management officials "asking why we couldn't have issued a warning," but this time the twister inflicted "no more than minor injuries from flying glass." No less enthusiastic was the Houston office's science and operations officer John Livingston, whose article "NEXRAD Now!" in NOAA's *Mariner's Weather Log* predicted an increase in the tornado warning time from 2 to 20 minutes as well as improved surveillance of deadly line squalls in the "data sparse" Gulf of Mexico.[27] NEXRAD's unprecedented foresight was equally impressive on the Pacific Coast, where forecasters at Oxnard, California, could now issue 45-minute warnings whenever heavy rain and mud slides threatened Malibu's fire-stripped hillsides.[28]

Although NEXRAD not only worked but worked well, its areal coverage was inherently unequal. With each site costing approximately $4 million, federal planners had designed a widely protective network with minimal overlap.[29] Understandable gaps occurred in remote parts of the Rocky Mountains, where troublesome terrain and sparse population reduced reliability and need. Elsewhere, though, communities a hundred miles or more from the nearest radar felt threatened by undetectable tornadoes and flash floods. More troubling for NOAA administrators was the poor performance of the military, which had been assigned 22 of the 138 NEXRAD sites in the conterminous United States. During 1994 and early 1995, for instance, some air force sites were out of service for several weeks, and on a monthly basis between 10 and 62 percent of air force radars were down more than 4 percent of the time.[30] The GAO, which reported these statistics, voiced dismay that the National Weather Service was not monitoring availability at individual sites.

Another critic, the National Weather Service Employees Organization, asked how far NEXRAD could see and volunteered the troubling answer: not far enough.[31] To which a National Research Council (NRC) panel retorted: it all depends on what you're looking at.[32] The fact is there's no easy answer—or at least no single answer—because the depth of NEXRAD's vision varies widely with the type of weather. For a broad, slow-moving phenomenon like the eye of a hurricane, WSR-88D has a generous reach of about 250 miles—not much better, though, than the 220-mile reach of WSR-74.[33] Even so, NEXRAD's range for wind velocity in a hurricane (which older radars could not measure) is an impressive 140 miles (officially, 230 km or 124 nautical miles), and for supercell thunderstorms, convective rain, and other "tall storms" that present a prominent target, NEXRAD has 20/20 vision at least as far as 120 miles—more than double the distance from most populated places to their nearest radar site. For tornadoes and microburst thunderstorms, though, WSR-88D is comparatively myopic. Areas 50 miles or more from the antenna are at a distinct disadvantage because beyond that distance, the lower edge of the radar beam overshoots tornado hooks and other disturbances occurring largely below 10,000 feet. And for microburst thunderstorms, which threaten aircraft with strong winds near the ground, the range is a depressing 25 miles. In addition, there's a "cone of silence" above the antenna site because NEXRAD cannot aim its beam higher than 19.5 degrees: for weather targets 10,000 and 40,000 feet above the antenna site the corresponding blind spots have radii of 5 and 24 miles, respectively. Small wonder, then, that the NRC panel addressed the coverage question with 22 separate maps.

The most telling of these maps (figure 8.5), which compares NEXRAD with its predecessor, uses plus-signs and Xs to distinguish National Weather Service sites from Defense Department sites, and shows areas with effective surveillance for supercell thunderstorms. Described by the NRC panel as "potentially the most dangerous convective storm type," the typical supercell is about 5 miles wide, travels with the prevailing wind, lasts several hours, and produces heavy rain, high winds, large hailstones, and deadly tornadoes.[34] The wider area shaded in light gray outlines the extent of WSR-88D coverage, which is nearly complete east of the Rocky Mountains and along the Pacific coast. By contrast, darker gray circles with minimal overlap

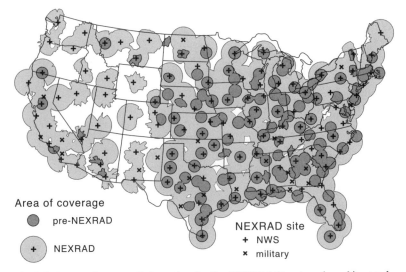

Area of coverage
- ⬤ pre-NEXRAD
- ⊕ NEXRAD

NEXRAD site
- + NWS
- × military

Fig. 8.5. Areas of supercell detection by the NEXRAD network and its predecessor. Point symbols differentiate NEXRAD sites operated by the National Weather Service (NWS) and the military. Compiled from NEXRAD Panel, National Weather Service Modernization Committee, Commission on Engineering and Technical Systems, National Research Council, *Toward a New National Weather Service: Assessment of NEXRAD Coverage and Associated Weather Services* (Washington, D.C.: National Academy Press, 1995), fig. 2-3a on p. 21.

describe the spotty coverage of the earlier network. Although 28 percent of the conterminous United States appears vulnerable, the blank areas are sparsely inhabited, experience few if any tornadoes, and are generally too rugged for cost-effective radar surveillance. Erratic and lopsided coverage zones in the Rockies and the Appalachians affirm the difficulty of looking around in mountainous terrain.

To see NEXRAD in action I drove 70 miles south from Syracuse on a snowy January morning. My destination was the Weather Forecast Office (WFO) near Binghamton, New York. Opened in 1993 under the National Weather Service's two billion dollar Modernization project, the new one-story building at the Binghamton Regional Airport is one of 118 WFOs equipped with the latest technology.[35]

In the restructuring that accompanied NEXRAD, Binghamton was a winner and Syracuse a loser. Under the old two-tiered system, both cities hosted a Weather Service Office (WSO), where personnel

took measurements and adapted forecasts prepared at Weather Service Forecast Offices (WSFOs) in Buffalo and Albany.[36] Like many
of the 250 former WSOs, the Syracuse office never had its own radar.
It didn't need one, weather officials figured, because the Syracuse airport was under the umbrella of a military radar station at Griffiss Air
Force Base, 35 miles east. Local weathercasters and air traffic controllers grumbled a bit when NOAA closed the local eight-person
WSO in 1995, and replaced it with one of 900 ASOS (Automatic Surface Observing System) units designed for continuous, 24-hour measurement of temperature, pressure, wind, precipitation, humidity, visibility, and cloud height.[37] Similar closures around the country
enabled the weather service to take fuller advantage of GOES,
NEXRAD, and ensemble forecasting at a smaller number of better-
equipped, better-staffed field offices.

Whether Modernization closed or promoted an office depended
largely on its location, its neighbors, and its experience with weather
radar. The WSO at Binghamton, conveniently situated between existing radar-equipped forecast offices at Albany and Buffalo, was an
obvious candidate for elevation. A node in the old WSR-74 network, the Binghamton office was neither too far from nor too close
to NEXRAD sites proposed for New York City; Philadelphia; State
College, Pennsylvania; and West Leyden, New York, 40 miles northeast of Syracuse. (West Leyden is the antenna site for Fort Drum,
near Watertown, where the army inherited NEXRAD responsibilities intended for Griffiss Air Force Base, axed by the Base Closing
Commission in the early 1990s.) Closure of nonradar offices in Syracuse; Rochester; and Scranton, Pennsylvania, enlarged Binghamton's service area from 10 to 24 counties, mostly within an 80-mile
circle. (West Leyden is closer to a few of these counties, but the military provides only data, not forecasts and warnings.) New duties
brought a larger staff, but not instantly. With 22 forecasters, techni-
cians, and administrators on its roster at the time of my visit, the
Binghamton forecast office was still a couple positions short of
being a full-fledged WFO.[38]

Jeff Waldstreicher, whose job title is Science and Operations Officer, is one of the office's four administrators.[39] Less a paper pusher
than a professional scientist, Jeff is both technical guru and scientific
mentor. His introduction to NEXRAD focused on the twin-screen
color graphics workstation (plate 7) called a PUP, for principal user
processor.[40] One of three key subsystems, the PUP interacts directly
with the RPG (radar product generator), which manages data storage and archiving and interacts in turn with the RDA (radar data ac-

quisition) subsystem, which includes the transmitter, receiver, and antenna. The brain of the system is the RPG's library of algorithms (computer programs) for extracting from radar data a variety of specialized maps and profile plots.[41] Six families of algorithms offer a diverse selection of "analysis products," including wind-velocity profiles and maps of accumulated precipitation, storm tracks, and wind sheer. A flexible research tool, the RPG lets meteorologists alter existing algorithms as well as write their own.

Jeff slid another chair in front of the workstation and offered me a seat. The right-hand monitor displayed the current view from Binghamton while its companion screen presented a comparable scan centered north of Buffalo. Using the mouse and a two-dimensional menu tablet about 15 inches square, he set the dual displays in motion, looping through the 12 most recent images, which looked back a little over an hour. I recognized immediately the twin bands of lake-effect snow that reached southeastward from Lake Ontario to coat 10-mile stretches of Interstate 81 with a freezing fluff, which had cut the safe speed from 65 to 40 mph and sent a few less fortunate drivers into a ditch. Vivid blues and greens highlighted areas where a slight-to-moderate radar reflection revealed this light but hazardous snowfall. Jeff's dynamic radar map also showed a wider zone of precipitation, now well to the southeast, accompanying the cold front that had passed through Syracuse shortly after midnight.

NEXRAD was running in "precipitation mode," with the antenna turning full circle every six minutes and simultaneously scanning upward at nine different elevation angles. For comparison, Jeff switched to "clear-air mode," whereby a slower, 10-minute rotation scanning only five elevation angles allowed a longer, more sensitive look at the weaker signals from very light precipitation as well as from clouds, dust, and the boundaries between relatively moist and dry air. Slowly the right-hand screen revealed a richer, more intricate pattern of snow and clouds, described in yellow and orange as well as blue and green. Clear-air mode is especially useful in central New York, Jeff explained, because cold temperatures often produce hard, tiny snowflakes with minimal reflectivity. On an especially cold day the previous year, the prodigious Lake Ontario snow machine had covered parts of the region with as much as 16 inches of white powder visible on radar only in clear-air mode. On sultry summer afternoons, by comparison, forecasters can explore vertical contrasts in moist, unstable air by electing "severe weather mode," which scans through 14 elevation angles in a five-minute rotation. As Jeff demonstrated, the twin screens let the operator monitor the entire forecast area while watching a profile along a selected azimuth for signs of hail or tornadic winds.

With twin screens, high-interaction graphic software, and a powerful array of analysis products, the PUP workstation is a lavish meteorological smorgasbord. In addition to calling up neighboring NEXRAD sites, the operator can focus the WFO's map on specific elevation angles, center the display anywhere within the region, and zoom in for a closer, more detailed look, as in plate 9, which depicts the violent tornado that killed three people in western Massachusetts on Memorial Day 1995.[42] Particularly revealing are specialized analysis products like the "storm-relative-velocity image" in plate 10, which describes winds relative to the radar by subtracting out the storm's forward motion. Contrasting colors—green for air moving toward the radar, red for air moving away—reflect locally intense contrasts of tornadic winds, centered on Great Barrington, just north of Sheffield, and moving east at 33 knots. Alerted by NEXRAD to an intensifying thunderstorm advancing on the town, forecasters at Albany notified local officials and the media 21 minutes before the tornado struck.[43]

Interactive symbols make NEXRAD's varied maps even more informative. The user interested in local effects or communities at risk can add principal highways, rivers, and major towns, while the analyst curious about extreme conditions can (temporarily) suppress echoes showing precipitation or wind speed above or below a specified level. The operator can also change the background from white to a less harsh light brown or customize the map's colors to satisfy personal taste or accommodate red-green colorblindness. Despite this flexibility, most forecasters prefer a conventional map key with variations in darkness embedded in a thermal-spectral sequence of hues running from cool blue to warm red. Plate 11, for instance, uses multiple shades of blue and green to describe an intense band of lake-effect snow extending from Lake Ontario to Utica. In the middle of the snow band medium green reflects a snowfall rate of 3–5 inches per hour, accompanied by thunder and lightning, whereas along the periphery light blue attests to less intense precipitation. Standardized colors obviate frequent references to the map key as well as confusion among workers sharing the same workstation. As plate 12 demonstrates, the conventional color sequence applied to a different scale of echo intensities in clear-air mode signals the map's heightened sensitivity with a markedly more vivid image.

Electronic archiving makes NEXRAD useful as a training tool and research database.[44] Stored images help interns and forecasters new to the area develop a fuller understanding of local influences as well as a deeper appreciation of the radar's capabilities. In addition to providing data for atmospheric research, archived images serve as "game films" with which forecasters can critique their own responses to perceived emergencies—as when Jeff and two colleagues examined their timely NEXRAD-assisted forecast of severe flooding in Wilkes Barre, Pennsylvania, following heavy rain and rapid snowmelt in January 1996.[45] And as plates 9–12 cogently confirm, a forecast office's NEXRAD archives can also accommodate an author eager for dramatically informative weather graphics.

NEXRAD is the latest, but not the last, milestone in NOAA's technological assault on meteorological uncertainty. The next step is the Advanced Weather Interactive Processing System (AWIPS), which promises full integration of Doppler radar, automatic surface measurement, satellite imagery, and computer forecast models.[46] Under development since the mid-1980s and touted as the "centerpiece" or "central nervous system" of a modernized National Weather Service, this new technology consists of a high-speed data network and highly interactive graphic workstations (plate 8).[47] When fully operational, AWIPS will link forecast offices with supercomputers as well as provide hardware and software for viewing, superimposing, and interpreting a variety of images, measurements, and predictions. In NOAA's view, the $500 million system will more than pay for itself through the more complete use of existing information. By integrating NEXRAD products with numerical models and automatic ground stations, for instance, AWIPS will promote geographically customized, locally specific flood and tornado warnings—valuable forecasts heretofore delayed, if not denied, by technological segregation.

As I write, AWIPS is experiencing delays and cost overruns similar to the growing pains that plagued NEXRAD in the late 1980s and early 1990s. Foremost among the program's critics is the U.S. General Accounting Office, which not only deplored NOAA's handling of the contract but questioned the relevance of many AWIPS features to the agency's professed goals of more accurate forecasts, fewer field offices, and a smaller workforce.[48] Although the GAO is no doubt justified in fretting over price and performance, system integration is a technological imperative with compelling promises, however vague. Having come this far, meteorological cartography cannot turn back.

Spreading the News

No other news map is as prominent and persistent as the weather map. I know this from everyday experience as well as from my research for *Maps with the News,* a history of American journalistic cartography.[1] Early in the project I saw the wisdom of noting but not counting weather maps, which for almost a hundred years were more numerous in the daily press than maps illustrating stories about war, crime, and public affairs. Surprisingly, the electronic media had no adverse effect on newspaper weather maps, which survived the spectacular rise of radio broadcasting in the 1930s and played an important role in combating competition from color television in the 1980s. Because weather is both news and drama, citizens seem as eager to read today's forecast as they are to listen to the latest update.

The history of the newspaper weather map is really the nexus of two histories, one treating print journalism as an information industry and the other addressing weather warnings as timely public information. Although a few American newspapers carried a daily weather map in the late nineteenth century, the press played a minimal role in disseminating forecasts and storm warnings until 1910, when the Weather Bureau realized that newspaper weather maps could help the agency not only reach a far larger (and presumably more appreciative) audience but also reduce substantially its increasingly burdensome cost of printing and posting—an expensive enterprise that diverted funds from data gathering and forecasting.

Technology is a key element of both histories. Development of photographic engraving in the late nineteenth century enabled newspapers to increase their use of photographs and other illustrations, including weather maps. Within two hours after the local forecaster finished his detailed national weather chart, photoengravers at a nearby newspaper could send their press room a printing plate showing the generalized isobars, isotherms, and wind arrows an assistant observer at the downtown weather office had copied in India ink onto a preprinted small-scale outline map. This example also highlights the institutional imperatives and bureaucratic inertia that can delay adoption of relevant technology: newspapers had been using photographic engraving decades before a stingy Congress, alarmed at the growth of the weather service budget, forced officials to treat their hallowed weather map as a press release.

The story of the newspaper weather map begins in England, where the press was an early collaborator, if not a precursor, of government meteorology. On August 31, 1848, the London *Daily News* published what British meteorologist William Marriott (1848–1916) called the first "telegraphic daily weather report."[2] Initiated in cooperation with the Royal Observatory's superintendent for magnetism and meteorology, James Glaisher, the two-month pilot project compiled data for 29 English cities. Volunteer observers noted sky conditions and wind direction at 9 A.M., Greenwich time, and sent their reports by telegraph to London, where the newspaper published a table of the previous morning's weather.

At the urging of the Astronomer Royal, the *Daily News* resumed the service the following year. Railways now carried for free reports contributed by a network of 50 observers, many of whom were stationmasters. Glaisher selected observers, inspected their wind vanes to verify compass points, and offered guidance on assigning winds to one of six categories: calm, gentle breeze, strong breeze, hard wind, storm, and heavy gale. No longer burdened by telegraphic charges, the newspaper paid a messenger to make the rounds of railway terminals at 2 A.M. Addition of wind-speed categories led Marriott to credit the *Daily News* with publishing the "first synchronous meteorological observations" on June 14, 1849. Glaisher also used the data to prepare a daily manuscript weather map—never published, though—for his own edification. In another famous first, the British government's "first daily weather report," prepared by Adm. Robert FitzRoy, was "issued to principal newspapers" on September 3, 1860.

Government's central role in data collection was well established by April 1, 1875, when the London *Times* published the first daily

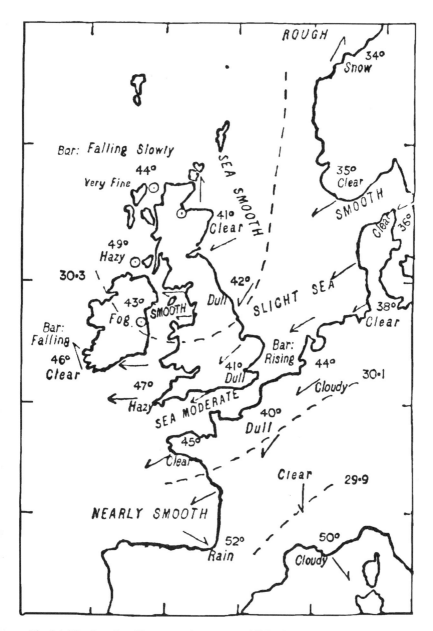

Fig. 9.1. The London *Times* weather map published on the morning of April 14, 1875, describes conditions for 8 A.M., April 13. From "The 'Times' Weather Chart," *Nature* 11 (1875): 473.

newspaper weather map.[3] Figure 9.1, the installment for April 14, demonstrates the importance of international cooperation and submarine cables, which delivered data from Ireland and the Continent. In addition to covering a wider area than the pioneering tabular reports in the *Daily News*, the *Times* weather map is meteorologically richer in its descriptions of temperature, pressure, and sea conditions. Three isobars (dashed lines representing barometric pressures of 30.3, 30.1, and 29.9 inches of mercury) describe a center of high pressure just off the north coast of Scotland, and simple arrows portray clockwise winds. A snapshot of conditions at 8 A.M. the previous day, the map describes a pleasant spring morning with temperatures in the 40s, clear or hazy skies throughout the British Isles, and smooth to moderate seas. Change was imminent, though: falling pressures in western Ireland and northern Scotland and rising pressure along the Channel suggested that the high was moving eastward, as the following day's map confirmed it had. Several months later another London newspaper, the *Hour*, also picked up the daily weather map.

Britain's Meteorological Office welcomed this wider dissemination. Its director, Robert Scott, willingly conceded that his bureau's own "circulation of 500 copies daily is absolutely nothing as compared with the number of readers of any of our leading papers."[4] In an article that July in the *Journal of the Society of Arts*, Scott acknowledged the persistence and technical ingenuity of Francis Galton (1822–1911), a prodigious British experimenter whose eclectic interests included meteorological charts.[5] Galton not only convinced the *Times*'s editors that weather maps were newsworthy but designed a drill-pantograph for transferring lines from a hand-drawn weather map to a printing plate.[6]

Rapid perhaps by nineteenth-century standards, the *Times* weather chart was more a recapitulation than a forecast. Observations taken at 8 A.M. at 50 stations in Britain and western Europe arrived in London by telegraph before 10 A.M. A private wire transmitted these data to the Meteorological Office, which prepared detailed weather charts for its own use as well as a simplified version for the media. A messenger took the newspaper edition to the Patent Type Founding Company, where an engraver traced isobars, isotherms, and wind arrows across the manuscript map while Galton's machine drilled corresponding grooves into a metal plate at a reduced scale. Cast from a mold representing the base map, the metal plate already contained grooves describing coastlines and grid ticks. Using templates and a drill to add words and numbers reflecting the day's weather, the artisan completed the engraving and from it cast several metal printing plates with raised lines, type, and point symbols. Another messenger carried these stereotype plates around to the *Times*, which incorporated the weather chart into the next morning's edition.

Scott was well aware that the *Times* weather chart was a bit stale: "It is simply impossible that reports taken once a day can give an account of all the changes which supervene during the twenty-four hours. It is equally clear that to see at breakfast time a chart for the preceding morning is far less useful to us than a chart for the previous evening, not to say, even for midnight, would be. The attainment of improved rapidity of publication is merely a matter of £. s. d. Given unlimited money and the thing can be done."[7] But whether the Crown should pay telegraphic charges was another matter. "The work could be done," he argued, "but it is hardly reasonable to ask the Government to pay such an amount for the ostensible purpose of furnishing recent weather information for the public press." If the media moguls wanted a more timely map, they could jolly well pay for it.

Scott's suggestion brought results. On January 1, 1876, the morning edition of the *Times* initiated a daily weather chart based on observations taken at 6 P.M. the previous evening, with the publisher paying additional telegraphic charges as well as the salaries of staff who kept the Meteorological Office open late to prepare a map for the newspaper's exclusive use.[8] In addition, the *Times*'s second edition, published in the afternoon and reaching a smaller readership, continued to carry a simplified weather chart based on the 8 A.M. data required for the government's own weather maps and storm warnings.[9] In 1879, two other London dailies, the *Standard* and the *Daily News*, confirmed the 6 P.M. chart's newsworthiness by picking up the daily weather map and sharing its cost with the *Times*.

The first American newspaper weather map was less a regular feature than a gimmick. The occasion was the International Exposition held in Philadelphia, in 1876, to commemorate the hundredth anniversary of the Declaration of Independence. One of the country's most influential dailies, the *New York Herald*, had set up shop on the exposition grounds to print a special edition as well as collect news for its New York readers. Sketchy accounts reveal little about the map's appearance, content, and transmission. The Signal Office reported using a system of "autographic telegraphy" to reproduce in

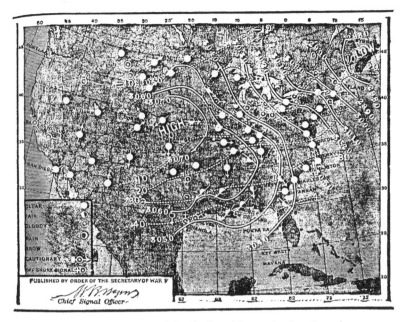

Fig. 9.2. Although the *New York Daily Graphic* was the first American newspaper to offer regularly a daily weather chart, this reduced facsimile reflects the crude letterpress reproduction of its original two-column white-on-black map. Reduced from the *New York Daily Graphic* 24 (January 27, 1881): 645.

Philadelphia a crude facsimile of the Washington weather map.[10] A special receiving instrument at the agency's exhibit reproduced the shapes and positions of lines and other symbols traced across the weather map by an operator in Washington. Although a forecaster on site might have plotted a weather chart directly from telegraphic observations, the Signal Office seemed eager both to demonstrate autotelegraphy and publicize its Washington weather map. In addition to posting a large copy for public inspection, weather service officials at the exhibition provided the *Herald* with a simplified version for the newspaper's on-site edition, starting on May 12, 1876. Although the weather map appeared "every day," its publication ceased several months later, when the exhibition closed.

The first regular American newspaper weather map dates from May 9, 1879, and ran for over three years. It appeared in the *New York Daily Graphic*, the world's first heavily illustrated daily newspaper.[11] No ordinary scandal sheet, the *Daily Graphic* charged five cents, about twice the price of its competitors, for an eight-page edition printed on high-quality rag paper, impressed on one side by conventional high-speed letterpress and on the other by lithography. However newsworthy, the daily weather map was apparently too aesthetically pedestrian to warrant the expense and precision of lithographic reproduction. As the example in figure 9.2 demonstrates, the *Daily Graphic*'s letterpress weather map is remarkable largely for the somber contrast of its white symbols and labels on a poorly inked dark background.

By its own accounts, the Signal Office was responding to "popular demand" for weather maps similar to those published in prominent European dailies.[12] America would have responded earlier, according to the 1881 Report of the Chief Signal Officer, but the nation's vast size posed "unusual difficulties." Unlike their English and French counterparts, the country's largest newspapers were not located in Washington, where the map was produced. And despite the weather map's apparent success at the centennial exhibition, autotelegraphy was "not effective at great distances." The solution was a cipher, telegraphed to the weather service's New York observer, who then prepared for the *Daily Graphic*'s engravers a "fac-simile" of the weather chart plotted that morning in Washington.[13]

The Signal Office cipher was a simple grid of horizontal and vertical lines identified by letters and numbers. A clerk in Washington placed the grid over the map, encoded the isotherms and isobars as strings of Bingo-like letter-number pairs, and telegraphed the disguised coordinates to New York. Although the decoded lines no doubt lost precision when plotted and smoothed, the cipher minimized telegraphy charges and satisfied the military imperative for command and control. "At that time," according to a short article published years later, after the weather service was a civilian agency, "no one outside of the forecast official in the Washington office was permitted to draw barometer and temperature lines and to locate highs and lows."[14]

In early 1882, the Signal Office bit the bullet and let its New York observer prepare his own weather map directly from data received by telegraph from selected stations.[15] This change was timely if not overdue: weather service officials were planning to include locally compiled weather maps in farmers' bulletins printed at various cities, and requests from numerous newspapers outside New York underscored the inefficiency and needless expense of reconstructing an imperfect facsimile of the Washington map. Ironically, this heightened interest in the daily weather map contributed to the feature's demise in the *Daily Graphic*, which received $10 per day to print the map. The weather service could ill afford similar subventions for other publishers, and in September 1882, when forced by Congress to cut its budget, the Signal Office canceled the subsidy and the *Graphic* dropped the map.[16]

Demise of the *Daily Graphic*'s weather map proved more portentous for the publisher than for the map—the *Graphic* folded two years later, while the newspaper weather map survived for brief periods in widely scattered American cities. In an 1894 article in the *American Meteorological Journal*, Harvard professor Robert De Courcy Ward chronicled the map's persistence.[17] The *Cincinnati Commercial* (later the *Commercial Gazette*) upheld continuity by initiating a daily weather map on April 12, 1881, with metal printing plates prepared by a Mr. S. S. Bassler, "formerly of the United States Signal Service." The map ran until November 1892, when Bassler left the paper to become the forecaster at the local Weather Bureau office. Newspapers in Atlanta, Buffalo, Cleveland, Detroit, Galveston, Kearney (Nebraska), Louisville, Memphis, Minneapolis, New Haven, Pittsburgh, St. Louis, and Springfield (Illinois) also carried a daily weather map for periods ranging from a mere month to several years. Editors dropped the map for a variety of reasons: their readers didn't understand it, they needed the space for more important news, or the information was too stale to justify the cost of making the plate. Even so, the *Atlanta Constitution* and the *Louisville Courier-Journal* still used a weather map to report the threat or aftermath of a severe storm.

Ward offered examples (figure 9.3) from the four newspapers running a daily weather map in early 1894. In all four cases content and symbology reflect conventions of the Weather Bureau, whose personnel drew the maps. Solid isobars and large labels for highs and lows focus on features of prime importance to meteorologists, while comparatively recessive dashed or dotted isotherms address professional and lay interest in outdoor temperatures. Point symbols at key cities conveniently combine the forecaster's appreciation of wind direction with a wider attention to cloudiness and precipitation. A century later journalistic weather maps drawn by news artists would suppress local details of pressure and wind, which might clutter the map as well as confuse readers interested largely in temperature and precipitation.

Equally significant are differences in artistic style and geographic focus. Each map bears the surname or initials of its preparer as well as his style of lettering and technique for differentiating land and water. Close inspection reveals other quirks. All but the Boston cartographer embellished station observations with spot temperatures, for instance, while the New Orleans artist identified rain with "R" and snow with "S," and the San Francisco artist highlighted his hometown in bold letters.

The *Examiner*'s less subtle symbols reflect the use of a map drawn at the forecast office in India ink on a small outline map provided by the newspaper and photographically reduced one-third during photoengraving. By contrast, the other three newspapers supplied the local Weather Bureau office with a steel plate coated with an eighth of an inch of soft chalk, into which an assistant observer used a sharply pointed stylus to engrave lines, labels, and other symbols.[18] Specially manufactured for each newspaper, the chalk plate already contained incised representations of the map's coastlines and boundaries. Artisans at the newspaper's engraving department placed the plate face-up on a stereotyping table and poured on molten metal, which seeped into the cuts. The metal slowly cooled and hardened, leaving the raised lines and labels of a printing plate.

Most prominent, though, are differences among the maps in geographic scope, which reflects a selection of stations relevant to the region if not to the local forecast. The San Francisco map thus affords the only coverage of the Pacific Coast, whereas the New Orleans map shows Tampa, and the Boston map includes Quebec City and Eastport, Maine. The Cincinnati version is also appropriately framed, with its city of publication just to the right of the map's center.

As testimony to the efficiency of late-nineteenth-century telegraphy and printing, these atmospheric snapshots reached readers only a half day old, or less. With the Weather Bureau taking nationwide observations at eight o'clock, eastern time, each morning and evening, the three morning dailies (the *Cincinnati Tribune*, the *New Orleans Times-Democrat*, and the *San Francisco Examiner*) carried the map for the previous evening, while the afternoon edition of the *Boston Herald* ran the morning map for same day, every day but Sunday.

In assessing the maps' impact, Ward compared these four newspapers' combined circulation with the Weather Bureau's total annual distribution of over a million and a half weather maps. Newspaper weather maps in just four big city dailies raised the total to nearly 50,000,000, but as "large as this number is it could be larger still."[19] Among the public benefits of prominent and readily available newspaper weather maps was a greater appreciation of the inherent un-

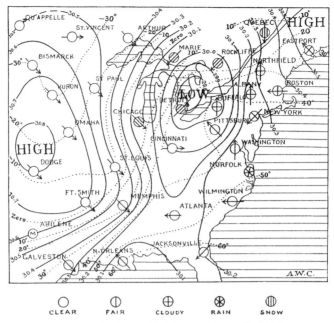

CLEAR   FAIR   CLOUDY   RAIN   SNOW

THE *BOSTON HERALD* WEATHER MAP

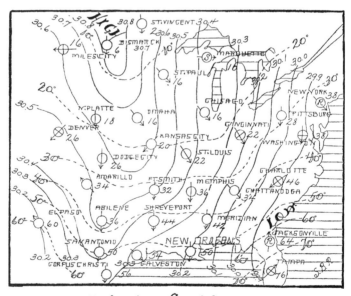

Explanatory Symbols.

◯ *Clear,* ◕ *Partly Cloudy.* ⊕ *Cloudy;* Ⓡ *Rain;* Ⓢ *Snow.*

THE *NEW ORLEANS TIMES-DEMOCRAT* WEATHER MAP

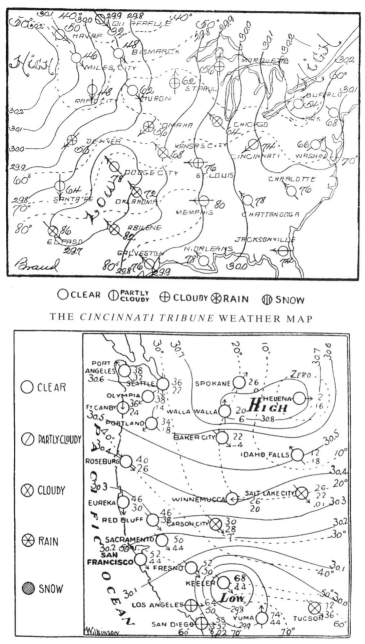

Fig. 9.3. Weather maps carried by daily newspapers in 1894. The *Boston Herald* map (upper left) is for 8 A.M., eastern time, January 24; the *New Orleans Times-Democrat* map (lower left) is for 8 P.M., eastern time, March 25; the *Cincinnati Tribune* map (upper right) is for 8 P.M., eastern time, May 9; and the *San Francisco Examiner* map (lower right) is for 5 P.M., Pacific time, February 21. From Robert De Courcy Ward, "The Newspaper Weather Maps of the United States," *American Meteorological Journal* 11 (1894): plate between pp. 100 and 101.

certainty of meteorological projection. Instead of blindly condemning flawed forecasts, the reader aware of the facts used in making predictions could become an "intelligent and reasonable critic of the forecaster."

Marked increases in cartographic dissemination followed the weather service's transfer to the Department of Agriculture. During 1891, the first year of civilian operation, weather stations outside Washington issued a total of 1,198,899 daily weather maps, principally to "business houses, public offices, and schools."[20] By 1898, the annual count more than quadrupled, to 5,239,800, and over the next 10 years distribution of station weather maps doubled, to 10,500,000—23 times the 449,530 copies of the Washington weather map distributed in the same year.[21] Yearly totals like these, rather than daily averages, seem intended to impress a Congress dismayed by repeated requests for additional printers and better equipment. As bureau chief Willis Moore argued in 1903, because "weather maps afford the only effective means possessed by the Weather Bureau for promptly placing before the public its daily observations and summaries, the improvement and extension of the maps . . . is urgently recommended."[22]

Equally impressive was the growth in number of stations printing their own weather maps from 52 in 1891 to 112 in 1909.[23] Of the latter, only 33 stations actually printed weather maps using lithographed outline maps, letterpress, and chalk-plate engraving, which the Weather Bureau adopted for its own editions in 1896.[24] Budget constraints or small demand forced the other 79 stations to rely on the "milliograph" (mimeograph) process, which could reproduce at best a couple hundred crude but legible copies on preprinted outline maps. Prints were made individually, with a roller, by forcing ink through tiny holes in a wax stencil.[25] An assistant observer drew isobars and isotherms onto the stencil freehand and used a special typewriter to cut symbols and labels.[26] Because repeated impressions weakened the stencil and widened the holes, after about a hundred copies the maps became blurry.

Moore persistently lobbied Congress to expand the number of stations with printing presses. In 1903, for instance, he proposed equipping 20 "of the more important stations" to print a larger, more detailed chalk-plate map and transferring their present equipment "to smaller stations, there to replace the milliograph process."[27] Although legislators apparently found his argument less than convinc-

ing, the number of stations issuing printed (rather than merely du-
plicated) maps crept upward from 23 in 1903 to 33 in 1909, with the
addition of new equipment and at least 20 employees. As Moore tes-
tified in 1907 at a congressional review of the Department of Agriculture budget, upgrading from milliograph to printing involved hiring not only a printer but one or two messengers to fold the maps and
post them at "the principal boards of trade and other important
places in the city."[28]

A budget hearing in 1908 provides an insight to Congressional
resistance.[29] In requesting 12 more printers, Moore had held up a
printed station weather map, which he called "a very neat publication [with] great value." Asked whether he had any examples of a
milliographed map, Moore presented "one of the best samples of a
map produced by the duplicating process that I have ever seen."
"That is pretty good," a committee member observed. "This is the
best sample," the chief retorted. "Copies produced by this process
become more and more illegible, until after 100 or 200 copies are
printed they are very bad." The committee chairman then asked,
"The only practical advantage of the printing over the duplicating
process is merely in the matter of neatness, is it not, and appearance?" Politely but firmly, Moore replied, "In legibility and cred-
itable appearance." Undaunted, he was back a year later, with a similar request, "largely for the printing in proper form of a better
weather map."[30]

Shortly thereafter Moore backed down, perhaps because of
strained relations with the secretary of agriculture.[31] But as his report
for 1909–10 made clear, backing down did not mean giving up:

> Early in 1910 the policy was adopted of discontinuing the station
> weather maps wherever the newspapers would undertake their publi-
> cation. The announcement of this purpose met with a cordial response
> from the press. The first "commercial weather map," as it has been
> called, was published in the Minneapolis Journal on March 1. Within
> the four months following its publication has been extended to 65
> morning and evening papers in 45 cities, while 55 additional stations
> will introduce the method as soon as suitable outfits can be supplied,
> which will probably be during the coming August. By this plan of pub-
> lication the weather chart is placed before the public twice daily, and
> reaches millions of people where it reached thousands before.[32]

In addition to far wider dissemination, the new policy promised cost
savings to which only whiners or freeloaders might object:

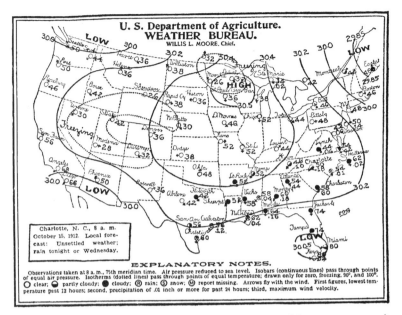

Fig. 9.4. A typical commercial weather map, prepared for a newspaper in Charlotte, North Carolina. From Henry L. Heiskell, "The Commercial Weather Map of the United States Weather Bureau," *Yearbook of the Department of Agriculture: 1912* (Washington, D.C., 1912), 538.

Some opposition to the plan has naturally been experienced in a few quarters where special purposes were served by the somewhat more elaborate charts formerly issued, but for purposes of study or for permanent file, the large Washington map remains available to those willing to pay its subscription price. I have no doubt that the change has resulted in vastly increasing the benefits of the map to the public in general, while the saving effected by discontinuing the printing at Government expense will permit the extension of the work of the bureau along other lines of usefulness.[33]

As figure 9.4 illustrates, the new map was a straightforward compilation of solid isobars, dotted isotherms, and station symbols describing wind direction and sky conditions. Isotherms were used sparingly, to highlight areas with severe cold (below 0°F or 32°F) or extreme heat (above 90°F or 100°F). Depending on the newspaper's preference, the local weather station provided either a clear, legibly inked copy for photographic engraving or an engraved chalk plate, to which coastlines and state boundaries had been transferred by pantograph from a master outline map. Explanatory notes described the map's symbols, and a box with the date provided a concise local forecast. Lest readers forget the source of the information, prominent labels above the map credited both the Weather Bureau's parent agency as well as its politically ambitious chief, Willis Moore, who hoped to become the next secretary of agriculture.[34]

News publishers eagerly adopted the new daily weather map. By July 1, 1912, a little over two years after its initial appearance, 147 newspapers with a combined daily circulation of 3,036,000 were publishing the commercial weather map in 91 cities.[35] Although I could find no count of daily newspapers with a weather map immediately before 1910—indeed nothing since 1897, when the number was estimated at fewer than 24—over 80 percent must have been new adopters.[36] This impressive response allowed the Weather Bureau to cut the number of stations printing their own weather maps nearly in half, from 112 in 1909 to 59 in 1912.[37] Even so, the cost of supplying castings or pen drawings devoured much if not all of the promised savings. In December 1911, Moore was again before Congress with outstretched hand, this time requesting $25,000 to set up two or three new stations as well as add "a little increase to our working force to keep pace with the demands of the newspapers who wish to print the daily weather map."[38] Because the first adoption in a city often sparked requests from competing publishers, a few stations were struggling to serve as many as six different newspapers. "The result now," he argued, "is that the map has become very popular, and our people are really overworked."

This popularity was short-lived. Moore's campaign to join the Cabinet proved offensive to President Wilson, who in April 1913 fired the weather map's key supporter.[39] The new chief, Charles Marvin, was less enthusiastic about publishing weather maps, in-house or in the press. His annual reports contained no cartographic statistics, and in 1917, when a paper shortage followed America's involvement in World War I, he replaced the station weather map at 50 cities with cards containing a weather summary, a local forecast, and an abbreviated table of observations.[40] Although many newspapers had dropped the daily weather map during the previous five years, wartime personnel shortages and a postwar recession forced the Weather Bureau to withdraw the service from still willing news publishers, most of whom did not reinstate the weather map for several decades.[41] But as figure 9.5 attests, in the early 1920s the regional forecast office in Boston warned of dangerous roads by printing its own cartographically illustrated highway weather bulletins.

Marvin's appeals to restore publication were marginally effective

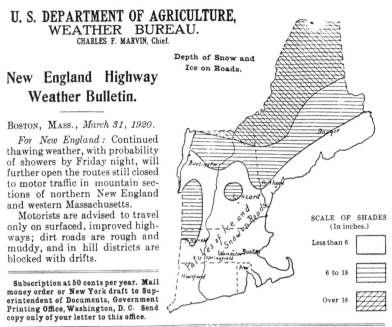

U. S. DEPARTMENT OF AGRICULTURE,
WEATHER BUREAU.
CHARLES F. MARVIN, Chief.

Depth of Snow and
Ice on Roads.

New England Highway
Weather Bulletin.

BOSTON, MASS., *March 31, 1920.*

For New England: Continued thawing weather, with probability of showers by Friday night, will further open the routes still closed to motor traffic in mountain sections of northern New England and western Massachusetts.

Motorists are advised to travel only on surfaced, improved highways; dirt roads are rough and muddy, and in hill districts are blocked with drifts.

SCALE OF SHADES
(In inches.)

Less than 6

6 to 18

Over 18

Subscription at 50 cents per year. Mail money order or New York draft to Superintendent of Documents, Government Printing Office, Washington, D. C. Send copy only of your letter to this office.

Symbols.—Points of the compass n., ne., etc.; ex., excellent; g, good; f, fair; p, poor or passable; imp., impassable; d, detour; mi-, miles; r, rough.
Note.—Information contained herein applies to the state of roads as affected by weather. Unimportant detours, construction obstacles, and local repair needs are not within the scope of this bulletin.

Fig. 9.5. This map of snow and ice summarizes conditions accounting for treacherous sections of main and state highways, which were described in greater detail on a list below. Upper half of U.S. Weather Bureau's New England Highway Weather Bulletin no. 18, March 31, 1920.

at best. His rhetoric was reverential—"Weather maps," his annual report for 1919–20 asserted ". . . are the most effective means of distributing weather information, and the educational, scientific, agricultural, commercial, and navigation interests of the country appreciate their value and expect the service in this form."—but Congress never provided the funds to restore the station weather map to its prewar level.[42] A 1926 memo in the collection of the NOAA Library lists only 23 stations printing a page-size station map, another 7 issuing a smaller, half-page version, and 17 producing only a comparatively crude milliograph edition.[43] Meanwhile, the Weather Bureau acquired a new ally, commercial broadcasting, and by 1924, over a hundred radio stations were broadcasting weather forecasts.[44]

A decade later electronic telecommunications revived the newspaper weather map. On January 1, 1935, the Associated Press inaugurated its Wirephoto network, a facsimile system for delivering timely news photographs to syndicate members willing to pay for the extra service.[45] Along with event-driven news and human-interest photos, the 39 initial subscribers received two daily weather maps—one for morning papers, another for afternoon dailies—redrawn by the AP's Washington staff from a larger, more detailed Weather Bureau product. Similar in symbology, content, and size to Willis Moore's commercial weather map, the Wirephoto map was a familiar collection of isobars, spot temperatures, and point symbols showing wind direction, sky conditions, and precipitation for selected cities.[46] With lines and type blurred by transmission noise and halftone reproduction, the Wirephoto weather map was aesthetically challenged, but as a single weather graphic disseminated in widely scattered cities, it was also truly revolutionary. In 1948, the Weather Bureau inaugurated its own facsimile network, to deliver centrally produced weather maps to forecast offices throughout the country.[47]

For a decade and a half the Wirephoto weather map was largely a feature of metropolitan dailies, for which photowire service was affordable as well as competitively advantageous.[48] In the early 1950s the network grew substantially, as small and medium-size newspapers reaped the benefits of the Fairchild Scan-A-Graver, which produced halftones without an expensive engraving plant, and the Photofax receiver, which was less expensive and more reliable than the original AP fax receiver. By 1956, the AP Wirephoto was serving 500 subscribers, while United Press International, an AP competitor with its own photofax system, offered centrally produced weather maps to 142 clients. Further improvements fueling additional expansion include the AP's AutoPhoto receiver in the mid-1960s, the UPI's fully automatic Unifax II receiver in 1975, and the AP's high-resolution Laserphoto II system in the later 1970s. In the early 1980s, satellite transmission eliminated the need for expensive dedicated telephone lines, and by 1985, the AP and the UPI each reached over 2,000 receiving dishes. By the mid-1980s, most daily newspapers, even small ones, ran an AP or UPI weather map, and many also included a black-and-white satellite photo showing cloud cover over the continental United States.

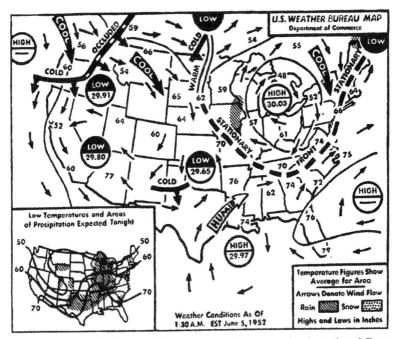

Fig. 9.6. In redesigning its Wirephoto weather map, the Associated Press added a forecast inset and replaced the isobars with point symbols describing pressure centers. From the *Syracuse Herald-Journal*, June 5, 1952, 56.

Media control of artwork and content led to simpler weather maps less likely to confuse lay readers and better able to survive photowire transmission. In 1946, for instance, the addition of fronts to the Wirephoto weather map resulted in less prominent isobars.[49] In 1950, after member newspapers asked for a forecast map, the Associated Press added a small, highly generalized inset map showing patterns of extreme temperatures and expected precipitation for the next 12 hours.[50] Figure 9.6 illustrates the redesign that replaced the map's isobars with prominent circles representing storm centers. Divided into an upper half labeled "high" or "low" and a lower half reporting sea-level pressure in inches, the circles were visually impressive, but as an anonymous AP source later observed, "the more striking the design is, the less information the map carries."[51]

Redesign proved contagious. Responding to further suggestions from members, the wire service overhauled its weather map in 1975, and again in the early 1980s.[52] As the "LaserGraphic" morning fore-

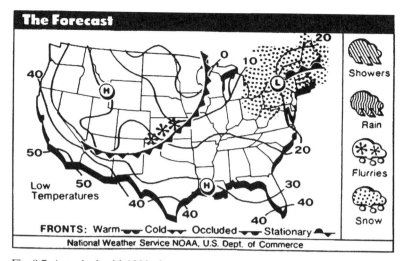

Fig. 9.7. A typical mid-1980s forecast map from the Associated Press. From the *Baltimore Evening Sun*, February 12, 1985, A2.

cast map in figure 9.7 illustrates, this latter makeover introduced a somber drop shadow along the eastern coasts and southern border—a functionless pseudo-3-D fad that infected many mid-1980s news maps. By contrast, United Press International simplified the key of its "Weather Fotocast" (figure 9.8) by labeling fronts directly on the map. Although the UPI weather map is clear and straightforward, its pandering to antimetric prejudice by converting the Weather Bureau's millibars to inches yielded isobars awkwardly labeled 29.77. Despite graphics generally more visually effective and appealing than their AP equivalents, the financially strapped UPI lost clients and clout in the late 1980s.[53]

No other newspaper weather map has had the impact of the large, visually distinctive map introduced in 1982 by *USA Today*.[54] As plate 13 illustrates, the map dominates the paper's revolutionary full-color weather package. Prominent bands of spectral hues present a generalized geography of expected high temperatures, while visually recessive numbers and letters below the names of selected cities predict local precipitation and sky conditions as well as the day's high and the following morning's low, and black boxes with white labels present an array of concise regional weather forecasts. This vivid display quickly became a symbol of Gannett Corporation's new national daily newspaper, printed in full run-of-press color at numerous sites throughout the nation. Threatened by this new and markedly more lively competitor, many local papers responded by buying color presses and overhauling their weather

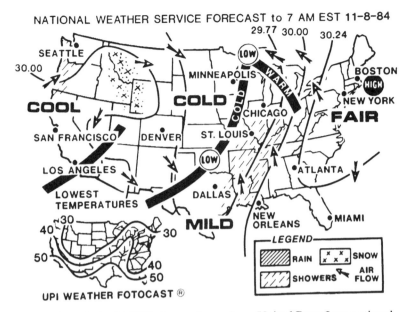

NATIONAL WEATHER SERVICE FORECAST to 7 AM EST 11-8-84

Fig. 9.8. A typical mid-1980s forecast map from United Press International. Courtesy of United Press International.

packages. In the late-1980s, the Associated Press again revamped its weather graphics and added a selection of regional weather maps, which allow most local newspapers to surpass *USA Today*'s treatment of regionally important places.

USA Today's weather map has changed subtly over time. An overprint of patterned black-and-white area symbols once described the predicted extent of snow, thunderstorms, and showers, but this information now appears below the main map, in two smaller displays, one for the day of publication and the other for the day following. In its early, pre-NAFTA years the large temperature map had a prominently precipitous drop shadow and an outwardly bowed outline map that omitted Canada and Mexico.[55] By contrast, today's less parochial view includes Vancouver, Montreal, and Mexico City. More persistent is the blatant omission of fronts, pressure cells, and air masses—meteorologically meaningful features most black-and-white newspaper weather maps present, along with temperature, in less than a quarter of the space. Ironically, a small customized weather graphic to the left

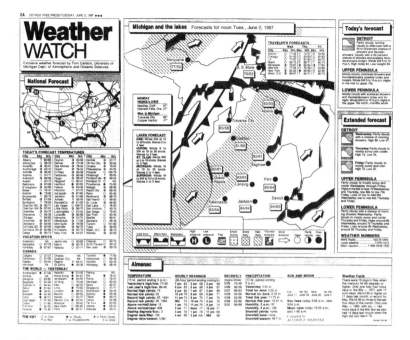

Fig. 9.9. A regional map highlighting Michigan weather was the dominant element of the *Detroit Free Press*'s half-page weather package. From the *Detroit Free Press*, June 2, 1987, 2A.

of the precipitation maps occasionally offers a clear, insightful interpretation of an otherwise unenlightening forecast.[56]

Threatened by television as well as *USA Today*, most newspapers redesigned their weather packages as the print media scrambled toward lively graphics, carefully organized news and features sections, and reader-friendly page layout.[57] Although comparatively few dailies initiated a color weather map, many introduced regional forecast maps like figure 9.9 or adopted national maps strikingly different from AP or UPI offerings.[58] The early 1980s were heady times in news graphics: an art department could use its new Macintosh computers to modify the AP Laserphoto map, downloaded as a Macintosh graphic from the wire-service satellite, or buy a color or black-and-white *USA Today*–like product from a graphics syndicate or weather-forecasting service.[59] An ironic casualty of this electronics-inspired news graphics revolution was the black-and-white cloud-cover map: an impressive innovation eagerly embraced two decades earlier, the photowire satellite map had become an unwelcome reminder of TV's visual superiority.[60]

Present-day newspaper weather maps only faintly resemble their turn-of-the-century cousins with which the Weather Bureau hoped to educate citizens, justify its existence, and explain failed forecasts. And there's more afoot here than improvements in instrumentation, data processing, and printing. No longer does the name of the weather service, much less that of its chief, adorn the daily weather map: in a symbiotic accommodation of free enterprise, the National Weather Service collects data, runs computer models, and issues storm warnings while commercial intermediaries market weather maps, forecasts, and other "value-added" information products. For the most part this limited privatization works well, stimulating innovation and satisfying consumer wants. But when the intermediary is a newspaper, meteorological cartography is embedded in the publication's overall design, where it serves as both news and packaging.

In this dual role, the newspaper weather map is, in the lingo of cultural studies, "a text" that holds different messages for different readers. For readers attracted to the colorful hype of *USA Today* a spiffy weather map becomes a symbol of the hipness of McPaper's journalists, designers, and readers, who need only glance at the map to know it's chic. For the presumably less hurried, more thoughtful readers of the *New York Times* and other prestigious newspapers, the map is a conveniently organized source of geographic information, concisely explained by supplementary weather maps like the examples in figure 9.10—nontechnical descriptions designed to edify casual readers as well as the meteorologically literate.[61] For readers interested only in the latest guestimate of weekend weather, the weather map is graphic confirmation of the weather package's scientific sophistication. For this audience, a regional view like the *Detroit Free Press* map of Michigan (figure 9.9) provides an emblem of the publisher's commitment to a state or region. (The Persian Gulf War of 1991 witnessed a different form of geographic solidarity whereby weather maps of southwest Asia not only revealed where clear skies might allow daytime air strikes against Iraq but also signified support of the American-led campaign to free Kuwait from Saddam Hussein.)[62] And for weather junkies who view the atmosphere as one of nature's great dramas, the weather package and its intriguing maps are akin (in a small dose, at least) to the sports section's rich array of box scores and information graphics.[63]

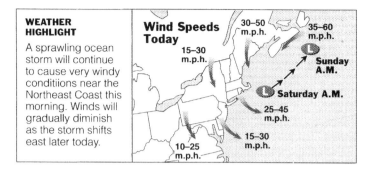

WEATHER HIGHLIGHT

A sprawling ocean storm will continue to cause very windy conditiions near the Northeast Coast this morning. Winds will gradually diminish as the storm shifts east later today.

Wind Speeds Today

30–50 m.p.h.

35–60 m.p.h.

15–30 m.p.h.

Sunday A.M.

Saturday A.M.

25–45 m.p.h.

15–30 m.p.h.

10–25 m.p.h.

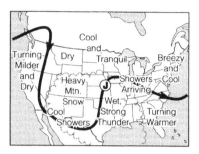

FRIDAY, SATURDAY A sprawling storm system will bring heavy mountain snow to parts of the southern Rockies while potent thunderstorms plague the southern Plains and Mississippi valley. Dry and tranquil conditions will prevail from the northern Rockies to western Great Lakes. The Northeast will be breezy and cool, while the coastal Southeast has a warming trend. Meanwhile, the West Coast States will be mainly dry and mild.

Fig. 9.10. Supplementary weather maps in the *New York Times* weather package describe in greater detail, but often without explanation, important aspects of the daily forecast. From the *New York Times*, April 19, 1997, 30 (above); and April 21, 1997, B11 (below).

The most promising of these roles by far is education. If designed not merely to describe but to explain, the newspaper weather map can reinforce an understanding of atmospheric behavior acquired in high school, college, or later in life from television. Not many newspaper weather maps do, though, and those that come close often fall short because the map author is reluctant to present a more than cursory description some readers might find too technical. Consider, for example, the two maps in figure 9.10: representative specimens of the skillful expository cartography with which the *New York Times* enriches its daily forecast. The upper map, which shows the winds expected when a large storm was to move up the coast later in the day, makes no mention of the roles of Coriolis force and counterclockwise (cyclonic) circulation in this typical northeaster. Nor does the caption for the lower map, which predicts foul weather for Colorado and New Mexico four and five days hence, mention the blocking high over northern Canada that will keep the jet stream and associated storm tracks well to the south.[64] As observed in the next chapter, television is less inhibited.[65] Doubly advantaged by animated graphics and personable weathercasters, the electronic equivalent of the newspaper weather package is better able to entertain, engage, and educate viewers with more thorough, less scientifically timid tutorials.

Weather Channels and Web Sites

Aside from a shared reliance on electronic screens, television and the Internet foster radically different approaches to meteorological cartography. Televised weathercasts, for example, offer a fully narrated sequence of maps, almost always with live, on-camera interpretation, but with content and scheduling controlled by the local station or cable channel. By contrast, the World Wide Web affords nearly instantaneous access, on demand, to a much broader range of weather graphics, including current satellite images and highly technical medium-range upper-air forecasts. There's something on the Web for nearly everyone: tutorial pages and detailed leading-edge treatments at university Web sites as well as maps with familiar themes and concise interpretation at Web sites advertising cable channels, local stations, and national newspapers. Ironically, the Internet fails weather enthusiasts when they need it most—when bad weather overwhelms Web servers with a flood of queries.

Despite fundamental differences in content and scheduling, the electronic media share a key cartographic advantage: symbols that move. Because display screens are dynamic, television and networked computers allow engagingly realistic reconstructions of the weather drama as well as plausible projections into the near future—representations that make storms and other atmospheric phenomena comprehensible by shrinking time as well as space. Moreover, the screen can switch in an instant from a sequence of GOES or NEXRAD images to a collage of moving fronts, pressure cells, and jet streams, and then to a regional map on which flashes portraying lightning strikes explain a flurry of fires and power outages. Almost as important is the nearly universal use of color in television and on the Internet. A versatile ingredient in meteorological cartography, color mimics temperature, signals danger, and promotes contrast among varied map symbols.

Equally significant are the shared limitations of low resolution and awkward labels. Computer screens have a resolution of about 72 dots per inch (dpi): adequate for word processing and simple graphics, but severely challenged by full-screen views of large national and hemispheric maps with intricate symbols, tiny type, and nonhorizontal labels—maps meteorologists are accustomed to drawing, printing, and faxing. Although zooming and panning can improve legibility, interactive close-ups risk losing the big picture: a serious loss professional forecasters escape with expensive twin-screen workstations.

Televised images are similarly constrained by the number of horizontal scan lines, currently fixed at 525 in North America and 625 in Europe.[1] Graphics must be simple, with labels not much smaller than 1/10 to 1/25 the height of the screen.[2] Unlike computer users, who can improve legibility by zooming or panning, TV viewers have little recourse: sitting close to the screen or buying a large, tavern-size receiver only makes the image grainier. What's more, the analog signal used in television transmission produces a generally fuzzier, less tolerant picture than the typical computer monitor, which reproduces a more visually reliable digital image. Because it's easy to change channels, broadcasters avoid complex displays likely to antagonize viewers and undermine ratings. Digital high-definition television promises more informative graphics by broadening the screen and doubling the number of scan lines, but even these enhancements cannot replicate *USA Today*'s wastefully simplistic temperature map.[3]

This chapter explores three fundamental issues in the electronic media's use of weather maps: the development of weathercasting as a key component of television news, the enhancement of weather information for delivery to cable channels and local stations, and the Internet's role in disseminating data and forecasts. Although I readily concede the underlying importance of computer graphics and electronic telecommunications, my treatment here of new technology highlights adoption and impact, not research and development.

American weathercasting began on October 14, 1941, on WNBT-TV, an experimental station (later WNBC) serving at best a few thousand viewers in New York City. Hawking Botany Wrinkle-Proof ties, a cartoon character named Woolly Lamb introduced the forecast by singing, "It's hot. It's cold. It's rain. It's fair. It's all mixed up together. But I, as Botany's Woolly Lamb, predict tomorrow's weather."[4] The forecast that followed consisted of a single screen with several lines of text, but no map. As far as I can tell, the distinction of telecasting the first weather map belongs to the British Broadcasting Corporation, which for a brief period in 1936 transmitted to far fewer viewers simple weather charts with isobars and pressure centers as well as temperature, sky conditions, and precipitation for a handful of unnamed cities in the British Isles and the adjacent continent.[5] As Woolly Lamb presaged the weathercaster's role as an entertainer, the BBC experiment demonstrated television's ability to deliver timely weather maps to a mass audience. Although World War II interrupted the new medium's development, postwar prosperity stimulated a new industry in which news programs with weather reports became a standard feature.

Weathercasting's first full decade could be called the turbulent '50s.[6] Annual sales of millions of new receivers made television attractive to advertisers, whose lucrative support triggered intense competition at local and national levels. With no immediate consensus on how to report weather, broadcasters experimented with formats ranging from terse Weather Bureau reports read by the station-break announcer to mini-lectures by a long-winded meteorology prof from the local university. Where weather was an especially fickle antagonist of farmers or travelers, larger stations hired their own full-time meteorologist, typically a weather service retiree or war veteran. National networks did likewise, following the lead of CBS, where John Clinton Youle reported the weather each weekday evening on the *Camel News Caravan*. More common, though, was the performer selected less for meteorological savvy than for an engaging personality, good looks, or fondness for comic costumes—an umbrella and slicker for rain, a bathing suit and beach ball for summer sun, and the mandatory parka and snowshoes for anything from light snow to a raging blizzard. Alarmed by ill-informed weather clowns and air-head weather girls, the American Meteorological Society established its Seal of Approval program in 1957 to promote understanding and reliability, if not decorum.

Since 1959, the AMS has certified over 900 weathercasters.[7] Requirements are rigorous—too rigorous for many wannabe's without a degree in meteorology or earth science. In addition to core courses in atmospheric science and mathematics, applicants must submit videotapes of on-camera reports for three consecutive days. The Board of Broadcast Meteorology—its members are working weathercasters—evaluates submissions in four areas: technical and scientific validity, informativeness, explanatory value, and communication skills. In addition to informing viewers about "recent, current, and anticipated weather conditions locally and nationally," seal holders must explain the processes that produced these conditions. Lapses following certification are taken seriously: if a seal holder "clearly fails to conduct himself or herself in a manner that reflects the dignity and honor of the profession or . . . fails repeatedly to adhere to the four criteria," certification can be revoked or suspended.[8]

For many stations, an AMS seal holder on the news team was once a mark of distinction, especially where competing stations lacked the society's implied endorsement. Now though, the specialized formal training required for certification is of little interest to a station manager who can pay a technically adequate but engagingly relaxed performer appreciably less than a card-carrying meteorologist. Small wonder then that a 1985 survey revealed that none of the managers of small-market stations considered the seal of approval "very important," while almost half rated it "not important."[9] Among respondents at large-market stations, which often face keen competition, 20 percent rated the seal very important, but 39 percent strongly disagreed. As evidence that stations managers don't confuse competence with certification, the qualifications most valued at all levels were, in order, knowledge of weather, personality, and broadcast experience.[10]

Many of today's weathercasters began their careers in college. Prestigious research-oriented meteorology departments at Penn State and Florida State University offer at least one course in broadcast meteorology, while smaller schools like Mississippi State University and Vermont's Lyndon State College run media-tailored meteorology programs for undergraduates eager to combine science and show biz.[11] At Mississippi State, for example, students in broadcast meteorology supplement an atmospheric science major with required courses in TV production, acting, and voice and articulation. At both MSU and Lyndon a well-equipped lab linked to the college television station helps students develop forecasting and presentation skills. Even so, the college laboratory is no substitute for the TV newsroom. A few lucky graduates land positions at the Weather Channel, but most fledgling weathercasters begin their careers in low-paying jobs at small-market stations, which have the greatest turnover.

These days competence in weathercasting implies skill in computer graphics as well as an understanding of atmospheric images.[12] It's been that way for nearly two decades—even longer at large-market stations better able to afford new technology. In the 1985 survey, for example, the most prevalent graphics were satellite images, used at 46, 51, and 96 percent of the small-, medium-, and large-market stations responding.[13] Color radar—pre-NEXRAD, of course—was a close second, with corresponding adoption rates of 40, 56, and 79 percent. Less common, except at large-market stations, were computer-generated maps, with rates of 29, 37, and 80 percent, respectively. By the late 1990s, though, any station wanting lively graphics had them: a gift of powerful, low-cost personal computers and fierce competition among software vendors and service bureaus eager to cover every corner of a specialized but profitable market.[14]

If today's TV weather maps are largely automatic, their counterparts in the turbulent '50s were purely manual.[15] With tiny labels and complex symbols, faxed images from the Weather Bureau were unsuitable even for close-ups, and had to be redrawn and simplified. As stage props with a sparse geographic framework of state boundaries, weather maps were huge—about four feet high by five feet wide, typically. The weathercaster stood in front of the map, just to the side of where he or she needed to add symbols or point out features. The more cartographically industrious presenter might draw up several charts in advance and clip them to an easel in sequence (first in front, last in back) so that the drawing just discussed could be quickly ripped off and discarded. Some stations attached maps to panels that could slide up and out of the way on vertical tracks, like windows.[16] (Sets were small, and unless the program director provided a second camera, space and equipment discouraged moving about.) Gary England, who broke into weathercasting in the early 1970s, described his first live broadcast at an Oklahoma station that mounted maps on miniature Ferris wheels:

> The metal weather maps on the large four-sided drums somehow looked larger that night. Each drum weighed 180 pounds but felt much heavier. Every time I turned a drum, some the letters and numbers would fall off or would assume a crazy tilt and have to be rearranged. It was frustrating those days, the norm.[17]

Some presenters drew in their map symbols on camera: an impressive act to a home audience unable to see the thin red lines penciled in as guides. Skill as an illustrator was a marketable asset for the weathercaster who could quickly sketch clouds, lightning bolts, or a radiant "Mr. Sun."

Competition for ratings inspired clever embellishments. Artistically challenged performers could polish their presentations with a metal-backed outline map and a set of magnetic stick-on symbols representing highs, lows, fronts, clouds, and suns. Moved about carefully, these symbols provided a crude but sometimes instructive animation. But when coated with material responsive to a rotating source of polarized light, apparent movement within selected symbols mimicked falling rain or diverging solar rays. Equally ingenious was the boundary map painted on a large sheet of plate glass or Plexiglas: a solution that not only intrigued viewers with the picture of a map drawn from behind but also got the performer's hand and arm out of the way of the map's most immediately relevant symbols. When I first saw this ploy, I wondered how the weathercaster trained herself to print letters backward on a map with New England on the left and California on the right. Quite a feat it seemed, until I realized how easily video electronics could reverse left and right.

Among the many electronic gimmicks that enliven today's weathercasts, none is more fundamental than *chroma key*, which integrates electronic maps, singly or in sequence, with a human interpreter. The concept is quite simple: the weathercaster stands in front of a large screen coated bright blue.[18] A camera trained on the screen captures an image, which the chroma key unit splits into two layers: bright blue and everything else. For the blue layer the system substitutes a radar or satellite image, a computer-drawn forecast map or animation, a "title graphic" with numbers and simple symbols describing temperatures and sky conditions for next few days, or an outside shot of flowering shrubs or falling snow. Assuming none of the weathercaster's clothing is the same vivid blue, the "everything else" is the map's narrator, whom viewers perceive as standing in front of the display, as in plate 14.

Chroma key requires careful staging. Marks taped or painted on the floor indicate preferred positions for camera and reporter. To avoid pointing at Minnesota when describing a storm centered over Kentucky, the weathercaster cautiously eyes a monitor just out of camera range. Just above the monitor a prompter screen displays the script the weathercaster composed earlier while analyzing the latest data and selecting appropriate images. Another video prompter mounted above the lens makes it easy to remember lines and look at the camera.

Weather graphics follow an established sequence, with tomorrow's weather as its climax. According to Robert Henson, author of a comprehensive history of television weathercasting, the succession of "current weather conditions, previous highs and lows, current map, forecast map, and local forecast" was not only well entrenched by the mid-1950s but "carried such strong inherent logic that only the most daring of programmers altered it."[19] Although recent modifications include a state or regional map that follows the national map and complementary views from GOES, NEXRAD, and fly-through animation software, the typical sequence remains a blend of chronology and cartographic scale: the coherent conflation of past-present-future with continent-region-place.

Geographer Jim Carter, who examined television weather presentations by local stations and national networks, noted the prevalence of forecast maps with three different themes: high temperatures, low temperatures, and the juxtaposition of fronts and areas of precipitation.[20] Equally common are dynamic maps called loops, in which a rapid sequence of snapshots pauses for a moment at the most recent frame before starting over again. Typically a "satellite loop" of GOES cloud-cover images recapitulates the past 12 to 24 hours at the national or continental level, and a "radar loop" of NEXRAD images covering the last four to six hours describes developing thunderstorms or passing fronts. To make certain viewers see an important feature or relationship, the weathercaster can cycle through the sequence several times as well as pause at an especially revealing frame. Local stations make effective use of regional radar loops, while weathercasts from the major networks as well as the Weather Channel and CNN (Cable News Network) favor the national radar loop, a composite of smaller-scale images from individual radars.

According to Carter, television weather reports provide a "unique viewing environment" in which familiar base maps and predictable graphic sequences as well as a personable narrator make complex information intelligible to nontechnical viewers.[21] Equally important is the diversity of formats, tailored to time of day as well as to market size. Morning weathercasts, for example, are shorter and less detailed than their evening counterparts. According to Tal White, the morning and noon weather anchor at WWBT-TV in Richmond, Virginia, morning viewers have a "fast-food mentality": eager to eat and run, they "listen rather than watch."[22] By contrast, evening viewers have the patience and leisure to watch three to four minutes of weather—even more when a storm threatens or severe recent weather warrants a fuller explanation. Because morning viewers tune in at different times, White's a busy guy: "From 5:30–7 a.m., I do five full [90-second] forecasts, five 45-second quick forecasts, [and] four weather teases . . . From 7–9 a.m., I do four live Today cut-ins [and] four prerecorded Today cut-ins."[23] Cut-ins are important to local viewers because network morning shows like NBC's *Today* offer little more than a short satellite loop and a fleeting glimpse of the national forecast map—less than a quarter of a weather slot in which the amiable Al Roker walks outside, looks at the sky, and chats with tourists.

The Weather Channel (TWC), a 24-hour cable-only service, must meet a different challenge: grabbing "grazers" who flip channels when the networks break for commercials. Because viewers are most likely to graze three minutes before and after the hour and half hour, TWC runs a six-minute, uninterrupted summary of national and regional weather in these slots.[24] Summaries include a satellite loop, which the weathercaster can halt—often with a close-up—to point out a significant feature, as well as a national-radar loop covering the past 90 minutes.[25] Forecast maps show weather predictions for the next three to five days, while carefully tailored graphics warn of severe weather (plate 15) and recapitulate notable storms (plate 16). Although the six-minute roundup uses a generally familiar sequence of maps, TWC revises the content as conditions change and breaks the monotony by rotating throughout the hour among three or four reporters with similar presentations but different styles. Backing up a staff of 30 weathercasters are 50 off-camera meteorologists, who analyze data and work with TWC's graphics specialists to generate a thousand unique images every day.[26]

Founded in 1982 by Landmark Communications, Inc., in partnership with John Coleman, formerly the weather reporter on ABC's *Good Morning America,* the Weather Channel survived a shaky start-up by offering itself free to major cable systems for the first few years: a gamble needed to build a subscriber base attractive to advertisers, who now account for 40 percent of its revenue.[27] Available in 97 percent of all cable TV households, TWC attracts a wide range of advertisers, including Michelin Tires, sponsor of the *Michelin Travel Watch.*[28] According to company officials, the Weather Channel pulls in three distinct groups of viewers: weather enthusiasts, who watch regularly; people planning trips, work, or recreation;

and severe weather viewers, who tune in when storms threaten or strike.[29]

Jim Carter, who figures the ads consume a fifth of TWC's total airtime, noted the added variety of "international weather" segments focused on Europe as well as frequent local weather reports, which account for another fifth of total programming.[30] Little longer than a minute, these local segments appear several times an hour, usually in a predictable "local on the eights" time slot. To compile customized packages for local cable markets, a Weather Channel computer canvasses current National Weather Service forecasts for more than 800 local zones, extracts information for each cable system's central city and its neighbors, and transmits the modules by satellite to the entire country. Location codes allow the local cable system's computer to grab the right module for automatic insertion in the prescribed time slot.[31] In addition to a concise text forecast of the area's weather for the next 36 hours, an extended three-day outlook, and a few statistical and astronomical facts, each report includes a regional forecast map and a radar loop recapitulating precipitation throughout the area over the last 90 minutes. Accompanied by background music or an automatic voice-over, these customized local updates are a key attraction for viewers, who stay tuned an average of nearly 14 minutes.[32] Cable viewers can catch weather maps on two other 24-hour information channels: Cable News Network (CNN), founded by Ted Turner in 1980, two years before TWC, and MSNBC, a joint venture of Microsoft Corporation and NBC that went on the air and on the Internet in July 1996.[33]

<hr />

Cable channels and local stations depend heavily on specialized commercial intermediaries for sophisticated graphics software as well as for NEXRAD and satellite images. A consequence of the increased complexity of computer technology, these intermediaries are comparatively recent. In the 1950s and 1960s, TV weather reporters relied on government facsimile charts in working up locally tailored forecasts and crude supporting graphics. The picture changed in the early 1980s, when digital electronics created a demand for animated forecast maps and other engaging weather graphics.[34] Intermediaries acquired an even larger role when the Reagan administration pushed privatization and free-market competition. A single tax-supported agency still gathers raw data and makes predictions, but today's National Weather Service lacks the budget if not the inclination to satisfy viewers' demand for vivid and lively weather maps. Free enterprise kicks in as various private firms disseminate the information in raw and enhanced form to television, newspapers, and thousands of other consumers with specialized needs.[35] Among the few exceptions are dial-up telephone forecasts and NOAA Weather Radio, which broadcasts local and regional summaries over seven high-band FM channels accessible only to receivers with a "weather band."[36]

The lords of meteorological dissemination are the four "private-sector data providers" with wires running out the back of every NEXRAD receiver. If the NEXRAD Information Dissemination Service (NIDS) has a charter member, it's Unisys, the contractor that developed NEXRAD's Principal User Processor (PUP) and now delivers radar imagery to the government's principal users: the National Weather Service, the Federal Aviation Administration, and the military. In July 1990, nearly three years after the government advertised for pre-proposals, the Department of Commerce selected three additional NIDS contractors to service other government agencies as well as numerous external users: Alden Electronics Incorporated, Kavouras Incorporated, and WSI Corporation (Weather Services International).[37] An annual access fee of approximately $1,350 per NEXRAD site entitles NIDS data providers to sell radar data to whoever wants to buy it, including firms like Accu Weather and EarthWatch Communications, which market enhanced weather graphics to much the same clientele.[38] Restrained by federal antitrust regulations and NOAA guidelines, Adam Smith's invisible hand thus promotes wide dissemination at an arguably fair price.

Specialized vendors offer broadcasters and cable networks a full range of studio-ready weather products, including GOES imagery from NOAA's National Environmental Satellite, Data, and Information Service (NESDIS), which remains in government hands. (In the early 1980s, the government tried to sell the nation's weather satellites but backed off when commercial weather services and other users complained loudly.)[39] Key attractions include integrated data and graphics software, which allow the operational convenience of one-stop shopping and guaranteed compatibility as well as a broad range of products and services readily tailored to clients' needs.[40] Experience in developing weather software and computer workstations for television stations and various government and commercial customers accounts for the eagerness of Alden, Kavouras, and WSI to become NIDS providers: they knew their market and its potential for profit.[41] In mid-1997, for instance, WSI's Web site boasted of a healthy list of media clients, including the Weather Channel, CNN, NBC's *Today Show,* and over 325 U.S. television stations.[42]

Attesting to television's addiction to trendy technology, most of these customers eagerly bought into three-dimensional animation in the mid-1990s, when WSI and its competitors took advantage of faster, less costly computers to implement earlier advances in digital cartography.[43] Now widely affordable in large and medium-size markets, 3-D weather systems enliven radar and cloud-cover presentations with spectacular fly-throughs and dynamic simulations of falling rain. Although the rapid succession of images can be overwhelming, animated graphics provide a vivid integration of continental weather systems with local landmarks. With 3-D—actually 4-D, if we recognize time as a dimension—a weather reporter can move seamlessly from a bird's-eye view of cyclonic circulation to a cross-sectional below-the-clouds description of an approaching cold front (plate 17). The result is radical if not revolutionary, with abstract map symbols attaining a "virtual reality" of sorts as the barbed blue line tracking a cold front's forward progress triggers thunderstorms, which highway symbols and labels then relate to recent reports of flooding and traffic accidents. This flair for engaging the local audience is 3-D's strongest selling point.

Hyperreality took the Syracuse weathercasting market by storm in October 1995, when two of the three stations affiliated with a major network went 3-D.[44] My graduate cartography seminar had just visited WIXT-TV, where head meteorologist Dave Eichorn (plate 17) enthusiastically described his new system, recently purchased from WSI for around $200,000 by a supportive management, which further affirmed its commitment to weather reporting by hiring a third meteorologist. With "3-D SkyTrack," WIXT's weather team could design its own fly-throughs and make changes minutes before going on the air—a distinct advantage over its rival's 3-D simulations, composed hours earlier by a weather graphics service in California.

A tape I made a few weeks later captured Eichorn's appreciation of coherence: an otherwise baffling SkyTrack segment added a different but welcome perspective when preceded by the more familiar national satellite and regional radar loops.[45] Having presented an overview of recent weather, Eichorn explained his forecast by taking viewers beneath the clouds and moving westward through rain and snow falling over Lake Ontario, southern Canada, and

Michigan. "It's kind of nice," he observed, "to get into a storm in 3-D like that, and take it apart, and see what's coming our way." Yes, indeed—as long as simpler 2-D sequences provide an introduction.

Like many television stations, WIXT supplements its NEXRAD loop with a lightning-stroke display. Blips on this dynamic map record the locations of short but powerful cloud-to-ground discharges, often but not necessarily associated with thunderstorms. As a complement to radar and satellite imagery, lightning-detection systems help the Forest Service and electric utilities monitor the likely locations of forest fires and power outages.[46] WIXT's display taps into the high-resolution lightning detection network developed by the Niagara Mohawk Power Corporation to help dispatch repair crews as well as redesign transmission and distribution lines. Receiving directional data from 20 lightning detectors covering 24,000 square miles, the system correlates tripped circuit breakers with lightning discharges and triangulates damage locations.[47] Because electrical storms threaten many other industries, including air-traffic management, casualty insurance, and construction, television broadcasters comprise a small part of the market for lightning data offered by Accu Weather, Alden, and similar weather services.

There's little limit to the weather images and programming a television station can buy. As I write, Syracuse's NBC affiliate rents production facilities to the area's Fox Network station, which broadcasts a local news program at 10 P.M., an hour before the three major-network stations go head-to-head with half-hour late-evening newscasts. The Fox outlet provides its own news anchors, while the electronic host supplies video footage, production staff, and some abbreviated, almost-as-nice weather graphics as well as its well-known weather personality. There's another option, though: small stations can avoid the expense of generating their own weather reports by buying a complete localized weather segment from the National Weather Network, whose five weathercasters serve over 50 small stations throughout the country from a satellite uplink in Jackson, Mississippi.[48] Broadcasters willing to pay extra for a special feed can even warn their viewers of approaching tornadoes or severe thunderstorms.

Among our four local stations, I most enjoy watching weather on WIXT. In addition to impressive graphics and generally accurate forecasts, Eichorn and his colleagues faithfully identify sources of uncertainty—valuable information for folks like me, who believe that understanding weather is the best alternative to controlling it. Complementing this naturally triggered discourse on jet streams, blocking highs, and lake-effect snow is WIXT's "Weather School," which challenges and entertains viewers with jargon-free multiple-choice questions about atmospheric physics, synoptic meteorology, and weather maps. Most appealing, though, is Eichorn's ill-disguised lust for tumultuous days with "lots of lines on the weather map."

Need anyone be told what the World Wide Web is, or how it has revolutionized information retrieval? Although many readers might not enjoy direct access at home, most will have seen (and probably used) the Web at work, a local library, or a friend's house. Developed in 1989 at CERN, the European nuclear research institute, the World Wide Web is an enhancement of the Internet, which the Defense Department initiated in the early 1980s.[49] Hyped by the White House and Congress as the "Information Superhighway," the Web exploded in the early 1990s with exponential growth in numbers of account holders, service providers, and Web sites offering text, pictures, and other propaganda. To gain access, one need only buy a personal computer, subscribe to an Internet service provider like America Online or Microsoft Net, and install a computer program called a browser.

Designed to help users navigate the Web, point-and-click browsers like Netscape Navigator and Microsoft Explorer reconstruct the layout of Web pages identified in cyberspace by URLs (uniform resource locators) like <http://www.usatoday.com>, which accesses *USA Today*'s home page.[50] "Web site" and "home page" refer interchangeably to one or more screens of text and graphics designed or commissioned by an "information provider." Although books and technical journals now include URLs in footnotes and bibliographies, many users uncover the addresses of promising Web sites by querying a content index like AltaVista and Yahoo!, which provide lists of links for topics as general as "weather" or as precise as "jet stream."[51] Indexed sites are identified on the screen by their names, usually in brilliant blue type. Although the browser simplifies the index screen by hiding URLs, the user can link instantly (well almost, if there's not too much traffic) to a desired Web site by pointing at the site's name and clicking. In addition, page designers provide internal links, identified on the screen with names or small labeled icons called buttons, for jumping to specific parts of a rich but well-organized Web site. Names or buttons in a Web site's menu can also link to pages stored on another computer, elsewhere in the same building or halfway around the world. Cyberspace has its own geography, which defies conventional notions of location and proximity.

As several weather pages illustrate, menus can take the form of a map on which area symbols or names represent links. At the University of Michigan's *Weather Underground* (Appendix: Underground), for instance, a map of current temperatures provides a menu for several hundred city forecasts. Clicking within the border of New York State jumps the user to a table listing current temperature, humidity, pressure, and sky conditions for 18 cities throughout the state. By clicking on the label "Syracuse," I can retrieve a more detailed local report with a four-day forecast. In addition to checking the relevant city forecast, a traveler concerned about what to pack might also check out the site's WeatherCam menu, with links to live video pictures of weather conditions at over a hundred cities and tourist spots throughout North America.[52] Although I find WeatherCam images a trifling bore, the *Weather Underground* and its Syracuse forecast are two of many Web pages I have "bookmarked" by adding their names (and sub-rosa URLs) to the list of noteworthy pages to which my browser can quickly return. Web users already know this, of course, and nonusers can learn far more satisfactorily by sitting down at the computer, tapping into the Web, and browsing serendipitously.

Unlike *Weather Underground,* many weather-page providers need no introduction. In addition to various branches and regional offices of the National Weather Service, the list of Web sites offering maps and other weather information includes *USA Today*, the *Weather Channel, CNN, MSNBC,* the meteorology departments at Penn State and Lyndon State College, and even Syracuse's WIXT-TV, which offers forecast maps, satellite and radar images, and climate statistics at its *Channel 9 WeatherNet.*[53] In addition, on-line weather graphics services not only advertise their wares on the Web but even provide free samples. For example, Accu Weather, an internationally prominent weather service with headquarters in State College, Pennsylvania, offers samples of "more than 35,000 different types of products" at its home page, *Get Weather* (Appendix: Accu Weather). Plate 18 affords a taste of the site's "Feature Graphic," a simple but colorful explanatory updated daily by the firm's staff of over a hundred meteorologists and graphic artists.

Among my favorite university sites is the *Weather Visualizer* (Appendix: Visualizer) at the University of Illinois.[54] A checklist menu lets users compose highly customized weather maps for cur-

rent weather as well as a variety of forecast models for the continental United States or its principal regions. Choices include forecast times reaching up to five days into the future as well as a selection of forecast models, atmospheric layers, and weather
parameters. Maps may cover the entire continental United States or
a geographic subdivision. The menu lists many of its choices in blue
type indicating "hot keys," which link to an explanation of the option and its cartographic representation. When finished selecting,
the user clicks the "Submit Query" button and waits a minute or so
while the server in Urbana, Illinois, tailors a map to match the specifications. Plate 19, the result of my request for a current surface
map with isobars, fronts, and a radar summary, shows a broad band
of precipitation moving eastward across the central states.

Among the Web's attractions is a wealth of highly specialized
maps, including satellite and NEXRAD images, upper-air snapshots, and colorful maps of short-, medium-, and long-range and
forecasts. The *Space Science and Engineering Center* (Appendix:
Wisconsin SSEC) at the University of Wisconsin offers a wide variety of satellite snapshots as well as derived maps such as the GOES-
8 Sounder image in plate 6, while Web sites at NOAA's *National
Centers for Environmental Prediction* (Appendix: NCEP) make
available numerous surface and upper-air forecast maps. And a direct link to the *Center of Land-Ocean-Atmospheric Studies* offers
similar maps for Europe, East Asia, and other regions. There's far
more out there, though, including time-lapse cloud-cover animations similar to their TV counterparts and made-for-Internet
movies, which users can download for very-small-window replays of
Hurricane Andrew and other notable storms. Although constant
turnover in Web sites, URLs, content, and page design precludes a
fuller account of what's available, look for significant advances in
three-dimensional weather maps encoded in VRML, the Virtual
Reality Mark-up Language. Interactive viewing, especially the ability to stop, back up, and replay, might well make 3-D more a tool
than a glitzy gimmick.[55] In a different vein, "push technology" allows Web sites to automatically deliver the latest forecast and cur-
rent-weather maps as well as warn users—assuming they are on-
line—whenever storms threaten.[56]

I once believed that the growth of the Internet and the wealth of
weather data on the World Wide Web, most of it free, would eventually eliminate noninteractive services like the Weather Channel.[57]

But despite the undisputed boon to scientists and weather enthusiasts, Web sites and on-line services show little inclination to provide the truly user-friendly interpretation of the television weathercaster. In this sense, the Weather Channel succeeds because, like CNN, WIXT, and weather telecasting in general, it employs two channels of communication, one visual and the other aural. Even so, future refinements of the Web might prove equally adept in demonstrating the value of the knowledgeable human interpreter who selects as well as decodes relevant but complex maps.

II

Hindsight as Insight

Climate, I tell my students, is average weather, which it is, sort of, if a broad notion of "average" acknowledges seasonal factors that make mean annual temperature and similar measures myopic if taken alone.[1] Equally significant are temperature extremes, rainfall variations, and the ratio of actual precipitation to potential evapotranspiration—just a few of many statistics portrayed in comprehensive climate atlases. In addition to these largely descriptive treatments, climate has important applied dimensions, often measured and mapped with indexes representing weather's effects on human physiology, energy consumption, or specific crops. Especially relevant to geographers and tourists are climatic classifications that identify similarities sufficiently salient to constitute "a climate." Besides these traditional views of climate, this chapter looks at El Niño, ozone depletion, and global warming, which have given maps new relevance in discovering threats and informing debate.

I could begin with Edmund Halley's pioneer 1686 map of trade winds (figure 2.5) or perhaps with Alexander von Humboldt's widely revered 1817 map of average temperatures (figure 2.4)—both are milestones in the history of cartography as well as the history of meteorology—but I won't. However seminal, these famous firsts are less representative of climatological mapping than the 12 charts in Lorin Blodget's (1823–1901) comparatively obscure 536-page 1857 magnum opus *Climatology of the United States and of the Temperate Latitudes of the North American Continent.*[2] Based largely on data compiled by Joseph Henry's network of Smithsonian observers, Blodget's maps treat climate's two most basic variables, temperature and precipitation. And they do, indeed, portray averages: separate annual averages for the northern hemisphere and North America as well as three-month continental averages for spring, summer, autumn, and winter. Juxtaposed and interpreted in a single volume, these now mundane charts were not only a welcome addition to mid-nineteenth-century American science but an enduring legacy of their author. In an article commemorating the hundredth anniversary of the book's publication, a Weather Bureau scientist canonized Blodget as the "Father of American Climatology."[3]

Who was Lorin Blodget? Certainly not a Halley or a Humboldt, according to historian of science James Rodger Fleming, whose account of "the Blodget Affair" portrays him more like the Deadbeat Dad of American Climatology.[4] In 1851, after two years as one of Henry's cooperative observers, Blodget went to Washington and with his congressman's help landed a temporary job at the Smithsonian drawing maps and reducing meteorological measurements to sea level. Quitting eight months later in a salary dispute, Henry's "disgruntled clerk" found similar work preparing meteorological tables for the census and writing an essay on the climate of the West for the army medical department. Lobbying Congress to transfer the Smithsonian meteorology project to the army with himself as director, Blodget further angered his former boss by publishing an article on agricultural climatology in the Patent Office Report for 1853. Much of the heretofore unpublished data belonged to the project, not its former employee, Henry argued.

Equally perturbed was surgeon general Thomas Lawson, who criticized Henry for letting an underling publish army data without proper acknowledgment. Henry's real battle, it seems, was with Lawson, who used the occasion to remind Congress of the army's important and considerably older role in compiling weather observations. Although a surprisingly conciliatory Henry rehired Blodget in late 1853, the arrangement lasted less than a year—long enough, though, for the ambiguously ethical climatologist to acquire the maps and data for his landmark text. Before the book's publication in 1857, Blodget had left Washington for Philadelphia and abandoned climatology for a second, eminently successful career as a publisher, business journalist, and customs official.[5]

Blodget's treatise proved remarkably durable. In an obituary published a month after his death in 1901, Cleveland Abbe noted that *Climatology* was "still often quoted, although its different sections are of very unequal value."[6] A dozen years later a rereading of Blodget inspired a laudatory essay in the *Monthly Weather Review* by Harvard professor Robert De Courcy Ward, who had taken the 56-year-old book along on a vacation in the New Hampshire mountains.[7] "More than ever before," Ward wrote, "I am impressed by the labor involved in the preparation of this book; by the author's broad and clear view of his subject; and by the practical application of the facts."[8] Particularly his application of cartography: "Much of Lorin Blodget's best pioneer work was done in the construction and discussion of his seasonal and annual temperature and rainfall charts, in which he took pardonable pride."[9] Through maps, "this almost forgotten author" had found order among "the confused mass of scattered observations which had been accumulating from different sources at the Smithsonian Institution and the Office of the Surgeon General."[10] Further praise came from the prominent Weather Bureau climatologist Helmut Landsberg, who in a short 1964 article on the history of American climatology opined that Blodget's volume "is even today a very readable book and, while superseded by more and better data, [its author] had certainly even then a clear grasp of many fundamental facts about the climate of the United States."[11]

It's hard to quibble with Abbe, Ward, and Landsberg about the effort and originality of Blodget's cartography. Yet the frustration of selecting a representative and reliably reproducible facsimile convinced me that his maps are far larger in size and scale than their climatological contents warrant. Indeed, their merit—the "fundamental facts" of which Blodget had "a clear grasp"—lies principally in smooth, highly generalized isolines that convey a clear picture of salient trends like comparatively moderate temperatures near the coasts and an enormous rain shadow east of the Cascades. By contrast, much of the detail consists of tiny type and fragile landform symbols, skillfully engraved by a Philadelphia lithographer.[12]

In figure 11.1, an enlarged portion of the map showing the "mean distribution of heat for the autumn" illustrates my point. As on his other seasonal maps, Blodget's isotherms swing southward around the comparatively chilly mountains of Vermont and New Hampshire and then turn north to parallel the New England coast. Printed in orange-red on the original, his temperature contours stand out from the tiny hachures portraying terrain. Unencumbered by current cartographic conventions, Blodget cleverly communicated uncertainty with isotherms that bifurcate and stop abruptly. To his further credit, he augmented the thick isotherms, drawn at an interval of 5°, with thin, supplementary isotherms representing intermediate tempera-

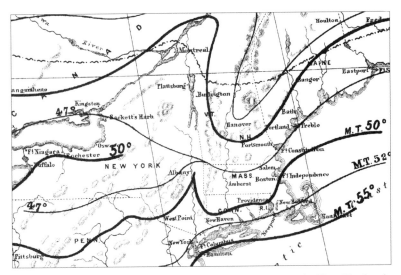

Fig. 11.1. Blodget's portrayal of average fall temperatures in New England. Excerpt enlarged from Lorin Blodget, *Climatology of the United States and the Temperate Latitudes of the North American Continent* (Philadelphia: J. B. Lippincott and Co., 1857), map facing p. 286.

tures: 47°F and 52°F in this instance. Even so, his 20-inch-long fold-out map is hardly more meteorologically informative than similar, much smaller national maps in the equally classic 1941 Yearbook of Agriculture, *Climate and Man*.[13] By omitting place names and woolly caterpillar mountains, the yearbook's authors were able to fit their climatic portraits into a space only one-sixth as large.

I wanted to include a comparable map showing *Climate and Man*'s portrayal of average autumn temperatures, but couldn't—like many climatic compendia, the yearbook ignores averaged seasonal temperatures to focus instead on January and July, which provide a sharper depiction of extremes than three-month averages. As these and its other 44 national maps demonstrate, variety and specificity more than compensate for the absence of terrain symbols. In treating temperature, for example, the book complements cartographic portraits averaged for January, July, and the entire year with a pair of maps portraying average annual minimums and maximums, another pair of maps showing the highest and lowest temperatures ever observed during a 40-year period, and several maps describing agriculturally relevant indexes like the average length of the frost-free period (the growing season), shown in figure 11.2. Closely related maps displaying average dates of the first frost in autumn and the last frost

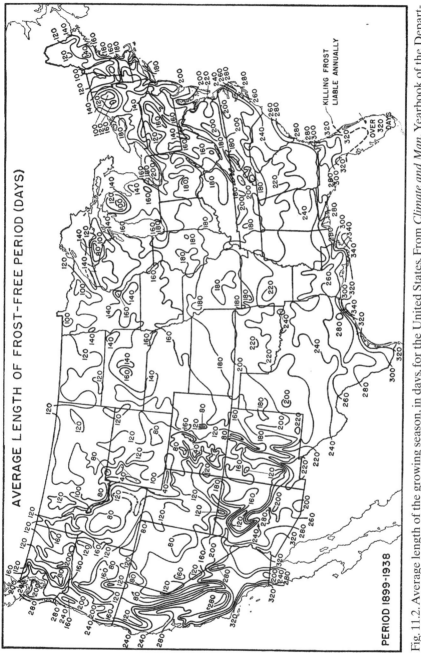

AVERAGE LENGTH OF FROST-FREE PERIOD (DAYS)

PERIOD 1899-1938

KILLING FROST LIABLE ANNUALLY

OVER 320 DAYS

Fig. 11.2. Average length of the growing season, in days, for the United States. From *Climate and Man*, Yearbook of the Department of Agriculture (Washington, D.C., 1941), 746.

in spring are typical of the meaningful measures with which applied climatology serves farmers, builders, and highway departments.

Plotting isotherms and other climatic isolines is both science and art. Map authors must choose an interval (5°? 10°? 1°?), decide where (if at all) to place supplementary intermediate contours, determine which data points (stations) to use and how heavily to weight them, and decide whether and how to incorporate effects of mountains, large lakes, and other terrain features. Blodget made these decisions, consciously or unconsciously, but declined to discuss his cartographic choices.

The authors of *Climate and Man*, equally close-mouthed about how they placed their isolines, supplemented their national overviews with state maps drawn by local experts. Individual state essays include seven maps tailored to farming: average January and July temperatures, average dates of the last killing frost in spring and the first killing frost in the fall, average number of days without a killing frost, and average precipitation throughout the year as well as during the growing season. Comparisons with the equivalent portion of the national map underscore the importance of scale and local knowledge. Consider, for example, the New York growing-season map in figure 11.3. With more sinuous contours than its national counterpart and an interval of 10 days, not 20, this regional map shows the moderating effects of the larger Finger Lakes and other influences too specific or subtle for the corresponding national overview (figure 11.2). Although the countrywide portrait captures some effects of the Great Lakes, the Atlantic Ocean, the Catskills, and the Adirondacks, its coarse delineations contrast markedly with the state map's geographic framework of county boundaries.

However useful as background information, boundary lines and other reference features invite misinterpretation by the cartographically naïve. Because generalized isolines based on widely separated observations cannot capture local variations in terrain and other factors affecting the microclimate of a farm or neighborhood, conscientious authors of small-scale climate maps now caution users not to take their depictions too literally. For example, the *Climatic Atlas of the United States*, published in 1968, includes specific caveats on all its contoured maps of temperature and precipitation. A typical warning reads:

NOTE.—CAUTION SHOULD BE USED IN INTERPOLATING ON THESE GENERALIZED MAPS. SHARP CHANGES MAY OCCUR IN SHORT DISTANCES, PARTICULARLY IN MOUNTAINOUS AREAS, DUE TO DIFFERENCES IN ALTITUDE, SLOPE OF LAND, TYPE OF SOIL, VEGETATIVE COVER, BODIES OF WATER, AIR DRAINAGE, URBAN HEAT EFFECTS, ETC.[14]

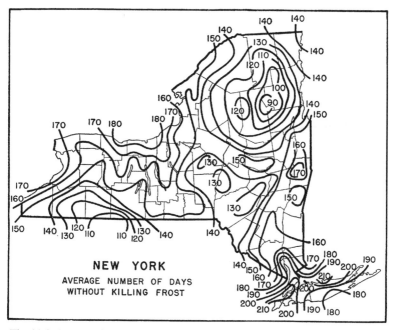

Fig. 11.3. Average length of the growing season, in days, for New York. From *Climate and Man*, Yearbook of the Department of Agriculture (Washington, D.C., 1941), 1032.

This or a similar disclaimer appears on 93 of the atlas's 271 climate charts.[15]

A different kind of misinterpretation arises from the widespread use of mean temperature, mean precipitation, and other averaged measurements assumed to represent what is normal.[16] Readily computed by adding a column of numbers and dividing by the number of items summed, the arithmetic mean is familiar to most readers. Less common, I suspect, is an appreciation of the ease with which abnormally high or low values can distort the mean. Temperature, which varies continuously over time, is less vulnerable to distorted means than precipitation, which can fluctuate wildly in dry areas as well as for short time periods like the week or the month. Although maps based on median rainfall provide a convenient supplement, if not a substitute, for maps of mean rainfall, climatologists prefer to map means tailored to specific concerns. In this tradition, the authors of the *Climatic Atlas of the United States* balanced their maps of mean monthly precipitation with monthly maps showing "mean number of days with 0.01 inch or more of precipitation"—a complementary measure useful in assessing the need to carry an umbrella.[17]

Another source of uncertainty is long-term climatic change, which can undermine older maps as well as newer ones based on too many decades of data. Because nearby industrial and commercial development can alter the microclimate of observation stations, climatologists are wary of local as well as global change and prefer to calculate means for a fixed interval of 30 years—the period the World Meteorological Organization uses for climatic "norms." (Thirty years also provides a reasonable number of stations that were neither moved nor discontinued.) In addition to following the 30-year guideline, the *Climatic Atlas of the United States* demonstrates a further fondness for round numbers with a period (1931–60) that ends almost eight years before its publication.

My copy of the *Climatic Atlas* might soon become a collector's item. NOAA has not updated the volume since 1968, and its National Climatic Data Center publishes a far wider range of maps and data sets on magnetic tape and CD-ROM. Moreover, the center's Web site (Appendix: NCDC) offers much the same information, including some measures less than two hours old and others reaching back to 1895.[18] What's more, Internet users have ready access to on-line information at the six regional climate centers affiliated with universities and partly supported by NOAA.[19] Instead of merely archiving data, each center monitors weather patterns and conducts applied research for clients throughout its region. Climatic data thus provide a tool for assessing the significance and likely impact of unusual weather.

The drought map in figure 11.4, a black-and-white rendering of a color graphic once available on the World Wide Web, exemplifies the centers' focus on moisture. Produced near the height of the 1995 drought by the Northeast Regional Climate Center, at Cornell University, the map reveals substantial intraregional variation in the Palmer Drought Severity Index. One of four drought measures used by NOAA, the Palmer index compares cumulative differences between actual precipitation and the hypothetical amount required to sustain evapotranspiration, runoff, and groundwater recharge.[20] The areal units are state climate divisions, defined largely by drainage basins and key crops, for which the National Weather Service has tabulated average temperature, precipitation, and drought severity back to 1895.[21] Shown to public officials in several states, this map proved helpful in deciding to restrict water usage.

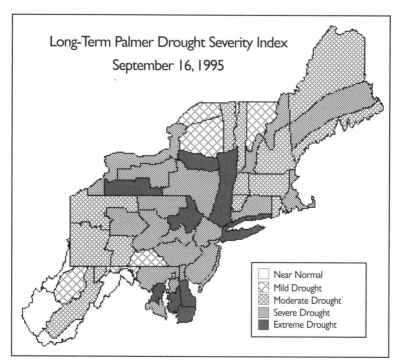

Fig. 11.4. Relative severity of the Northeast drought, as measured for September 16, 1995, by the Palmer Drought Severity Index. Modified from a color image on the Web site (Appendix: Northeast Climate) of the Northeast Regional Climate Center.

Despite evolving databases that make the Web a living atlas, climatology continues to humor geographic educators eager to point to places on a single-map overview and ask students "What's the climate like there?" The ancient Greeks had such a map, but it ignored crucial humid-arid and coastal-interior distinctions by dividing the world into only three, largely latitudinal zones: tropical, temperate, and polar.[22] Needed refinements awaited the concept of the climatic province, proposed by Alexander Supan (1847–1920) in 1884, and quantitative estimates of potential evapotranspiration, advanced notably by C. Warren Thornthwaite (1899–1963) in the 1930s and 1940s.[23] Although Supan, Thornthwaite, and many others proposed systematic classifications of the world's climates, almost all physical geography textbooks endorse a single scheme for small-scale world maps: the "Köppen system," which Wladimir Köppen (1846–1940) presented in 1900 and modified several times, more notably in 1918 and 1936.[24] Based on annual and monthly means for temperature and precipitation, Köppen's system is quantitative and hierarchical. It begins by assigning a capital letter that describes a climate as humid and hot (A), dry (B), humid and warm (C), humid and cool (D), or polar (E).[25] Intending to replicate vegetation boundaries, Köppen drew a line around B climates with the formula

$$R = .44T - N,$$

where T is mean annual temperature in Fahrenheit degrees, N is a constant (14 for dry summers, 3.5 for dry summers, and 8.5 otherwise), and R is mean annual precipitation in inches.[26] If a station's precipitation is less then R, it has a B climate; otherwise its climate is A, C, or D, depending upon the temperature of the coolest or warmest month. (E climates fall out even earlier, a reflection of no month warmer than 50°F.) Köppen's map for the United States, shown in figure 11.5, illustrates the complexity of subsequent breakdowns, which add one or more lowercase letters. The key, were I to include it, would consume more space than the map.

Climatologists have offered numerous modifications, which move boundaries, collapse or expand categories, or reflect the comfort of humans, not plants.[27] Though well intentioned, these latter alterations have not caught on.[28] Humans vary in tolerance to temperature and humidity, and air conditioning is readily available in homes, cars, and offices. If anything, these attempts underscore the value of maps depicting a single theme like heating degree-days, which estimates demand for energy by summing across an average year all negative departures of the daily mean temperatures from a standard of 65°F—a rough approximation at best, given variations in insulation and the cost of fuel, not to mention my British-born colleagues who consider 65°F balmy. Although Köppen warrants half a lecture in my introductory physical geography course, climate is too complex and intriguing for simplistic single-map regionalizations.

Outside in the garden it's a different matter. I enjoy perennials and flowering shrubs, but am wary of ornamentals too fragile for northern New York winters. When the arrival of nursery catalogs suggests that there will be a spring after all, Marge and I eagerly scan the pretty pictures. We're careful, though, to compare the zone number with the plant hardiness zone map, which most catalogs contain. According to the map, we live in zone 5, where the average annual min-

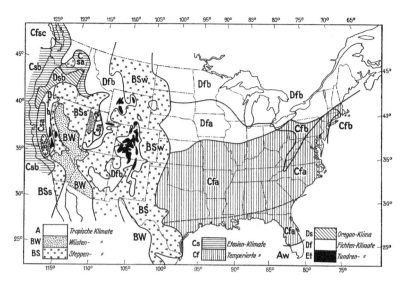

Fig. 11.5. Climatic regions of the United States, according to the Köppen system. From W. Köppen, "Die Klimagebiete nach Köppens Klassifikation," in *Handbuch der Klimatologie* (Berlin: Gebrüder Borntraeger, 1936), 2 (pt. J): map p. 194.

imum temperature ranges from –20° to –10°F. No matter how enticing, a plant vetted only for zone 6 or higher might not survive.

The Department of Agriculture revised the hardiness zone map in 1990. The previous version, compiled in 1965, reflected observations taken between 1931 and 1952 in some parts of the country, and between 1899 and 1938 in others.[29] More recent data from 14,500 stations throughout Canada, Mexico, and the United States—more than twice the number for the earlier edition—informed the update, which adds Mexico and accommodates Hawaii and the Florida Keys with a new, 11th zone, where annual minimum temperatures remain above 40°F. Colder conditions in the Southeast but slightly warmer temperatures in parts of the Northeast and Midwest confirm the need for revision. Although rumored effects of global warming are difficult to discern, the data indicated significant local warming in cities sufficiently large to create an "urban heat island"—features too complex and uncertain, though, for a small-scale climate map.

Gardeners, farmers, and others at the mercy of real weather, not annual minimums, should check out the Web site of *NOAA's Climate*

Prediction Center (Appendix: Climate Prediction). One of several National Centers for Environmental Prediction, the Climate Prediction Center analyzes trends and applies numerical models to predict "climate events" two weeks to several years in the future. Its principal near-range product is a 6-to-10-day outlook for temperature, precipitation, and the height of the 500-mb surface. Each theme occupies a separate map, the result of running standard forecast models beyond the range of the five-day forecast, averaging over days 6 through 10, and comparing this result with the corresponding 30-year mean. When the models improve, maps averaging days 8 through 14 will provide a true second-week forecast.[30]

Usually presented in tandem, the temperature and precipitation outlooks (plate 20) represent "normal" in white and highlight departures in contrasting hues: red and blue for temperature, green and brown for precipitation. A less-is-less strategy helps viewers distinguish "below" and "much below" normal temperatures portrayed in light blue and deeper blue, respectively, while depictions of "above" and "much above" normal temperatures take advantage of traditional associations of yellow with sunshine and red with danger. On the precipitation chart, lack of "much below" and "much above" categories reflects the increased challenge of forecasting rain or snow more than a week ahead. In addition to drawing on emotive associations of brown with parched and green with lush, the map uses a distinctly different, zigzag brown pattern to identify but not emphasize zones with no precipitation in the forecast—an appropriate distinction in the Southwest, where dry weeks are the norm for all or part of the year.

For fire officials, highway superintendents, and others who need to plan for bad weather months in advance, the Climate Prediction Center also offers "long-lead outlooks" for one- and three-month periods extending over the next year.[31] Updated in the middle of every month, each long-lead prediction consists of two series of temperature and pressure maps. Supplementing 13 monthly maps with lead times ranging from 0.5 to 12.5 months are 12 overlapping three-month, seasonal maps. Because the latter predictions include neither the current month nor the month following, a few of these seasonal maps actually look a bit farther than a year into the future. For example, the outlooks issued in mid-May begin with a prediction for July, August, and September of the current year and end with an outlook for June, July, and August of the following year. For the convenience of Web users, the center provides separate maps for North America and the conterminous United States.

Temperature   Precipitation

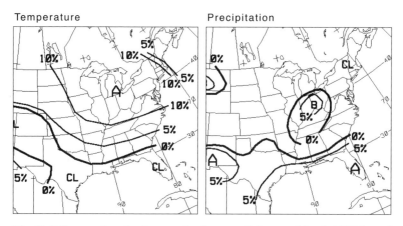

Fig. 11.6. Seasonal outlook maps for January-February-March 1998, issued the previous May. NOAA Climate Prediction Center images reformatted for Web-site distribution by the Northeast Regional Climate Center (Appendix: Northeast Climate).

The Northeast Regional Climate Center reformats the maps into smaller, slightly overlapping eastern and western views better tailored to low-resolution monitors. As figure 11.6 illustrates for the eastern states, symbols describing the prediction cover neither Canada nor Mexico, and the symbology is cryptic, at least at first glance. Issued in mid-May 1997 for January, February, and March 1998, the left-hand map predicts above-average temperatures for most of the East, whereas its companion calls for below-average precipitation within an oval-shaped region extending from Ohio into northeastern Arkansas. I can tell because "A" and "B" represent above and below normal conditions, respectively. Elsewhere the code "CL," for "climatology," identifies areas where above-normal, near-normal, and below-normal conditions are equally likely. For these zones there's no better guess than the distribution of seasonal temperatures for the reference period 1961–90.

What then does an above-normal temperature outlook mean for the area between New York and Iowa? To interpret this outlook, the viewer must first understand a few definitions:

1. The outlook map predicts seasonal mean temperature. For a given station the seasonal mean for January, February, and March is calculated by (1) adding the high and low temperatures for each day and dividing by two to find the *daily mean*, and (2) adding the daily means for the three-month period and dividing by 90 (91 for a leap year) to find the *seasonal mean*.

2. The distribution of 30 seasonal means for the period 1961 through 1990 can be ranked from highest to lowest, and divided into three equal-size groups of ten: above normal, normal, and below normal. In a "climatology" forecast, the seasonal mean has an equal likelihood (33.3%) of falling in each group.[32]

3. Percentages like the 10% on the temperature outlook map in figure 11.6 are *anomaly probabilities* indicating the excess likelihood that the seasonal mean will be above or below the middle third, or normal group. A station directly on the 10% line in the "A" region thus has a 43.3% likelihood (33.3% + 10%) that its seasonal mean temperature will fall in the above-average third of the 30-year distribution.[33] Because percentage probabilities must sum to 100%, the likelihood of a seasonal mean in the below-average third is only 23.3%.[34]

Because the map's contours describe a broad dome, stretching from New York to Iowa, with probability anomalies greater than 10% but not as high as 15% (which would require another isoline), the likelihood here of an above-average seasonal mean is greater than 43.3% but not as high as 48.3% (33.3% + 15%). As of the preceding May, the odds looked quite good for a winter-month mean *at least* in the normal range. And as my heating bills for early 1998 confirm, the map was right on target with its forecast of a mild winter.

How reliable are long-lead seasonal outlooks? It all depends, experts say, on the weather element predicted, the amount of lead time, the season, and the region of the country.[35] Skill scores are lower for precipitation than for temperature and, as seems reasonable, lower for a season a year away than for the next three months. Especially problematic are the transitions between winter and summer: in particular, seasonal outlooks for January-February-March and July-August-September typically produce the highest skill, whereas forecasts for April-May-June and November-December-January are the least reliable. Spring more than fall is also a "barrier" to reliable long-lead forecasting insofar as predictions looking ahead eight months from January typically exhibit less skill than eight-month predictions made in July. In general, any forecast that reaches forward through spring tends to be less reliable than a forecast similar in length that does not look past spring. Skill scores also tend to be lower in the interior of the continent, where the oceans cannot dampen fluctuations in temperature, as well as in the Rocky Mountains, where short-term variations create "climatic noise."[36]

Ed O'Lenic, head of forecast operations at the Climate Prediction

Center, compares the skill and promise of NOAA's long-lead seasonal outlooks to batting averages in the major leagues, where .250 is moderately good these days. Right now, he says, the seasonal outlook is "batting in the .220–.230 range but steadily approaching .250." And look for improvements, O'Lenic predicts, "with averages above .250 in the next decade."[37]

Impressive results already achieved reflect improvements in data collection as well as the synergy of combining statistical analysis with dynamic computer models.[38] Especially important are the worldwide network of buoys for measuring tropical temperatures, winds, and currents, and coupled ocean-atmosphere models that relate Hadley cells and tropical thunderstorms to changes in sea-surface temperatures. Simulations that link ocean currents and atmospheric circulation are relevant because the equatorial Pacific Ocean plays a key role in North American weather. Every three to five years on average the surface waters of the equatorial central and eastern Pacific become a couple of degrees warmer: just enough to increase the incidence of tropical thunderstorms, intensify Hadley circulation, shift the subtropical jet stream farther east, and affect much of the world's weather. Lasting from 12 to 18 months, these changes typically include stormy, wet conditions across the southern states; warm, dry weather across the northern United States, especially the Pacific Northwest; and severe drought in Australia and Indonesia. Because this abnormal warming of the tropical Pacific typically appears in late December, Peruvian fisherman long ago commemorated the birth of Christ by calling the phenomenon El Niño (the Child).[39]

Meteorologists recognize El Niño as the warm phase of the Southern Oscillation, a multiseason seesaw phenomenon described in 1924 by Sir Gilbert Walker, director of the Indian meteorological service.[40] Sifting through climatic data for relationships useful in forecasting variations in the Asian monsoon, Walker noticed a negative correlation between sea-level pressure in Darwin, on the northwest coast of Australia, and Tahiti, in the eastern Pacific near the equator: higher-than-average mean pressure at Darwin meant a negative anomaly at Tahiti, and vice versa. Although he discovered correlations with American weather, he failed to find a reliable predictor of abnormal monsoons.[41]

Walker's discovery was of little interest until the late 1960s, when Jacob Bjerknes proposed a physical explanation for the eastward drift of the warm surface waters that trigger El Niño.[42] Subsequent analysis confirmed ENSO (El Niño and the Southern Oscillation) impacts on a variety of weather events in North America, including the drought of summer 1988 and the floods of 1993 and 1995.[43] A confident understanding of El Niño allowed the Climate Prediction Center to give further meaning to the map in figure 11.6, described on its Web site as "broadly consistent with past patterns associated with above normal [sea surface temperatures] in the central equatorial Pacific in that season."[44] An evolving understanding of other broadly periodic, "interannual" phenomena—none as blatant as El Niño, though—holds the promise of further improvement in predictive skill.[45]

More vexing than any other question in climatology is the threat and uncertainty of global warming.[46] The atmosphere is heating up, studies suggest, and there's a physical reason for it: increased concentrations of "greenhouse gasses," which absorb infrared (heat) radiation from the surface below. Carbon dioxide, water vapor, and ozone, to name a few, reradiate some of this heat into space and some of it back to the surface.[47] As these gasses become more abundant, more heat is trapped and surface temperatures rise. Although fluctuations in the annual mean temperature make this increase difficult to confirm, it's clear that carbon dioxide, methane, nitrous oxide, and chlorofluorocarbons are markedly more abundant in the atmosphere now than in the preindustrial early nineteenth century. Explanations, all plausible, include chemical manufacturing, the use of fossil fuels, and deforestation.

Controversy over global warming is inherently political.[48] Efforts to control the release of greenhouse gasses are unpopular in the United States and other developed nations, where cutbacks in manufacturing would diminish consumption and increase unemployment. Although we accept some restrictions to control air pollution (a more immediate problem with demonstrable health effects), some farmers might well prefer slightly warmer temperatures—even at the expense of coastal residents forced to move because of a rise in sea level. Still less enthusiastic are citizens of developing nations, asked to forgo improvements in their standard of living while the West consumes and pollutes at markedly higher levels.

Scientists remain skeptical about global warming for a variety of reasons.[49] We still have much to learn about linkages among human activity, atmospheric chemistry, and climate change. Particularly problematic is carbon dioxide, an abundant but comparatively inefficient greenhouse gas. Sea water, which absorbs carbon dioxide from the air, might mollify the harmful effects of burning coal, oil, and wood. By contrast, increased temperatures could produce further warming if shrinkage of the polar ice caps were to diminish the

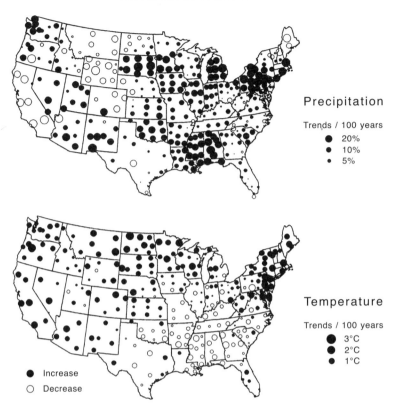

Precipitation

Trends / 100 years
● 20%
● 10%
· 5%

Temperature

Trends / 100 years
● 3°C
● 2°C
● 1°C

● Increase
○ Decrease

Fig. 11.7. Precipitation and temperature trends by climate division, 1900–1994, converted to per-century rates. From Thomas R. Karl and others, "Indices of Climate Change for the United States," *Bulletin of the American Meteorological Society* 77 (1996): 282.

planet's reflectivity. Because long-term interactions of the atmosphere, hydrosphere, and biosphere are not well understood, these effects and their impact are difficult to predict. Moreover, natural cycles of warming and cooling over the past million years raise the possibility of global cooling, perhaps in a mini Ice Age similar to the reduced temperatures that repeatedly clobbered Scottish agriculture between 1550 and 1850, when the Gulf Stream seems to have shifted southward. And as Mount Pinatubo demonstrated in 1991, volcanic eruptions can lower temperature for several seasons by ejecting ash particles, which tend to repel incoming solar radiation.

To further complicate matters, ambiguous data suggest that some regions have cooled while others warmed. In 1996, for instance, climatologists at NOAA's National Climatic Data Center published a study of year-to-year variations in precipitation and temperature between 1900 and 1994.[50] After diligently adjusting for changes in instrumentation, urban heat island effects, and other biases, Thomas Karl and his colleagues computed area-averaged trend indices for the 344 climate divisions in the conterminous United States. Reproduced in figure 11.7, their maps reveal coherent regions of increase and decrease for both precipitation and temperature. In particular, a warming trend dominates New Jersey, New York, New England, and the upper Midwest, while cooling is equally prominent in the Southeast and South Central regions. Although weighting for differences in area yielded an overall nationwide increase of 0.4°C per century, a time-series graph showing considerable annual fluctuation echoes the geographic contradictions of the lower map. Whether warming or cooling is the global trend, climate change is far from straightforward.

More immediately ominous is the hole in the earth's ozone layer, which few scientists dispute.[51] A pungent pollutant near the surface, ozone in the stratosphere shields life below from excessive ultraviolet radiation, a prime cause of skin cancer. A form of oxygen in which three rather than the normal two atoms bond together, ozone molecules (O_3) are vulnerable to chlorine oxides, which accumulate over the poles during winter and unleash a vicious attack in early spring. By one estimate, a single free radical of chlorine can collide with and destroy a hundred thousand ozone molecules.[52] The source of this ozone-destroying chlorine is a group of chemicals called chlorofluorocarbons (CFCs), developed in the 1930s and widely used until recently in refrigerators, air conditioners, aerosol sprays, and various foam products. Highly stable at lower altitudes, CFCs rise slowly into the stratosphere, decompose under intense ultraviolet radiation, and release highly reactive chlorine.

In *Mapping the Next Millennium*, science writer Stephen Hall describes the surprising and ironic discovery of the ozone hole.[53] In 1957, during the International Geophysical Year, meteorologist Joseph Farman installed ozone spectrometers at two of Britain's Antarctic ground stations. Estimating the amount of ozone from the intensity of ultraviolet light reaching the surface, his sensors detected a distinct decline in springtime (August-November) ozone levels between 1978 and 1982. The trend persisted in 1983, and after a new instrument installed in early 1984 confirmed his earlier measurements, Farman and his colleagues wrote an article, which *Nature* promptly published.[54] Serendipitously it seems, advance word of his findings encouraged NASA researchers at the Goddard Space Flight Center to exhume data archived from the Total Ozone Mapping Spectrometer (TOMS) on Nimbus 7, launched into a polar orbit in 1978. Because the instrument generated an overwhelming 190,000 ozone readings every day, NASA scientists had yet to look for trends. When they did, the TOMS data provided persuasive cartographic confirmation of the ozone hole and eventually convinced 149 nations to ratify the Montreal Protocol of 1987 and agree to cut the use of CFCs in half by 1999.[55] As NASA official Robert Watson recalled, "As soon as we made images, we could see it wasn't a small, isolated hole."[56]

Images of ozone abound in the late 1990s, to verify the threat as well as hint at the turnabout projected for early in the next century. A key relates the maps' vivid colors to a scale of Dobson Units (DU), based on the intensity of backscattered ultraviolet radiation and calibrated to reflect the gas's thickness, in hundredths of millimeters, if all the ozone molecules in a vertical column were gathered near the ground at sea-level pressure. Scales ranging from 150 to 450 DU suggest that 300 DU, representing 3 mm of "total ozone," is more or less typical. But spread upward through 30 miles of atmosphere a 3-mm column of ozone is less a "layer" than a mist. Even so, the maps can reveal frightening trends when December views of the south pole are arrayed side by side for adjoining years, as in Hall's book.[57]

For me, few media are as persuasive as film and no ozone propaganda are as instructive as the time-series animations that NASA distributes over the World Wide Web.[58] In a window consuming a tiny portion of the screen—much more might tax the Internet as well as strain hard drives and central processors—a rapid succession of highly generalized TOMS images can recreate the ozone layer's spectacular shrinkage in the austral spring, when disappearance of the black circle of polar night heralds a new assault of CFCs, followed by a slow but not fully complete recovery through summer and fall, when solar energy splits O_2 molecules into loose partners that combine with intact molecules to form O_3 threesomes. Although the annual recovery is promising, a month-by-month comparison of 1979 and 1992 (plate 21) verifies a frightening trend.

Our best defense against catastrophic climate change rests on a four-fold strategy of monitoring, modeling, archiving, and visualizing. In addition to revealing less subtle environmental hazards like volcanic eruptions and the ozone hole, networks of satellite and surface sensors provide a foundation for numerical models that test hypotheses about ocean-atmosphere interactions as well as predict the consequences of remediation tactics or new dangers. Archiving provides an information base essential for validating long-range models and assessing the significance of newly discovered anomalies. Visualization supports the other three strategies by helping researchers and policy makers cope with an increasingly overwhelming variety of intriguing, highly complex maps and by encouraging the data to speak for themselves.

Managed Myopia

Atmospheric cartography is part paradox, part enigma. No other scientific discipline or government enterprise is as pervasively cartographic as meteorology yet so disconnected from the community of professional cartographers. Maps abound in the meteorological literature, but rarely do its research studies address projections, symbols, and other overtly cartographic issues. The feeling seems mutual: cartography's journals and technical conferences have had little to say about the mapping of weather and climate—far less, to be sure, than the profession's attempts to improve maps portraying terrain, geology, soils, city streets, and property boundaries. The atmosphere, it seems, is a province apart, invisible and largely ignored by those who study the design and use of maps.[1] As if to retaliate, meteorologists go their own way, making many more maps in both number and variety than any other earth science, and doing so without explicit recourse to the professional cartographer's theories and practices.[2]

Equally perplexing is the scant attention accorded maps of the atmosphere by historians of cartography: a perverse ignorance matched only by the cartographic inattentiveness of historians of meteorology, who employ cartographic artifacts to decorate and authenticate yet pay scant attention to their design and construction. It's easy to do, though—until this chapter I've said little about symbols and projections—because maps of weather and climate can be intriguing eye candy as well as convincing testimony to the indisputable progress of atmospheric science.[3]

An explanation for this mutual unawareness lies partly in the distinction between mapping and map making. Weather science, it's clear, is immersed in mapping, that is, in the meaningful representation of data and spatial relationships in a two-dimensional plane, whereas cartography has traditionally focused on designing mass-produced atlases and topographic maps as well as one-of-a-kind, more or less handcrafted maps for books, scholarly journals, and the news media. In this sense, meteorological maps are akin to corporate memoranda, military orders, and other highly codified communications, whereas cartography's products are the graphic equivalents of novels, short stories, and news articles. It's a useful analogy because weather maps, like corporate memos and military orders, are not only fleeting and frequent but intended for rote and rapid use by specialized readers who understand their meaning. And like other highly coded institutional communications, weather maps can be straightforward or complex, clear or muddled.

With unchanging keys and ephemeral patterns, weather maps comprise a distinct cartographic genre serving unique user communities (professional meteorologists, pilots, weather enthusiasts), often in the supportive, map-rich environment of the forecast office, briefing room, or television weathercast.[4] Despite constraints of tradition and bureaucracy, the weather map is not only a product, more or less efficient in its design than plausible alternatives, but also a representation that favors a discernible viewpoint, however subtle. It's informative as well as proper, then, to examine the weather map's graphic integrity.

I'll begin with scale, the foremost of any map's three key properties: scale, projection, and symbolization. Defined as the ratio of map distance to ground distance, scale reflects both level of detail and geographic scope. When reported as a fraction with one in the numerator, scale affords a revealing distinction between large-scale and small-scale maps. In the cartographer's lexicon, fractions of 1/25,000 or greater are typical of *large-scale* maps, which can carry lots of local information, whereas fractions of 1/500,000 or less characterize *small-scale* maps, which have little room for local detail, except at isolated points like weather stations. (If this sounds confusing, consider that "large" and "small" refer to the whole fraction, not its denominator, in the sense that 1/2 pie is larger than 1/4 pie.) Because weather maps usually cover a country, continent, or hemisphere in a space no wider than two newspaper columns or a computer monitor, they're almost always small-scale maps. Complex point symbols encoding up to 18 separate observations for stations about 150 miles apart require somewhat larger formats, as do the surface-analysis charts worked up manually to inform forecasters at the National Meteorological Center, but these purposeful enlargements easily pass beneath the 1:500,000 threshold.[5] And even coarse, zoomed-in NEXRAD images, like the tornado close-up in plate 9, have scales in the intermediate range, well below 1:25,000. Rarely, it seems, do meteorologists present their maps at inappropriately large scales.

Closely related to scale is resolution. Whether generated with a numerical model or captured by a satellite or radar scanner, gridded data constrain the range of acceptable scales. At the lower limit are scales too small for the uncluttered display of interpolated isolines or individual pixels, which are a more significant limitation for ground-based radar systems like NEXRAD than for purposefully coarse numerical models and satellite imagery, constrained by computational complexity, bandwidth, and various trade-offs examined in chapters 5 and 7. In most instances, though, coastlines, boundaries, legible labels, and other frame-of-reference features set the display scale's lower limit.

Less obvious is the upper limit of meteorological map scales. Although needlessly large weather maps are easy to create, aesthetics and graphic efficiency discourage zooming in too closely on the myopic patterns revealed by imaging systems and sampling networks. Except for tornadic winds and other storms that merit close inspection, weather scientists prefer as wide a view as they can cram onto a display screen or printed page. In general, scales no larger than 1:10,000,000 are convenient for observing and forecasting "synoptic scale" phenomena like hurricanes, northeasters, and high-pressure systems, with diameters up to 3,000 miles, whereas scales between 1:2,000,000 and 1:100,000 are generally adequate for tracking "mesoscale" features, such as tornadoes and thunderstorms, with diameters between .5 and 50 miles.[6] Although turbulence and other "microscale" features are too fleeting to warrant detailed mapping over large areas, researchers occasionally use map scales greater than 1:25,000 to examine terrain's influence on climate and weather's impact on trees and soil.[7]

Explicit statements of scale are uncommon on weather maps, partly because scale varies considerably from place to place on maps of continents and hemispheres. When scale is noted, as on the National Weather Service's weekly series of Daily Weather Maps, the statement is graphical, not verbal or numerical.[8] Figure 12.1, taken from the largest of the four maps used to describe each day's weather, is a melange of five separate bar scales illustrating a steady

POLAR STEREOGRAPHIC PROJECTION TRUE AT LATITUDE 60°

SCALE OF NAUTICAL MILES AT VARIOUS LATITUDES

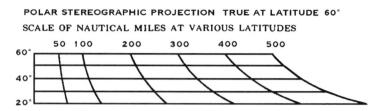

Fig. 12.1. Graphic scale (enlarged) from the 7 A.M. surface weather map, the largest of four maps describing one day's weather in NOAA's weekly series of Daily Weather Maps.

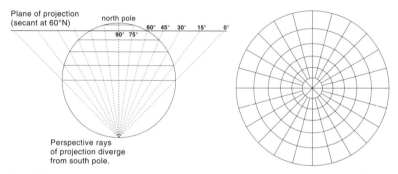

Fig. 12.2. A polar stereographic projection constructed from globe (left) secant to a plane at 60°N yields a grid (right) of converging meridians and concentric parallels.

poleward decline in scale. According to the scale diagram, an inch at 50°N, the approximate latitude of Winnipeg, represents a distance 18 percent longer than would an inch at 30°N, the latitude of New Orleans. Along a given parallel, though, scale remains constant.

Scale variation of some sort is not only common but essential: the only map on which scale is constant everywhere—and in all directions from every point—is a globe. While the stretching needed to flatten a sphere need not be apparent on a large-scale map of a tiny portion of the earth, distortion is at least noticeable, if not prominent, on small-scale maps of continents. On the map from which I extracted figure 12.1, the parallels for 25° and 30°N are 25 percent farther apart than the parallels for 50° and 55°N. However subtle, north-south scale on the Daily Weather Map rises steadily with distance from the north pole.

As the label above the graphic scale notes, NOAA plots the Daily

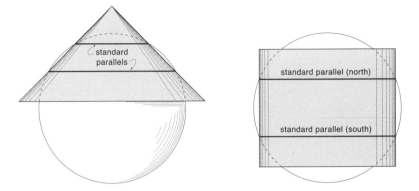

Fig. 12.3. Projecting from the globe onto a secant cone (left) or secant cylinder (right) yields two standard parallels and a broader low-distortion zone than if the projection were tangent (merely touching) at a single standard parallel.

Weather Map on a polar stereographic projection with true scale at latitude 60°N. The left half of figure 12.2 describes this projection's construction from a hypothetical globe, which defines the map's scale.[9] The plane of projection intersects the globe along the 60th parallel, which has the same scale on both projection and globe. Rays diverging from the south pole specify projected positions for other parallels, plotted 15° apart in the diagram to the right. The result is a grid of concentric parallels and evenly spaced meridians that meet at the north pole. Because the angles between converging meridians are true azimuths, the projection is called azimuthal. Unlike other azimuthal projections, the stereographic projection is also conformal, meaning that it represents true angles at all other points as well. Meteorologists consider conformality useful for plotting wind direction at stations and for describing relationships between winds and isobars.[10] As a bonus, stereographic projections portray all circles on the globe as either circles, circular arcs, or straight lines.[11]

In cartographic jargon, a projection that cuts the globe is *secant*, and the circle along which globe and plane intersect is a *standard parallel*. These terms are important because whatever distortions occur increase with distance from the standard parallel, which defines a belt of comparatively low distortion. Moreover, as figure 12.3 describes, projecting from a globe onto a secant cone or cylinder can yield an even larger zone of low distortion defined by two standard parallels, the positions of which the cartographer can adjust to lessen distortion within a specific region. In particular, a secant conic projection with both standard parallels in the same hemisphere is especially suited to mid-latitude regions, whereas a cylindrical projection with identical standard parallels in both hemispheres can define a broad low-distortion zone centered on the equator.

An early advocate of regionally tailored map projections was Vilhelm Bjerknes, patriarch of the Bergen school. In 1919, Bjerknes urged weather services to adopt three conformal projections: a stereographic projection secant at 75° for polar regions, a Lambert conic projection secant at 30° and 60° for middle latitudes, and a Mercator cylindrical projection secant at 15° for equatorial regions.[12] He also proposed four common scales: 1:2,500,000, 1:5,000,000, 1:10,000,000, and 1:20,000,000. Standardization, Bjerknes argued, would greatly benefit meteorology by allowing researchers to readily assemble detailed mosaics of synoptic charts from various countries.

Weather officials agreed in principle but took nearly two decades to refine a set of nine resolutions, approved in 1937 by the International Meteorological Committee, meeting in Salzburg. Guided by Bjerknes's principles, the Salzburg resolutions broadened the scope of the polar stereographic and Mercator projections by shifting their standard parallels to 60° and 22.5°, respectively.[13] Recognizing the trade-off between geometric accuracy and geographic scope, the resolutions' authors preferred the flexibility of a polar stereographic projection covering a full hemisphere and a Mercator projection accommodating somewhat higher latitudes.[14] This flexibility proved beneficial to the Daily Weather Map: although the map's focus on the conterminous United States and adjacent portions of Canada and Mexico might make the Lambert conformal conic projection a more appropriate choice, its preparation is closely linked to a hemispherical weather chart, which requires a polar stereographic perspective.[15]

The Salzburg resolutions also recognized the different context of climate maps, on which the relative areal extent of a drought, shortened growing season, or specific climate classification is more important than precise angles and wind directions. Because conformal projections can egregiously distort areal relationships, the International Meteorological Committee called for equal-area projections for "climatological uses."[16] And because equal-area projections can greatly distort shape and angles in regions well removed from a standard parallel, the resolutions' authors reiterated for climate maps their plea for regionally tailored projections based on a plane secant at 60°, a cone secant at 30° and 60°, or a cylinder secant at 22.5°.

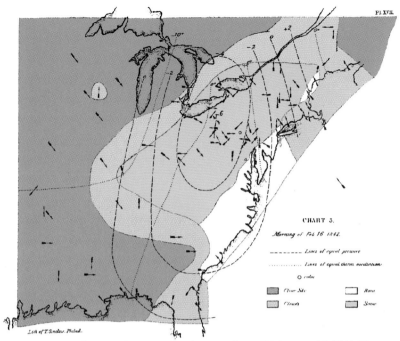

Plate 1. Loomis's weather map for the morning of February 16, 1842. From Elias Loomis, "On Two Storms Which Were Experienced Throughout the United States, in the Month of February, 1842," *Transactions of the American Philosophical Society* 9 (1846): plate XVII.

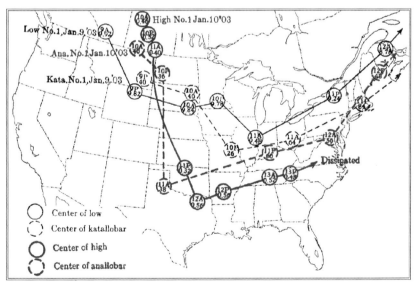

Plate 2. Allobar and pressure-cell tracks for a high formed on January 10, 1903, following a low formed on January 9, 1903. From Alfred J. Henry and others, *Weather Forecasting in the United States*, Weather Bureau publication no. 583 (Washington, D.C., 1916), fig. 28, facing p. 85.

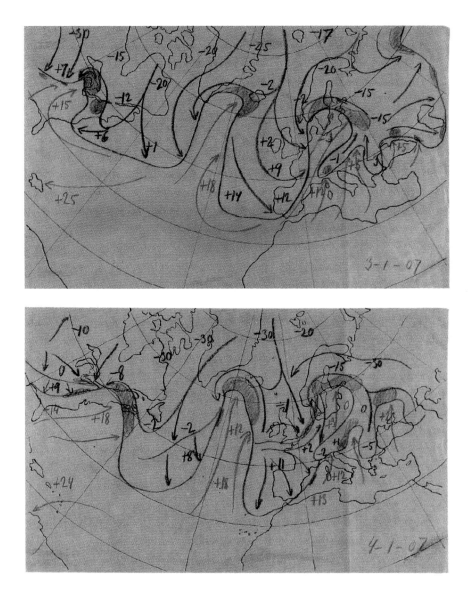

Plate 3. Solberg's drawing of a continuous circumpolar front for January 3, 1907. Courtesy of Ralph Jewell.

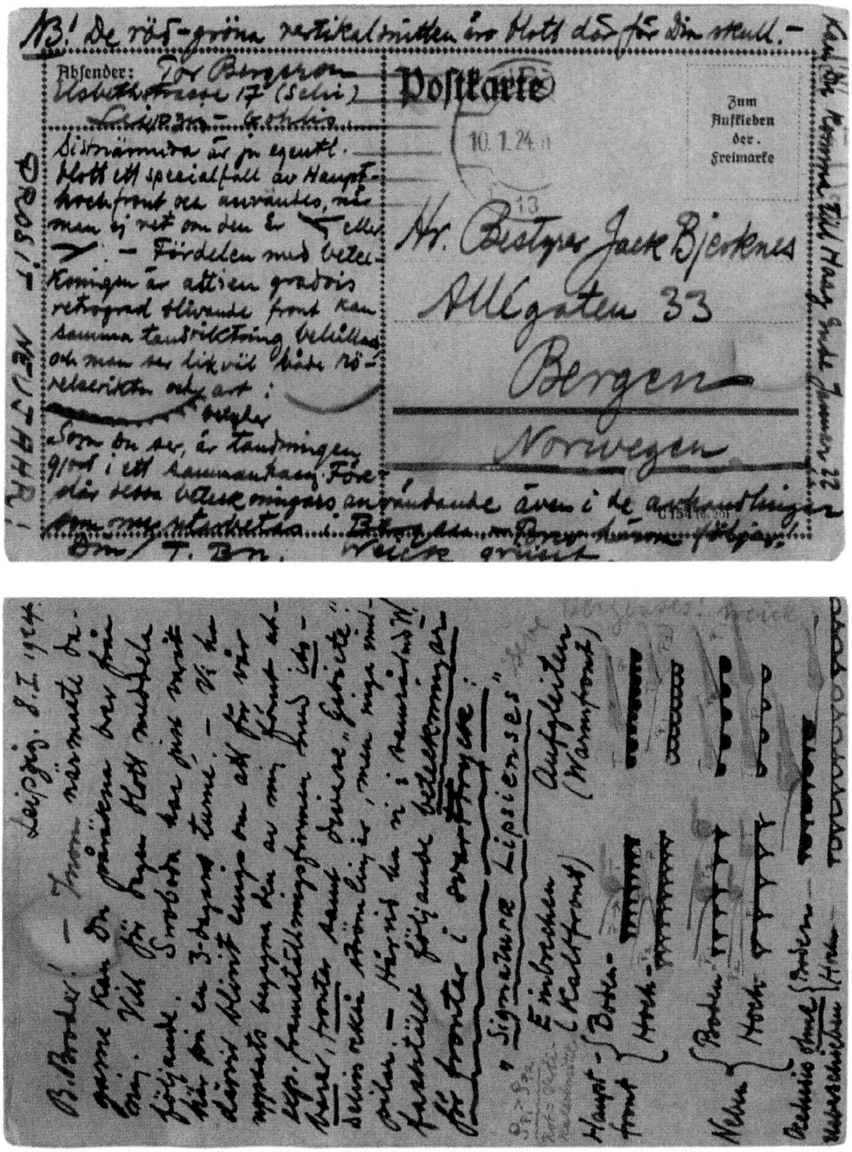

Plate 4. Bergeron's postcard to Bjerknes, dated January 8, 1924, describing a cartographic code for differentiating fronts without color. Courtesy of Ralph Jewell.

GOES-8 Color Enhanced IR
August 29, 1996 11:15 UTC

NOAA

Hurricane Edouard

Hurricane Fran

Tropical Storm Gustav

Plate 5. GOES-8 color-enhanced infrared image showing hurricanes Edouard and Fran, followed by tropical storm Gustav. The image was scanned on August 29, 1996, starting at 7:15 A.M., eastern daylight time. Red and yellow represent extraordinarily high amounts of moisture and cloud cover. Courtesy of the National Oceanic and Atmospheric Administration.

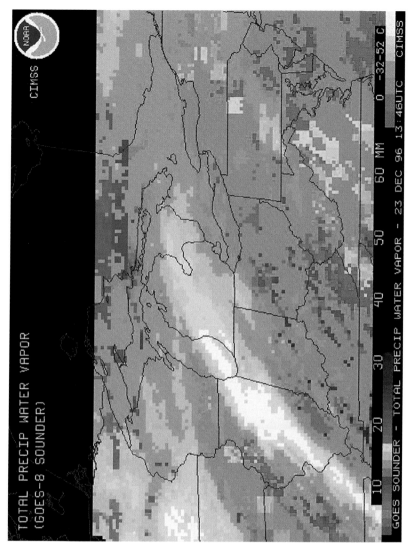

TOTAL PRECIP WATER VAPOR
(GOES-8 SOUNDER)

10   20   30   40   50   60 MM   0 -32-52 C

GOES SOUNDER - TOTAL PRECIP WATER VAPOR - 23 DEC 96 13:46UTC   CIMSS

Plate 6. Total precipitable moisture in the Great Lakes area, derived from GOES-8 Sounder measurements taken at approximately 9 A.M., eastern time, December 23, 1996. Measured in millimeters, precipitable moisture is the amount of liquid water if the vapor in a vertical column extending from the surface to the top of the atmosphere were condensed. As indicated by the color bar, the driest areas are brown and the most moist regions are red. The computer cannot reliably estimate precipitable moisture in areas with extensive cloud cover, shown in gray. Precipitable moisture is important because westerly winds passing over relatively warm lake surfaces pick up considerable moisture, which is deposited downwind over land as lake-effect snow. Available every hour, these maps are most informative when viewed as frames in a time-series animation. Courtesy of the Cooperative Institute for Meteorological Satellite Studies, Space Science and Engineering Center, University of Wisconsin–Madison.

Plate 7. Twin-screen NEXRAD workstation. Courtesy of the NOAA Central Library.

Plate 8. Interactive AWIPS workstation. Courtesy of the NOAA Central Library.

Plate 9. Zoom-magnified NEXRAD image of the F4 tornado (characterized by winds 207 to 260 miles per hour) that tracked across Berkshire County, Massachusetts, around 7:15 P.M. on May 29, 1995. The Albany radar, at Berne, New York, just off the upper left corner of the map, took the image at 7:16 P.M. Courtesy of the National Weather Service, Binghamton, New York.

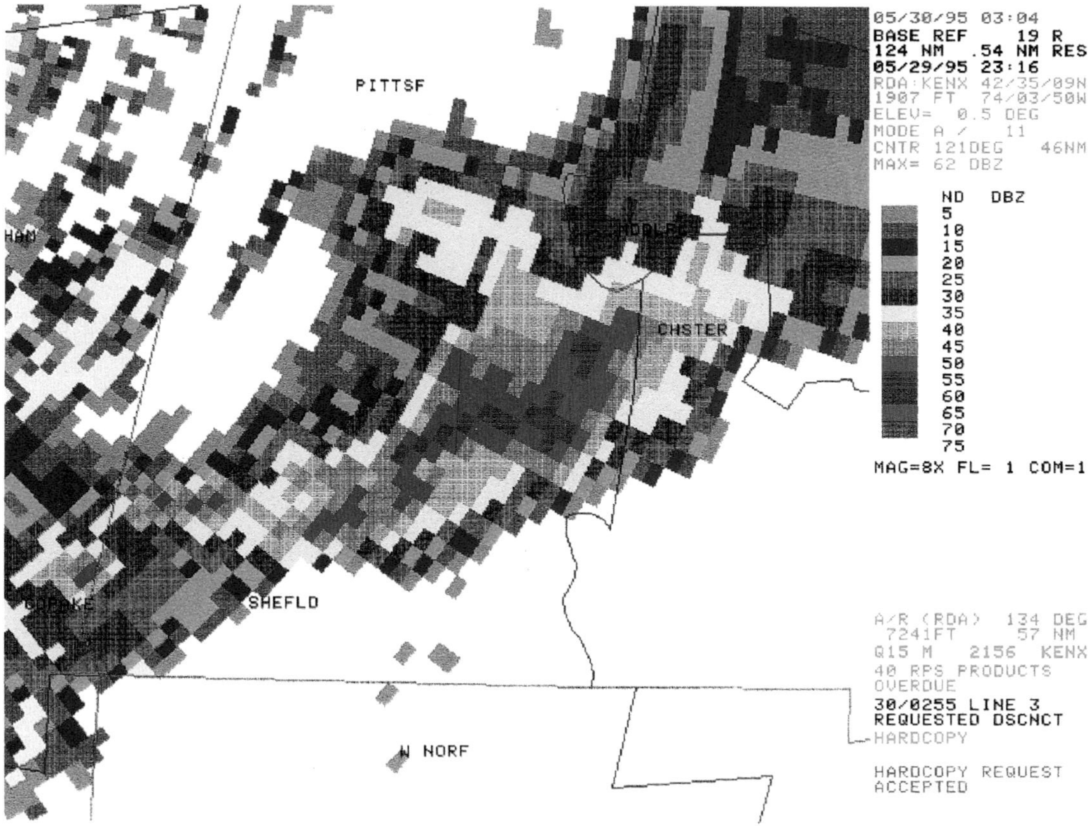

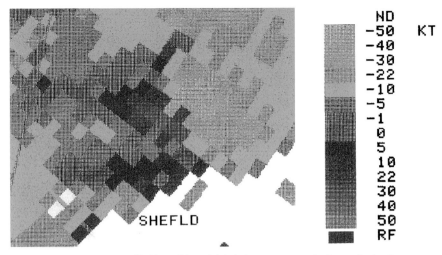

Plate 10. Excerpt (left) and key (right) from storm-relative-velocity image of the Memorial Day tornado in plate 9. Taken at 7:06 P.M., as the tornado rampaged through Great Barrington, the NEXRAD image shows winds moving toward the radar in green and away from the radar in red. Wind speed levels are in knots (KT). Courtesy of the National Weather Service, Binghamton, New York.

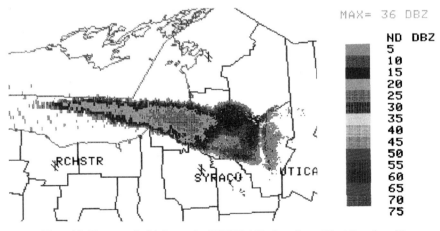

Plate 11. Excerpt (left) from the NEXRAD view from West Leyden, New York, of a band of lake-effect snow reaching eastward from Lake Ontario at 9:15 P.M. on January 19, 1994. Map key (right) illustrates how embedding three levels of intensity variations (light, medium, dark) in a sequence of spectral hues (blue, green, yellow-orange, red, and purple) yields a coherent scale with 14 distinct levels. Precipitation-intensity levels are in decibel reflectance units (DBZ). Courtesy of the National Weather Service, Binghamton, New York.

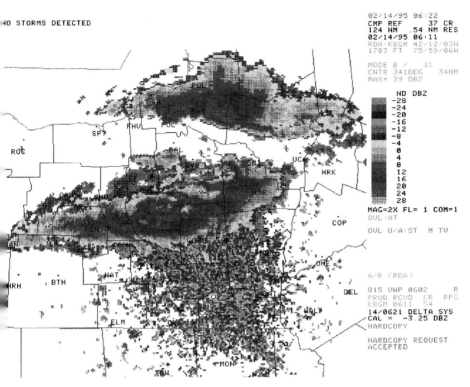

Plate 12. Operating in clear-air mode at 1:11 A.M. on February 14, 1995, the NEXRAD at Binghamton, New York, captured two bands of lake-effect snow stretching across upstate New York. The northern band off the east end of Lake Ontario is more typical than the southern band, extending more than 180 miles east from Lake Erie. Courtesy of the National Weather Service, Binghamton, New York.

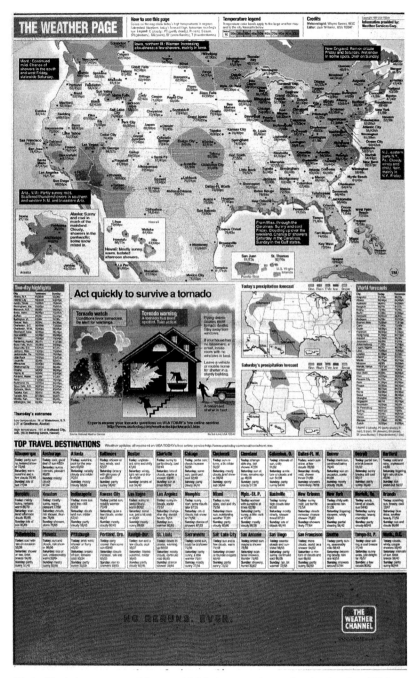

Plate 13. A large map of forecast high temperatures dominates the *USA Today* weather page. From *USA Today*, April 19, 1997, 16A.

Plate 14. The chroma-key graphics system adds text and logo after placing a satellite image behind the weathercaster. Courtesy of CNN.

Plate 15. Sea-Doo, a manufacturer of recreational watercraft, warns Weather Channel viewers of imminent thunderstorms.

Plate 16. The Weather Channel describes recent snowfall with red and white symbols on a shaded-relief base map.

Plate 17. Dave Eichorn interprets a "3-D SkyTrack" animation. Courtesy of WIXT-TV, Syracuse, New York.

Plate 18. Accu Weather's daily "feature graphic" retrieved with Netscape Navigator, a widely used Web browser.

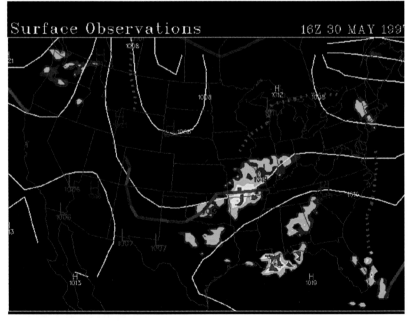

Plate 19. Surface weather map customized by the *Weather Visualizer* Web site at the University of Illinois.

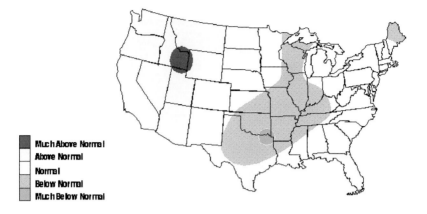

SIX TO TEN DAY MEAN TEMPERATURE FORECAST
June 26 - 30, 1997

Much Above Normal
Above Normal
Normal
Below Normal
Much Below Normal

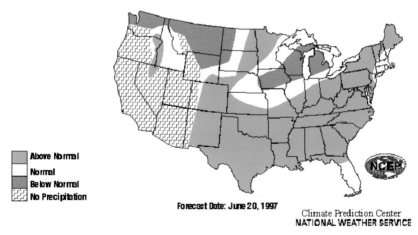

SIX TO TEN DAY MEAN PRECIPITATION FORECAST
June 26 - 30, 1997

Above Normal
Normal
Below Normal
No Precipitation

Forecast Date: June 20, 1997

Climate Prediction Center
NATIONAL WEATHER SERVICE

Plate 20. Six-to-ten-day forecast maps for mean temperature and precipitation from the *Climate Prediction Center* Web site (Appendix: Climate Prediction).

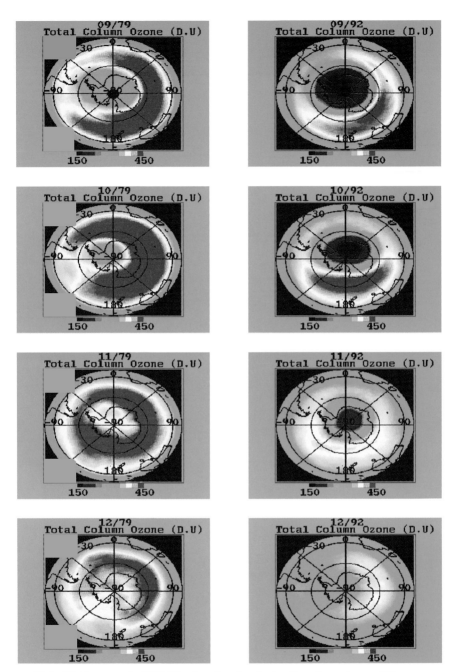

Plate 21. Frames of a TOMS animation for Antarctica show strong differences between 1979 (left) and 1992 (right) in mean monthly total ozone during fall and early summer. Lower DU values near the pole reflect the destruction of ozone. From "Monthly TOMS Data over Antarctica, 1978–1993," Quicktime movie distributed over the World Wide Web by the Goddard Space Flight Center (Appendix: Goddard DAAC).

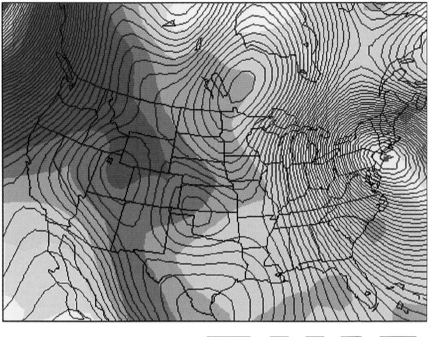

2 mb surface contours

models agree   models disagree

7 am Feb 4   [Begin] [R] [S] [F] [Play]

Plate 22. Frame from an animated pressure map computed by averaging three numerical forecast models. The key (inserted from a frame at the beginning of the animation) describes the use of color to communicate uncertainty. Five control buttons make the display interactive: "Begin" to display the map key, "Play" to run the loop, S to stop, and F and R to move forward or reverse one frame at a time. From on-line animation at Alan MacEachren's visualization Web site (Appendix: Geo Visualization).

More basic than standardized map projections are standardized map symbols. Efficient visual analysis demands detailed descriptions of diverse local weather conditions, encoded graphically on a single weather chart. But a code that allows effective communication within one nation's weather service might prove confusing, if not misleading, in countries with different linguistic and graphic conventions. Although straightforward metaphors like arrows portraying wind direction and a progression of open, split, and filled circles representing clear, partly cloudy, and cloudy skies afford common ground, alphabetic symbols such as R for rain and S for snow are inherently useless among nations with different vocabularies and alphabets. Equally troublesome are all-graphic codes of abstract symbols favoring one country's or region's unique weather. For these reasons the international meteorological congress held in Vienna in 1873 adopted a standard set of symbols broad in scope and largely pictorial or numeric.[17]

Standardized symbols, it turns out, predated the weather map. In a 1916 article in the *Monthly Weather Review*, Fitzhugh Talman, a professor at the U.S. Weather Bureau, attributed the first use of graphic meteorological symbols to the mathematician Johann Heinrich Lambert (1728–1755), coincidentally famous for the Lambert conformal conic projection endorsed by Vilhelm Bjerknes and the Salzburg resolutions. In 1771, Lambert proposed the small multielement symbols for clouds, rain, snow, fog, and thunder in the upper row of figure 12.4.[18] His goal was twofold: overcoming language barriers and saving space in printed records. The Meteorological Society of the Palatinate, which collected weather records from observers throughout Europe in the late eighteenth century, shared these goals. Notorious in the annals of meteorology for not mapping its data, the society devised an elaborate symbolic code for its correspondents and published records. As the middle and lower rows of figure 12.4 illustrate, the code included symbols for varying degrees of cloudiness and different forms of precipitation as well as for astronomical phenomena and atmospheric electricity. With no maps to provide a cartographic precedent, neither Lambert's nor the society's symbols had a significant influence on the international code.

Standardization of weather map symbols was neither instantaneous nor universal. The underlying terms and definitions—reflections of what meteorologists thought was worth observing or meas-

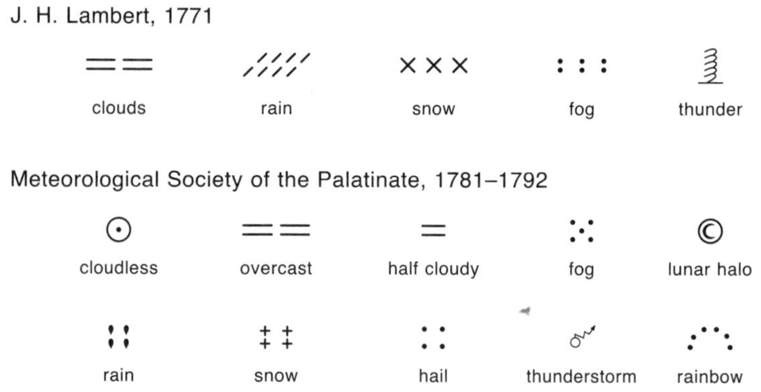

J. H. Lambert, 1771

═══	╱╱╱╱	✕ ✕ ✕	⦂ ⦂ ⦂	⹍
clouds	rain	snow	fog	thunder

Meteorological Society of the Palatinate, 1781–1792

☉	═══	═	⸫	Ⓒ
cloudless	overcast	half cloudy	fog	lunar halo

⁚ ⁚	⧻ ⧻	⁚	♂	⸫⸱
rain	snow	hail	thunderstorm	rainbow

Fig. 12.4. Weather glyphs devised by Lambert in 1771 (upper row) and the Meteorological Society of the Palatinate between 1781 and 1792 (lower rows). Compiled from C. Fitzhugh Talman, "Meteorological Symbols," *Monthly Weather Review* 44 (1916): 265.

uring—continued to evolve in the late nineteenth and early twentieth centuries.[19] With little international collaboration on chart production to guide them, government weather bureaus devised idiosyncratic symbols. Although an international conference held at Munich in 1891 expanded and modified the Vienna code largely by endorsing symbols used on German weather maps, the weather services in Britain and France continued to plot their own versions. Even in 1916, Talman reported, "the forms of these symbols are more or less flexible."[20] His table of 27 "international meteorological symbols" used by the United States Weather Bureau includes one to four "principal variants" for all but six of them.

Although some flexibility is still apparent, adherence to an international code is greater now than in Talman's time. It's too simplistic, though, to attribute wider acceptance of common map symbols to the International Meteorological Committee or its successor, the World Meteorological Organization.[21] The WMO supports the code with detailed lists of terms, definitions, and symbols, but increased standardization also reflects the rise of commercial aviation, the needs of NATO, increased international collaboration among atmospheric scientists, and larger, richer meteorological databases. The complexity and diversity of weather maps, especially maps of the upper atmosphere, is a strong incentive for international standardization.[22]

Continued flexibility is apparent in NOAA's Daily Weather Map, which employs an abridgment of the international code appropriate for educators, researchers, and weather enthusiasts. The specimen

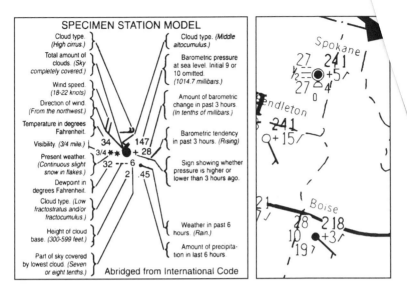

Fig. 12.5. A specimen station model (left) provides an abbreviated map key for decoding station symbols on the Daily Weather Map. As the enlarged excerpt (right) illustrates, not every station requires all 18 items.

station model on the left side of figure 12.5 shows the designated positions of 18 numbers and pictograms describing weather conditions at a single station. Measurements in feet, miles, and Fahrenheit degrees, rather than their metric equivalents, acknowledge American preferences—at least those of nonscientists. As the enlarged excerpt on the right illustrates, the map does not use all available symbols for every station. The station model and a large graphic key available from the Climate Prediction Center make decoding straightforward.[23] In this example Spokane, Washington, reports a cloudy sky, no wind, a temperature of 27°F, a dew point of 27°F, half-mile visibility in patchy fog, four-tenths of the sky covered by low clouds, and a barometric pressure of 1024.1 millibars: 0.5 millibars higher than three hours earlier as a result of pressure that rose then fell slightly. By contrast, Boise, Idaho, is a degree warmer, with greater visibility (10 miles) and a breeze out of the southeast at 3–7 knots (3–8 miles per hour.) Absence of a decimal number in the lower right indicates that neither city recorded measurable precipitation during the previous six hours.

Spokane's three pictograms are easy to spot in figure 12.6, a graphic array divided in two by a double vertical line. Immediately

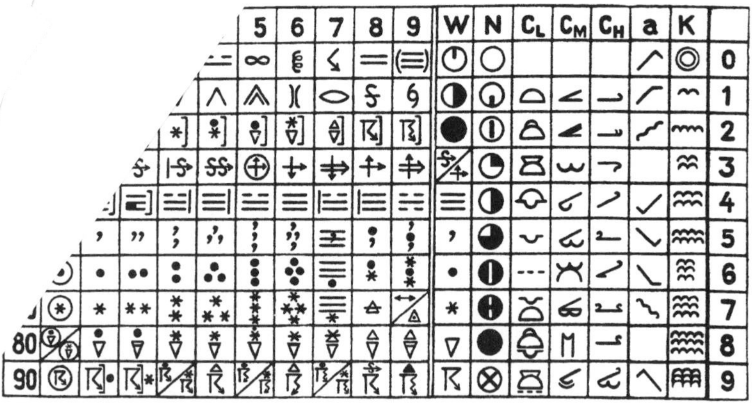

Fig. 12.6. Table of plotting symbols for observed data. From U.S. Weather Bureau, *Preparation of Weather Maps*, unnumbered circular dated July 1, 1942, 27.

to the left of this divider, grouped in the 40s row with symbols describing various foggy conditions, is pictogram 49, representing "fog in patches" with a solid horizontal line between two dashed lines. To the right of the divider, in the 10s row under the label C_L in a column of symbols describing low-level clouds, is the puffy, cloudlike semicircle on a flat base that represents cumulus clouds. And three columns to the right, in the 00s row below the label "a" in the column of pictograms for barometric tendency is the aptly chosen symbol for "rising, then falling." Ironically perhaps, the international standard that supports this graphic archaeology refers principally to the array's numbers, part of a precise system for exchanging data among government meteorological services. Required to conform to the World Weather Organization's rigorous definitions and formats for data exchange, weather bureaus are free to substitute their own symbols on charts for domestic use.[24]

Weather services usually employ the recommended plotting symbols, or at least most of them. The United States, as noted, uses a slightly different station model, while Britain's Meteorological Office incorporates a few of its own pictograms, most notably for "wet fog," "patches of shallow fog over land/sea," and "zodiacal light" (aurora borealis).[25] Wide acceptance of the standard symbols reflects both the inconvenience of maintaining one's own schema as well as the graphic logic embedded in the international code. Many symbols are suggestively pictorial, in particular the zigzag, lightninglike symbol for thunder and lightning (90s row), the counterclockwise whirl with the open

center for hurricane (pictogram 19), and most of the cloud symb
(under columns C_L, C_M, and C_H, for low-, medium-, and high-le\
clouds). Others, like the fog symbols in the 40s row, adopt a commc
element (horizontal lines) and vary their shape or number to connotc
change, coverage, or intensity.[26] Still others, like pictogram 85, representing "slight showers of rain and snow mixed," combine two or three
symbols, in this case, dots for rain, asterisks for snow, and downward
triangles for slight showers. In general, the same symbol repeated vertically indicates greater intensity, different symbols superimposed or
arranged vertically indicate coexistence, and different symbols
arranged horizontally indicate sequence. Hieroglyphics to the uniniti-
ated, the international code quickly becomes second nature to those
who use it regularly. In that sense, the graphic code of the weather map
is like the verbal languages of tribes and nations: adaptable to change
and understood within its community of users, it works.

While it's philosophically precarious, if not politically incorrect, to
question the efficiency of a natural language like English or Urdu, it's
hardly inappropriate to appraise the weather map's application of visual variables, a key concept in map design. Formally stated in the late
1960s by French semiologist Jacques Bertin, visual variables are what
the cartographer manipulates when designing a map.[27] Bertin identified eight of them: the grid or locational coordinates X and Y and
six retinal variables: shape, size, value, texture, hue, and orientation.
Of these, the typical weather map relies on *shape* to depict conditions
at weather stations and *orientation* to describe winds, advancing
fronts, and storm tracks. *Hue*, which differentiates among red, blue,
and other named colors, is also important, especially on radar and
satellite images, which encode wind speed and moisture with rain-
bowlike sequences. *Value*, which refers to the relative shade of gray
or a particular hue, occasionally portrays intensity or comparative
risk, and *size* operating as line width helps distinguish fronts (thick
lines) from isobars (thin lines). Although variation in *texture* (spacing) is rare, meteorological cartography invokes a ninth visual variable, *numerousness*, to encode both wind speed (with multiple barbs
on wind arrows, as in figure 12.5) and the intensity or continuity of
precipitation (as for sequences 61–66 and 71–76 in figure 12.6).[28]

In addition to identifying visual variables, Bertin placed restrictions on their use. Shape and hue, he argued, are inherently qualitative: ideal for representing differences in type or kind but poor if not
misleading for describing relative intensity or amount. By contrast,

insically suited to portraying quantities, and
y efficient for depicting direction: up or down,
rth or south. As embodied by wind arrows, orien-
the best graphic metaphor of all: learned in an in-
ᵖhor "arrows fly with the wind" obviates further need
ᴧap key.

or hue, meteorological cartography respects Bertin's pro-
s. Although some color symbols obey the canon—the red
ᴧe lines depicting the qualitative difference between warm and
ᴦronts on Internet maps like plate 19 are a good example—most
not. Radar and satellite images are blatantly deviant: I can envi-
ᴧon Bertin frowning at the contrast between a light drizzle coded in
green and the heavy downpour of an intense tropical storm coded in
orange. Intensity data, he would argue, call for an easily decoded sin-
gle-hue sequence with the darker-is-more logic embedded in the
Weather Channel's thunderstorm map (plate 15), not the wildly contrasting hues of a GOES or NEXRAD snapshot.

He'd be wrong, though, or at least only partly correct. Weather
forecasters make a fundamentally qualitative distinction between
light and heavy rain as well as between severe and light winds, and
radar imagery depends on vividly contrasting hues to differentiate
storm echoes from ground clutter.[29] Simply put, ease in spotting tiny
amounts of orange or red in a speckled mass of blues and greens out-
weighs other considerations. Visual variables are a useful construct,
but because tornadic hooks and sudden squalls demand immediate
attention, meteorological images must defer to a higher cartographic
principle: informative contrast.

Making contrasting colors widely and reliably informative is not
easy, especially when a single sequence must cover a wide range of intensities. As several of this book's color plates testify, long color keys,
with many categories, are common. For example, NEXRAD velocity images code winds with 14 colors running from light green
through red, whereas NEXRAD reflectivity images employ 15 colors ranging from light blue through white. In both instances, hues signify major differences while tonal variations within a hue portray less
substantial but nonetheless important differences. In plate 12, for example, various shades of red highlight areas of more intense lake-effect snowfall. But within bands of less intense snow more subtle differences between weaker and stronger greens and weaker and
stronger blues convey potentially meaningful information about
trends and intensity. Despite the complexity, mixing hue and value affords richer, more versatile color coding.

Hues also differentiate foreground information like velocity or flectivity from background features like coastlines and internation boundaries. Innately qualitative, these reference features compet with foreground data for a share of the color palette. Reserving brown and solid blue for terrain and water bodies, respectively, removes them from the list of contrasting hues that might portray moisture or wind velocity, and contributes indirectly to idiosyncratic symbols like green rain. Where possible, meteorological cartographers have adopted and reinforced conventional associations of red with danger, white with snow, red with heat, blue with cold (or water bodies), gray with clouds, and brown with land. Continued use has not only established green rain as a cartographic convention but anointed other associations, like purple areas for freezing temperatures and purple lines for occluded fronts. Because temperature maps rarely include fronts and many other feature combinations seldom coincide, a named color like red or purple can portray markedly different features on different maps with little risk of confusion.

Color conventions reflect in part the availability of printing inks and colored pencils. Although Elias Loomis's pink clouds and yellow rain (plate 1) never caught on, Tor Bergeron's pencil sketch of red and blue fronts (plate 4) proved as lasting as his pioneering black lines with triangular and hemispherical pips. Yet neither inks nor pencil leads proved as influential as cathode-ray tubes based on the red-green-blue (RGB) color model. These are the colors of the three sets of tiny phosphorescent dots lining the inside of a color monitor or television picture tube. The dots glow in color when excited by electrons fired from the back of the tube. A separate electron gun addresses each set of dots, so that an area bombarded by a single gun glows either red, green, or blue. Full-strength electron beams from all three guns yield white, full-intensity bombardment by just the red and green guns produces yellow, and adding full-strength red to half-strength green makes orange. An American Meteorology Society study published in 1993 identified 15 named colors used on weather maps.[30] As the percentage strengths in the table indicate, most of these standard colors rely on only one or two phosphors (table 12.1). Alarmed at a "proliferation of new applications of color" and eager to "unify" the variety of color schemes depicting weather information, the AMS study group "searched for commonality" and devised guidelines for software developers and the media.[31] For the most part, its survey identified an emerging consensus, with unique but consistent use of color for specific applications like radar and satellite imagery, aviation weather maps, and colors for weather information.

	.een	Blue	Name	Red	Green	Blue
	100	100	Tan	75	50	0
	75	75	Brown	50	25	0
	0	0	Green	0	100	0
,0	0	0	Light green	50	100	50
100	50	50	Blue	0	0	100
100	25	0	Light blue	0	50	100
100	50	0	Purple	50	0	100
100	100	0				

.: The numbers in the table are the percentage intensities of red, green, and blue ꞓd to produce the named colors. Tints (shades) of these colors are produced by mul-ꞇplying all three numbers for a color by the tint percentage. For example, 40% orange is the combination of 40% red (100 × 0.4), 20% green (50 × 0.4), and 0% blue.

various advisories, alerts, watches, and warnings. For example, aviation maps show levels of turbulence with a yellow-orange-red sequence relying on the easily remembered darker-is-worse principle. Less ergonomically correct are maps of air, soil, or sea surface temperature, which rely on the traditional spectral sequence blue-green-yellow-orange-red. Potentially confusing because perceived brightness is greatest in the middle (yellow) and lowest at both ends (blue, red), this cold-to-warm scheme is best reserved for maps of temperature—a theme reinforced by color-coded water faucets. Where a consensus was lacking—"a rare situation," we're told—the AMS committee applied human-factors principles.[32]

In appending a list of 12 recommendations for using its guidelines, the committee drew heavily on human-factors research. Suggestions include choosing colors that reflect familiar relationships, limiting their number, using them consistently, and allowing experienced users to manipulate colors on interactive workstations. Robert Hoffman, a psycholinguistics expert who advised the committee, is apprehensive about combining too many features on one map.[33] "The fundamental problem," as he sees it, "is that there will usually, if not always, be more things to depict in a display than there are clearly discernible colors."[34] Meteorologists are used to consulting numerous black-and-white maps, collected in clipboards organized in arrays on large walls, but recreating this environment on an electronic workstation requires not two color screens, but three or four. Moreover, users must be able to adjust category breaks as well as colors, and until problems of inattention and visual discrimination are better understood, "recommended" color schemes must remain tentative.[35]

Outnumbered by retinal variables, Bertin's positional variables and Y are no less important. Unlike coastlines and other compar‧ tively fixed features, isolines and air mass boundaries imply fa greater choice than their crisp symbols suggest. Moreover, uncertainty arises whether the lines are plotted manually or by machine. Computers can thread contours between observation points instantly and consistently, but the software user must either set distance weights and other interpolation specifications or default to the programmer's decisions, which are often arbitrary.[36] And if the interpolation software computes its distances on a flat map, rather than on a globe, the choice of projection can play a significant role in the location of isolines.[37] By contrast, manual contouring can accommodate terrain and other local influences, but only at the mercy of the mapmaker's concentration, innate dexterity, and geographic savvy. Although sufficiently robust collectively to reveal large storms and other macroscale features, the position and shape of individual lines on weather maps are at least slightly deficient in what cartographer Alan MacEachren calls "truth value."[38]

A key culprit is the front. Although most boundaries between air masses that differ in temperature are sufficiently sharp to warrant bold lines pipped with triangles or hemispheres, weather maps typically make no distinction between strong and weak fronts. More egregious, in the eyes of meteorology professor Clifford Mass, is the unthinking imposition of an imperfect model of cyclone structure and development.[39] "It is sobering to consider," he asserts, "that 70 years after the seminal work of the Bergen School, we still do not have a comprehensive understanding of the detailed air motions and evolution of midlatitude cyclones."[40] Familiar and firmly entrenched, Norwegian frontal theory resists promising attempts to modify its inadequate treatment of occlusions or replace it with radically different dynamic, three-dimensional concepts.[41]

Bergen concepts do indeed make a difference. In his 1934 classic *The Drama of Weather*, English meteorologist Sir William Napier Shaw juxtaposed two surface weather maps for the same day and time: one from the British Meteorological Office and the other from the Meteorological Institute at Bergen.[42] While the former anticipated change with labels like "barometer falling," the latter focused on fronts and air masses. As figure 12.7 indicates, the two maps were equally distinctive in their isobars: smoothly rounded on the British

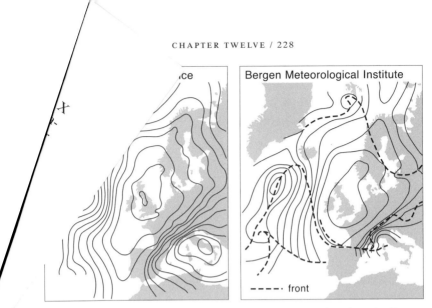

Fig. 12.7. Isobars interpolated from similar surface weather data, for 7 A.M., Greenwich time, February 28, 1929, by weather forecasters in Britain (left) and Norway (right) using intervals of 4 and 5 mb, respectively. Compiled from larger facsimiles in Napier Shaw, *The Drama of Weather* (Cambridge: Cambridge University Press, 1934), 250–1.

version but notably angular on its Norwegian counterpart. To explain these differences, Shaw invoked his theatrical analogy:

> We have seen too that in the Norwegian revision of the ideas of the play, the fronts which were brought face to face as discontinuities new or old became the leading actors. Pressure's isobars had to accommodate themselves to the suggestions of polar and tropical air on their march of adventure.[43]

This effect was hardly accidental: textbooks touting the Bergen line insisted on refractive kinks, reflecting a discontinuity in pressure, whenever isobars crossed air mass boundaries.[44] In 1944, for instance, in the first edition of the highly influential English-language *Introduction to Meteorology*, Bjerknes disciple Sverre Petterssen tied two of his five rules for plotting isobars to frontal theory:[45]

> 3. The isobars in the vicinity of fronts should be drawn so as to bring out the frontal discontinuity in the horizontal pressure gradient.
> 4. The isobars should be drawn first in those areas where the analysis is simplest and should then be prolonged into areas where the solution is more difficult.

Trained to see fronts, meteorologists drew maps celebrating the Norwegian view of storms.

Twister, a 1996 cinematic blockbuster about tornado chasers, introduced moviegoers in the Northeast to an unfamiliar term: dryline.[46]
As residents of Oklahoma and west Texas well know, the dryline is a
surface boundary between dry and moist air. In the spring and early
summer, a strong discontinuity often forms between dry, continental
air from the Rockies and moist, tropical air from the Gulf of Mexico.
Eastward movement of this narrow zone can trigger severe thunderstorms with huge hailstones and tornadoes.[47] Although High Plains
forecasters pay attention to the dryline, it appears on NOAA's Daily
Weather Map only when misdiagnosed as a cold front.[48] Missing
from the Bergen lexicon, the feature is not a part of the map's symbolic code. To remedy this deficiency, National Weather Service forecasters occasionally embellish their detailed in-house surface analy-
ses with a dashed symbol marked "DRYLINE" with a rubber stamp.[49]

Erratic movement and a need for mesoscale measurement accounts for the dryline's underrepresentation in weather cartography.
Figure 12.8, based on a comparatively fine-grained data-collection
network set up in southwestern Oklahoma by the National Severe
Storms Laboratory, describes a classic dryline.[50] Dots indicate measurement points, and isotherms describe trends in the dew point—the
temperature at which atmospheric moisture starts to condense. A
moisture boundary forms in the morning, when the sun heats the
arid, continental air to temperatures well above its dew point. (Dry
air cools down substantially at night, while the humid air to the east
remains warm and muggy, with air temperatures closer to the dew
point.) As the day progresses, dew-point temperatures differing by
several degrees within a few miles make the discontinuity more distinct. At 11 A.M. the dryline is west of the sampling network. Over the
next three hours advancing dry air lowers the dew point by more than
15°F, and by 2 P.M. a distinct dryline is apparent near the Texas-Okla-
homa border, where the 50° and 55°F dew-point isotherms are a
mere five miles apart. At 5 P.M., the discontinuity is farther eastward
and more distinct. With evening, though, the dryline begins to
weaken and retreat westward. Chances are high that the next day will
witness a similar scenario.

Present on more than 40 percent of spring days, the dryline can be
inferred from station plots of dew-point temperatures.[51] Observers
on the ground experience a passing dryline as a drop in humidity and
a shift in wind direction, typically from southeasterly to westerly. Although the temperature difference is minimal, the dryline can mimic

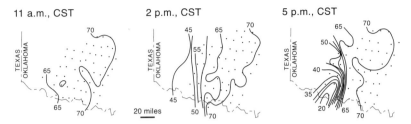

Fig. 12.8. Dew-point isotherms describe the eastward advance of a typical dryline on May 22, 1966. Dots indicate measurement points on a mesoscale network established by the National Severe Storms Laboratory. Compiled from Joseph T. Schaefer, "The Life Cycle of the Dryline," *Journal of Applied Meteorology* 13 (1974): 448.

a strong, belligerent cold front—in the spring continental air is commonly a bit colder, heavier, and more invasive than the unstable humid air to the east. Forced aloft by the invasion, tropical air surrenders its moisture rapidly, often at rates of an inch or more in several minutes. Occasionally violent, the storms are often welcome, according to George Bomar, author of *Texas Weather*.[52] "Blustery winds, vivid lightning—even a dash or two of small hail—seem to be a small price to pay to get the kind of worthwhile rain that the west Texas thunderstorm provides."[53]

Sir William Napier Shaw wrapped up *The Drama of Weather* by comparing the Victorian meteorology of the late nineteenth century with the then-revolutionary Bergen principles.[54] In focusing on surface pressure and its effect on wind, the Victorians had paid scant attention to temperature and humidity. By contrast, the Norwegians, in accounting for air movements underlying pressure differences, drew attention to shifting battle lines between tropical and polar air. Although skeptical of the comprehensiveness of frontal theory, Shaw valued upper air measurements and called for vertical cross sections and a renewed interest in storm centers—"a point of crucial importance because there the lines of the fronts join."[55] He concluded by reaffirming the crucial role of visual analysis: "The play is there for all the world to see and hear; and the action will be plain when our minds can get the true perspective into focus."[56]

My own forecast for future weather maps defies the rigid printed page. I see a flexible and intelligent graphic interface supporting many maps, some animated and most interactive. And not just maps: time-series plots and vertical cross sections are an important part of the mix. Some maps will reflect the recent past, while others reach into the future, perhaps as much as a month for detailed forecasts, maybe several years for more generalized climatic predictions. Hot keys will summon carefully tailored explanations of symbols or specific features—the system will build on the viewer's knowledge and experience—and visualization-support software will generate relevant sequences of maps and diagrams as well as monitor forecast models, radar, and satellite imagery for locally severe weather and anything else the user deems important.[57] Motion and change will be key visual variables, and the display will represent uncertainty overtly if requested. After all, the viewer who wants weather predictions for different places should also understand how the forecast's reliability varies geographically.

I've had a glimpse of this future in an experimental animated forecast map developed by Alan MacEachren's visualization research group at Penn State.[58] Isobars describe a pressure surface computed by averaging hourly predictions for three short-term forecast models. Beneath the isobars, colors running from light blue through orange red portray consensus among the models. Where agreement is high, the simulated isobars are highly visible against a light blue background. Where the models disagree, darker reddish colors partly obscure the comparatively uncertain pressure surface. The prototype, a 48-hour forecast launched at 7 A.M., February 2, 1995, describes the successful prediction of the severe blizzard that struck the east coast two days later. In plate 22, the animation's final frame, strong agreement among the models indicates a reliable forecast near the storm's center, identified by a vivid red dot. Control buttons that let the user replay the simulation or pause at individual frames promote a careful examination of this otherwise complex display.

Despite the likely misgivings of cartographic historians wary of unqualified notions of progress, the final decades of the twentieth century have witnessed massive and indisputable improvements in meteorological cartography. On average, atmospheric measurements are more comprehensive, forecasts are more reliable and can see further ahead, information is more accessible, and weather maps are, by almost any definition, more useful than ever. What's more, recent advances in theory and technology promise further improvements early in the new millennium. Less certain are the future pace and ultimate extent of progress and the impact of these improvements on society. I'd be surprised, though, if weather maps don't lead the way in setting design standards for cartographic forecasts of crime, disease, and other volatile phenomena that future generations will no doubt attempt to predict.

Appendix

WEB SITE ADDRESSES

The column on the right lists uniform resource locators (URLs) for weather Web sites identified in the text by the brief names on the left.

Accu Weather	http://www.accuweather.com
AMS Certificate	http://www.ametsoc.org/AMS/memb/approv.html
AWIPS 1	http://www.nws.noaa.gov/modernize/awiptech.htm
AWIPS 2	http://tgsv5.nws.noaa.gov/msm/awips/awipsmsm.htm
Climate Prediction	http://nic.fb4.noaa.gov
Coding Practices	http://www.nws.noaa.gov/metarex.shtml
CoVis Project	http://typhoon.covis.nwu.edu
Geo Visualization	http://www.gis.psu.edu/research.html
Goddard DAAC	http://daac.gsfc.nasa.gov
Goddard PAO	http://pao.gsfc.nasa.gov
Lyndon State	http://apollo.lsc.vsc.edu/courses/courses.html.
METAR	http://www.nws.noaa.gov/oso/oso1/oso12/fmh1.htm
Mississippi State	http://www.msstate.edu/dept/geosciences/bson.html
NCDC	http://www.ncdc.noaa.gov
NCEP	http://www.ncep.noaa.gov
NESDIS Satellite	http://www.nnic.noaa.gov/SOCC/SOCC_Home.html
Northeast Climate	http://met-www.cit.cornell.edu
Underground	http://wunderground.com
USA Today	http://www.usatoday.com
Visualizer	http://covis.atmos.uiuc.edu
Weather Channel	http://www.weather.com
Weather History	http://www.nws.noaa.gov/pa/history/history.htm
Wisconsin SSEC	http:// cimss.ssec.wisc.edu
WMO	http://www.wmo.ch

Note: As experienced Web users are well aware, organizations maintaining a Web site occasionally remove or rename files and even rename the server (computer) on which Web pages reside. Valid when this book was written, the URLs listed here might not apply several years after its publication. Equivalent or updated information might be found by browsing higher in a Web site's hierarchy, with a URL truncated at com, edu, gov, or org. For example to find a Web page maintained by the National Weather Service, begin browsing at <http://www.nws.noaa.gov>.

Notes

Chapter One

1. References for the Portland disaster include the *Boston Globe*; the *New York Times*; Cleveland Abbe, "The Aims and Methods of Meteorological Work, Especially as Conducted by National and State Weather Services," in *Maryland Weather Service* (Baltimore: Johns Hopkins Press, 1899), 1: 219–330; E. B. Garriott, "Forecasts and Warnings," *Monthly Weather Review* 26 (1898): 493–5; E. B. Rideout, "The Day the Weather Bureau Was Right," in *New England's Disastrous Weather* (Camden, Me.: Yankee Books, 1990), 57–60; and Edward Rowe Snow, "The Truth about the *Portland*," in *New England's Disastrous Weather* (Camden, Me.: Yankee Books, 1990), 61–4.

2. "Topics of the Times" [column], *New York Times*, December 1, 1889.

3. Contemporary newspaper accounts of the tragedy report that the *Portland* sank on Sunday morning. But elderly residents' recollections of sightings in late morning and early afternoon as well as the testimony of a diver who claimed to have found the wreck in the 1940s suggest that the ship sank on Sunday evening, after hitting a vessel carrying granite. See Snow, "The Truth about the *Portland*."

4. Citing the captain's reputation for caution, E. B. Rideout argued that Blanchard was merely obeying company orders. See Rideout, "The Day the Weather Bureau Was Right."

5. The Weather Bureau had already simplified the illustrations from which I traced figures 1.1 and 1.3 by eliminating, for clarity, the closely spaced vertical-line pattern showing areas with recent rain or snow—an important but occasionally overwhelming symbol on standard weather maps. See Abbe, "The Aims and Methods of Meteorological Work," 244, 248.

6. The effect of the Coriolis force on large wind systems is more complex than I have implied. In general, the earth's rotation deflects a moving object from its course, pulling to the right in the northern hemisphere and to the left in the southern hemisphere. This effect is strongest near the poles and nonexistent at the equator. In the upper air, around 20,000 feet, the Coriolis effect is strong enough to balance the force of the pressure gradient so that the wind—meteo-

rologists call this a geostrophic wind—moves parallel to the isobars rather than from an area of high pressure to an area of low pressure. The result is a marked circular flow: clockwise around a high and counterclockwise around a low in the northern hemisphere, but counterclockwise around a high and clockwise around a low in the southern hemisphere. Near the surface, by contrast, friction (also called drag) slows the air and lessens the Coriolis effect, so that radial movement in response to the pressure gradient—inward toward the center of a low, outward away from the center of a high—is also apparent.

7. For a fuller discussion of northeasters and their effects on the coast, see Robert E. Davis and Robert Dolan, "Nor'easters," *American Scientist* 81 (1993): 428–39. Perhaps the first person to recognize the northeaster was Benjamin Franklin, who in a May 12, 1760, letter to Alexander Small, described an informal analysis he carried out "twenty years ago, a few more or less," after correspondence with his brother in Boston, 400 miles away, suggested that large storms moved from southwest to northeast. See Nathan G. Goodman, ed., *The Ingenious Dr. Franklin: Selected Scientific Letters of Benjamin Franklin* (Philadelphia: University of Pennsylvania Press, 1931), 185–7; quotation on 185. Franklin's discovery seems to have gained wider circulation, unattributed though, in a book by Lewis Evans, which Franklin's firm printed in 1747. Evans's book, which described the British "middle colonies," included a map that contained several short notes, one of which read "All our great storms begin to leeward; thus a NE storm shall be a day sooner in Virginia than in Boston"—an observation that geographer William Morris Davis attributed to Franklin. See William Morris Davis, "Was Lewis Evans or Benjamin Franklin the First to Recognize That Our Northeast Storms Come from the Southwest?" *Proceedings of the American Philosophical Society* 45 (1906): 129–30. A recent popular account of hurricane research suggests that Franklin "conceived the idea of weather maps, of plotting all the storms that existed on a given day and following them on succeeding days." See David E. Fisher, *The Scariest Place on Earth: Eye to Eye with Hurricanes* (New York: Random House, 1994), 73. But this alleged discovery seems to have eluded scholars more familiar with the meteorological content of Franklin's writings. See, for examples, Davis, "Was Lewis Evans First?"; and Cleveland Abbe, "Benjamin Franklin as Meteorologist," *Proceedings of the American Philosophical Society* 45 (1906): 117–28.

8. E. B. Garriott, "The North Atlantic Coast Storm of November 26–27, 1898," *Monthly Weather Review* 26 (1898): 494–5.

9. Reports included recorded and reduced barometric pressures, dry-bulb and wet-bulb temperatures (used to calculate humidity), maximum and minimum temperatures, and amount of precipitation since the last report. For descriptions of Weather Bureau operations near the turn of the century, including data collection and the construction of weather charts, see Cleveland Abbe, "The Aims and Methods of Meteorological Work," 228–34, 240–3; Willis Isbister Milham, *Meteorology: A Text-Book on the Weather, the Causes of Its Changes, and Weather Forecasting* (New York: Macmillan, 1936), 360–8; Willis L. Moore, *Descriptive Meteorology* (New York: D. Appleton, 1910), 145–8, 217–9; Willis L. Moore, "Storms and Weather Forecasts," *National Geographic Magazine* 8 (1897): 65–82; Willis L. Moore, "Weather Forecasting," *The Forum* 25 (1898): 341–53; E. J. Prindle, "Weather Forecasts: The Manner of Making Them and Their

Practical Value," *Popular Science Monthly* 53 (1898): 307–33; J. Warren Smith, "The U.S. Signal Service Weather Forecasts," *American Meteorological Journal* 8 (1891): 49–52; and Robert De Courcy Ward, *Practical Exercises in Elementary Meteorology* (Boston: Ginn, 1899), 47–51.

10. Adjustment of barometric pressure to sea level was one of several corrections for local effects that, unless removed, would reduce the comparability of data and muddle the map. Other corrections included adjustments to pressure for differences in temperature and (after 1899) the force of gravity. See Abbe, "The Aims and Methods of Meteorological Work," 229–30. For a detailed discussion of the history, goals, and limitations of nineteenth-century barometric reduction, see Frank H. Bigelow, *Report on the Barometry of the United States, Canada, and the West Indies*, Report of the Chief of the Weather Bureau, 1900–1901, vol. 2 (Washington, D.C., 1902). For instructions on the adjustment of temperature and pressure measurements, see U.S. Weather Bureau, *Instructions for Voluntary Observers* (Washington, D.C., 1892), 12–15, 38–40. For information on turn-of-the-century weather telegraphy, see James Kenealy, "Weather Bureau Stations and Their Duties," *Yearbook of the Department of Agriculture: 1903* (Washington, D.C., 1904), 109–20, esp. 115–7. For an example of the compactness of the ciphers used for weather data, see J. Warren Smith, "The U.S. Signal Service Weather Forecasts," *American Meteorological Journal* 8 (1891): 49–52, esp. 50.

11. Aside from specifying values to be represented by isobars and isotherms, weather service publications offer little guidance on constructing the lines. A notable exception is a Signal Service manual that recommended starting with longer, presumably more salient isobars: "Ascertain by a general examination of the reports after they have been entered on the map whether any of the lines of equal pressure will, when drawn, traverse a large portion of the map, and if this is found to be the case these long lines will be drawn first." Written when the observer network was still quite sparse in the west, the manual also cautioned against extending isobars west of the hundredth meridian. See U.S. Army Signal Office, *Instructions to Observers of the Signal Service, United States Army* (Washington, D.C., 1881), 65. A pamphlet published ten years earlier provides the simple example illustrated in figure 3.3; see Office of the Chief Signal Officer, *The Practical Use of Meteorological Reports and Weather Maps* (Washington, D.C., 1871), 12.

12. For a detailed (but nongraphic) discussion of how to position isotherms between temperature points, see Ward, *Practical Exercises in Elementary Meteorology*, 51–60. Modern analysts recognize that linear interpolation of isolines, as described in figure 1.4, does not always offer the most locally appropriate portrayal of a temperature or pressure surface.

13. When millibars replaced inches on the Washington daily weather map in 1939, a 3-mb increment slightly increased the number of isobars (0.1 inch is approximately 3.3 mb). In 1954, the Weather Bureau experimented with a 5-mb increment before adopting an interval of 4 mb. For discussion of the advantages of metric conversion, see Charles F. Brooks, "Some Uses of the Metric System in Meteorology," *Bulletin of the American Meteorological Society* 19 (1938): 396–8.

14. Willis Moore, who joined the faculty of George Washington University after serving as chief of the Weather Bureau from 1895 to 1913, described his own experience in adapting the national weather map to a specific locality. In 1892,

while serving as the forecaster at the Milwaukee, Wisconsin, weather station, Moore developed five local forecasting guidelines, three of which concerned approaching storms described on weather maps. For example, "a Low from the northwest that reaches western Minnesota and western Iowa without precipitation or clouds will pass over Wisconsin as a dry Low, unless the isobars are closer than five eighths of an inch." See Willis Luther Moore, *The New Air World: The Science of Meteorology Simplified* (Boston: Little, Brown, 1922), 153–5; quotation 154.

15. As described in a 1906 scientific paper by a Weather Bureau official, a telegraphed report en route to Washington from Portland, Maine, would be copied for the local forecaster's use at Boston, Providence, Hartford, New Haven, New York, Philadelphia, and Baltimore, all located on the trunk line from Portland to Washington. (All telegraph offices on the same circuit received the message simultaneously.) Selected reports arriving in Washington from other parts of the country were then transmitted outward over the New England circuit to Portland and intermediate cities. See Alfred J. Henry, "Weather Forecasting from Synoptic Charts," *Journal of the Franklin Institute* 162 (1906): 297–316. For a slightly more informative explanation, circa 1920, see Gustavus A. Weber, *The Weather Bureau: Its History, Activities, and Organization* (New York: D. Appleton, 1922), 18–19. Weber described a system in which Washington received telegraphic reports from 175 stations, 140 of which received direct reports from 21 circuits.

16. George S. Bliss, "The Weather Business: A History of Weather Records, and the Work of the U.S. Weather Bureau," *Scientific American Supplement* 84 (1917): 110–11; quotation 110.

17. Frank H. Bigelow, *Storms, Storm Tracks, and Weather Forecasting*, U.S. Weather Bureau Bulletin no. 20 (Washington, D.C., 1897), 14.

18. Bigelow outlined his approach in straightforward but vague language: "The positions of these tracks have been determined carefully for the United States by studies made in the forecast division of the Bureau, on the long series of maps that have been made during the past twenty years. The line that the central point of a high area or that the center of a storm makes in passing over the country from west to east is laid down on individual charts, these are collected on a group chart, and from this the average line pursued can be readily described." Ibid., 10.

19. Use of a glass drafting table (or a smaller, table-top light box) is speculation. Electricity was available, though, and a light table would have been more efficient than a pantograph, a mechanical copying device available for changing scale.

20. Bigelow, *Storms, Storm Tracks, and Weather Forecasting*, 2.

21. Abbe, "The Aims and Methods of Meteorological Work," 249.

22. For the noon and 3 P.M. maps, see *Monthly Weather Review* 26 (November 1898): charts 10(b) and 11(a); or Abbe, "The Aims and Methods of Meteorological Work," 244–5.

23. For examples of what one advocate of weather typing considered similar patterns of precipitation, temperature change, and cold waves, see W. V. Brown, "A Proposed Classification and Index of Weather Maps as an Aid in Weather Forecasting," *Monthly Weather Review* 29 (1901): 547–8.

24. E. B. Garriott, "Classified Weather Types," *Monthly Weather Review* 29 (1901): 548–9; quotation on 549.

25. F. H. Brandenburg, "Facilities for Systematic Study of Corresponding Weather Types," *Monthly Weather Review* 29 (1901): 546–7.

26. Filing by district, not by state, avoided confusion over storms centered near a state border. Brandenburg was vague not only about the number of districts but also about how he dealt with pressure centers near district borders. Ibid.

27. Because "in general four [sheets] sufficed" for most days, Brandenburg's catalog, which covered a ten-year period, must have contained at least 15,000 separate entries. Ibid.

28. Three short articles in the December 1901 *Monthly Weather Review* marked a high point of interest in weather typing. But five years later an article presenting five empirically derived "specimen rules" and a huge table of related weather information demonstrated sustained interest in weather types. See J. L. Bartlett, "The Study of Practise Forecasting," *Monthly Weather Review* 34 (1906): 523–6. Frederik Nebeker noted that while weather typing was never prominent or successful, the approach received considerable attention during both world wars, when the military seemed eager to hedge its bets with a rhetorically plausible yet theoretically unsound forecasting strategy. See Frederik Nebeker, *Calculating the Weather: Meteorology in the Twentieth Century* (San Diego: Academic Press, 1995), 43. For an example of a climatological study of weather types, see Thomas A. Blair, "Weather Types and Pressure Anomalies," *Monthly Weather Review* 61 (1933): 196–7.

CHAPTER TWO

1. Instruments were standardized and calibrated. The number of stations varied, and not all reported continually. Initiated in 1780, the project ended in 1795, when the French army occupied Mannheim. David C. Cassidy, "Meteorology in Mannheim: The Palatine Meteorological Society, 1780–1795," *Sudhoffs Archiv* 69 (1985): 8–25.

2. Ibid., 22; and Malcolm Rigby, "The Evolution of International Cooperation in Meteorology (1654–1965)," *Bulletin of the American Meteorological Society* 46 (1965): 630–3.

3. Khrgian mentioned (but didn't cite) a paper by Steiglener, who compared barometric fluctuation in London, Regensburg, and St. Petersburg in 1776. See A. Kh. Khrgian, *Meteorology: A Historical Survey*, 2d ed., rev., trans. Kh. P. Pogosyan (Jerusalem: Israel Program for Scientific Translation, 1970), 1: 76.

4. "Aus einem Schreiben des Professor Brandes, meteorologischen Inhalts" (From Professor Brandes's letter of meteorological content), *Annalen der Physik* 55 (1817): 112–4. For an English translation of parts of Brandes's letter, see Karl Schneider-Carius, *Weather Science, Weather Research: History of Their Problems and Findings from Documents during Three Thousand Years* (New Delhi: Indian National Scientific Documentation Centre, 1975); originally published as *Wetterkunde, Wetterforschung: Geschichte ihrer Probleme und Erkenntnisse in Dokumenten aus drei Jahrtausenden* (Freiburg: Verlag Karl Albert, 1955), 179–80.

5. H. W. Brandes, "Einige Resultate aus der Witterungs-Geschichte des Jahres 1783, und Bitte um Nachrichten aus jener Zeit; aus einem Schreiben des Professor Brandes an Gilbert" (Some results from the weather history of the year 1783 and a request for information of that period; from a letter of Professor Brandes to Gilbert), *Annalen der Physik* 61 (1819): 421–6. Although Brandes's original

maps are yet to be found, the Palatine data were the basis for a series of daily European weather maps, published in 1988; see John Kington, *The Weather of the 1780s over Europe* (Cambridge: Cambridge University Press, 1988).

6. Cassidy, "Meteorology in Mannheim," 21-2.

7. Brandes, translated in Schneider-Carius, *Weather Science,* 179.

8. Ibid., 180.

9. Ibid.

10. Ibid., 181-2. Franekar is a small city in the Netherlands. "17 lines" probably refers to 17/12 inches of mercury. In eighteenth-century Europe, except in England, an inch typically was divided into 12 lines. Although inches, feet, and other measures varied significantly from country to country, the Mannheim network used comparable barometers graduated in Paris inches and lines. See W. E. Knowles Middleton, *The History of the Barometer* (Baltimore: Johns Hopkins Press, 1964), 134, 172-3. Although the "17 lines" refers to specific measurements at two places near the storm center, "lines, in which the barometric pressures were equally low" in the second quoted paragraph is an unambiguous reference to isobars.

11. From H. W. Brandes, *Beiträge zur Witterungskunde* (Leipzig: Johann Ambrosius Barth, 1820). Especially significant is the absence of a map by Brandes in Gustav Hellmann's historical survey, which includes facsimiles of maps by Edmund Halley, Alexander von Humboldt, Elias Loomis, Urbain LeVerrier, and Émilien Renou. Brandes's maps were so significant, and so widely mentioned, that Hellmann would surely have included one were an authentic print available. See G. Hellmann, *Meteorologische Karten,* Neudrucke von Schriften und Karten über Meteorologie und Erdmagnetismus, no. 8 (1897; reprint, Nendeln: Kraus Reprint, 1969).

12. Schneider-Carius, *Weather Science,* 180.

13. Khrgian, *Meteorology: A Historical Survey,* 138.

14. Ibid.

15. The absence of maps in *Beiträge* cannot be attributed to the publisher's inability to include graphics. The edition I examined at the NOAA Library, in Silver Spring, Md., near Washington, D.C., contains two fold-out pages with tables and seven fold-out pages with time-series charts on which readily differentiated hand-colored lines describe the passage of storms through 12 cities. Oddly, the volume has two title pages: one with *Beiträge zur Witterungskunde* as the title and the other with the much longer title *Untersuchungen über mittleren Gang der Wärme-Aendererungen durchs ganze Jahr....*

16. Heinrich Wilhelm Brandes, *Dissertatio physica de repentinis variationibus in pressione atmosphaerae observatis* (Lipsiae: Literis Staritzii, Typogr. Univers., 1826). The maps appear as unnumbered pages between pp. 44 and 45. Although the book examines storms occurring on December 24-26, 1821, and February 2-3, 1823, all four maps address the 1821 storm.

17. H. R. Scultetus, "Die erste Wetterkarte" (The first weather map), *Meteorologische Zeitschrift* 60 (1943): 356-9. To make a much reduced version of the map legible, Scultetus also removed the Latin title and relettered the city names with a simpler, more open typeface. I had originally thought Scultetus reconstructed, or faked, the map, because the maps were missing from the reprint edition I first examined. Though all four maps in the copy in the NOAA Library ad-

dress the 1821 storm, I would not be surprised to discover another copy with maps describing the 1823 storm.

18. Wilhelm Trabert, *Meteorologie und Klimatologie* (Leipzig: Franz Deuticke, 1905), 65.

19. Patrick Hughes, "The Great Leap Forward," *Weatherwise* 47 (October/November 1994): 22–7. The map appears with two other maps on pages 26–7. The relevant portion of the caption reads: "Heinrich Brandes pioneered the now-familiar isobars in 1820 when he analyzed the departures from normal pressures of a day 37 years before."

20. Albert Defant, *Wetter und Wettervorhersage* (Leipzig: Franz Deuticke, 1918), 7–8; and Hugo Hildebrand Hildebrandsson and Léon Teisserenc de Bort, *Les bases de la météorologie dynamique: Historique-état de nos connaissances* (The bases of dynamic meteorology: Historical review of our knowledge) (Paris: Gauthier-Villars et Fils, 1907), 46-47. Defant attributes the map to Hildebrandsson and de Bort, who conceded (p. 46), "we have reconstituted the synoptic map opposite [on p. 47]" (author's translation). For a more recent reconstruction, see Kington, *Weather of the 1780s over Europe.* The map for March 6, 1783, appears on p. 80, with 16 other daily maps. Although based on the same observations used by Brandes, and more recently by Hildebrandsson and de Bort, its isobars are similar but not identical to those in figure 2.2.

21. Helen M. Wallis and Arthur H. Robinson, *Cartographical Innovations: An International Handbook of Mapping Terms to 1900* (Tring, Herts., England: Map Collector Publications, 1982), 156–7, 223–4.

22. Arthur H. Robinson, "The Genealogy of the Isopleth," *Cartographic Journal* 8 (1971): 49–53. See also Werner Horn, "Die Geschichte der Isarithmenkarten" (The story of isarithmic maps), *Petermanns Geographische Mitteilungen* 103 (1959): 225–32.

23. For an examination of Halley's cartographic contributions, see Norman J. W. Thrower, "Edmond Halley as a Thematic Geo-Cartographer," *Annals of the Association of American Geographers* 59 (1969): 652–76. In addition to his isogonic maps, Halley's cartographic contributions include a map of tides, a map of the area affected by a solar eclipse, and a map of wind patterns. See Arthur H. Robinson, *Early Thematic Mapping in the History of Cartography* (Chicago: University of Chicago Press, 1982), 46–50, 69–71, 84–6.

24. Robinson, "The Genealogy of the Isopleth," 51.

25. The map's full title is *New and Correct Chart Showing the Variations of the Compass in the Western and Southern Oceans, as observed in ye year 1700 by His Majesty's Command by Edm. Halley.* For descriptions, see L. A. Bauer, "Halley's Earliest Equal Variation Chart," *Terrestrial Magnetism* 1 (1896): 28–31; S. Chapman, "Edmond Halley as Physical Geographer, and the Story of His Charts," *Occasional Notes,* Royal Astronomical Society, no. 9 (June 1941); Sidney Chapman and Julius Bartels, *Geomagnetism* (Oxford: Clarendon Press, 1940), 2: 911–4; and Thrower, "Edmond Halley as a Thematic Cartographer," 666–7.

26. Drawn on a Mercator projection, the copy of the map in the British Museum measures 48 by 57 cm (approximately 18.9 inches wide and 22.4 inches tall). See Bauer, "Halley's Equal Variation Chart," *Nature* 52 (1895): 196.

27. Halley's expedition had a practical goal: the possibility that regularity in magnetic declination might prove useful, in concert with latitude, in estimating

longitude. See Thrower, "Edmond Halley as a Thematic Cartographer," 663.

28. Printed as a separate sheet, rather than included in a scientific periodical, Halley's Atlantic chart was too easily overlooked, destroyed, or otherwise lost, so that by 1890 the British Museum held the only known copy. For contemporary comments on the discovery of a second copy, see L. A. Bauer, "Some Bibliographical Discoveries in Terrestrial Magnetism," *Nature* 52 (1895): 79–80; Thomas Ward, "Halley's Chart," *Nature* 52 (1895): 106; and L. A. Bauer, "Halley's Equal Variation Chart." To extend coverage, Halley compiled observations from "Several Journals of Voyages Lately Made in the India-Seas." See Thrower, "Edmond Halley as a Thematic Cartographer," 667–9.

29. Thrower, "Edmond Halley as a Thematic Cartographer," 669.

30. The impact was far wider: for the names and definitions of 90 isolines used in meteorology and related fields, see C. Fitzhugh Talman, "A List of Meteorological Isograms," *Monthly Weather Review* 43 (1915): 195–8. According to Talman, Humboldt had described (but not used) isothermal parallels as early as 1815. Ignoring Brandes, Talman attributed the isobaric line symbol to H. K. W. Berghaus's *Physikalischer Atlas* (1838) and the term *isobar* to A. K. Johnson's *Physical Atlas* (1849).

31. Alexandre de Humboldt, "Des lignes isothermes et de la distribution de la chaleur sur le globe" (On isotherms and the global distribution of heat), *Mémoires de physique et de chimie, de la Société d'Arcueil*, Paris 3 (1817): 462–602; quotation (in translation) from Schneider-Carius, *Weather Science*, 169. Humboldt described his intent to map temperature with isolines in an earlier paper; see Alexandre de Humboldt, "Sur les lois que l'on observe dans la distribution des formes vegetales," *Annales de chemie et de physique* 1 (1816): 225-39.

32. In dismissing a connection with elevation contours, Robinson and Wallis note that both Halley and Humboldt were concerned with linking points of equal value, not with describing a three-dimensional surface, such as the land surface, using contours formed by the intersections of the surface with evenly spaced horizontal planes. See A. H. Robinson and Helen M. Wallis, "Humboldt's Map of Isothermal Lines: A Milestone in Thematic Cartography," *Cartographic Journal* 4 (1967): 119–23.

33. Jan Munzar, "Alexander von Humboldt and His Isotherms: On the Occasion of the 150th Anniversary of the First Map of Isotherms," *Weather* 22 (1967): 360–3.

34. The map had an unusual publication history: Humboldt discussed his map at length in one journal (*Mémoires de physique et de chimie, de la Société d'Arcueil*) but published it in another (*Annales de chemie et de physique*). A folded copy of the map is bound between pages 112 and 113 of the latter, which summarizes a much longer paper, originally presented orally to the Paris Academy of Science, over four meetings during May and June 1817; see Alexandre de Humboldt, "Sur les lignes isothermes," *Annales de chemie et de physique* 5 (1817): 102–11. Perhaps, as Robinson and Wallis suggested, his friends who edited the *Annales* did him a favor by including copies printed for *Mémoires*, which was delayed for three years—hardly an unreasonable explanation for a map revolutionary in both message and design. See Robinson and Wallis, "Humboldt's Map of Isothermal Lines," 122, n. 5.

35. Robinson and Wallis, "Humboldt's Map of Isothermal Lines."

36. Chapman, "Edmond Halley as Physical Geographer," 2; H. Howard Frisinger, *The History of Meteorology to 1800* (New York: Science History Publications, 1977), 123–5; and Wallis and Robinson, *Cartographical Innovations*, 157–8. Like Halley's isogonic map, his winds chart became a classic. According to Edgar Woolard, further advances were not made "until well into the nineteenth century"; see Edgar W. Woolard, "Historical Note on Charts of the Distribution of Temperature, Pressure, Winds over the Surface of the Earth," *Monthly Weather Review* 48 (1920): 408–11; quotation on 410.

37. Thrower, "Edmond Halley as a Thematic Geo-Cartographer," 656–7.

38. E. Halley, "An Historical Account of the Trade Winds, and Monsoons, Observable in the Seas between and near the Tropicks, with an Attempt to Assign the Phisical Cause of the Said Winds," *Philosophical Transactions* 16 (1686): 153–68. The map is bound opposite p. 151.

39. Ibid., 162–3.

40. Ibid., 163.

41. Harold L. Burstyn, "Early Explanations of the Role of the Earth's Rotation," *Isis* 57 (1966): 167–87.

42. George Hadley, "Concerning the Cause of the General Trade-Winds," *Philosophical Transactions* 34 (1735): 58–62.

43. See Burstyn, "Early Explanations of the Role of the Earth's Rotation," 183–7; C. L. Jordan, "On Coriolis and the Deflective Force," *Bulletin of the American Meteorological Society* 47 (1966): 401–3; and H. E. Landsberg, "Why Indeed Coriolis?" *Bulletin of the American Meteorological Society* 47 (1966): 887–9.

44. H. W. Dove, "Ueber barometrische Minima" (On barometric minimums) *Annalen der Physik und Chemie* 13 (1828): 596–613.

45. Gisela Kutzbach, *The Thermal Theory of Cyclones: A History of Meteorological Thought in the Nineteenth Century* (Boston: American Meteorological Society, 1979), 10–16.

46. Dove first presented his theory, without illustrations, in 1827; see H. W. Dove, "Einige meteorologische Untersuchungen über den Wind" (A meteorological investigation of the wind), *Annalen der Physik und Chemie* 11 (1827): 545–90; quotation from translation by Schneider-Carius, *Weather Science*, 218.

47. The paper was derived from Dove's doctoral dissertation; see H. W. Dove, "Ueber barometrische Minima." Hans Scultetus not only accepted Dove's figures 8 and 9 as maps but cited them as milestones in the development of synoptic meteorology. See H. R. Scultetus, "Dove und Loomis als Wegbereiter der Synopsis" (Dove and Loomis as pioneers of synopsis), *Meteorologische Zeitschrift* 60 (1943): 419–22. Dove, who became professor of meteorology at the University of Berlin in 1830 and directed the Prussian Meteorological Institute from 1849 to 1879, was the world's preeminent meteorologist for several decades. In addition to influential contributions to meteorology and climatology, he published the first world maps of monthly temperature, using isotherms, around 1848; see H. W. Dove, "Remarks by Professor Dove on His Recently Constructed Maps of the Monthly Isothermal Lines of the Globe, and on Some of the Principal Conclusions in Regard to Climatology Deducible from Them," *Report of the British Association for the Advancement of Science*, 18 (1848): 85–97.

48. Kutzbach, *The Thermal Theory of Cyclones*, 15.

49. Khrgian, *Meteorology: A Historical Survey*, 170.

50. William Morris Davis noted the significance of America's size in an essay on weather forecasting in Europe. See W. M. Davis, "European Weather Predictions," *American Meteorological Journal* 8 (1891): 53–8, esp. 55.

51. The American storms controversy involved many more people and theories than discussed here. For a fuller account, see James Rodger Fleming, *Meteorology in America, 1800–1870* (Baltimore: Johns Hopkins University Press, 1990), 23–73. For a chronology that includes numerous other contributors to weather mapping and the theory of storms, see Jacques Dettwiller, *Chronologie de quelques evenements meteorologiques, en France et ailleurs* (Boulogne-Billancourt, France: Ministère des transports, direction de la météorologie, 1982).

52. Maxime Bôcher, "The Meteorological Labors of Dove, Redfield, and Espy," *American Meteorological Journal* 5 (1888): 1–13; and Walter E. Gross, "The American Philosophical Society and the Growth of Meteorology in the United States: 1835–1850," *Annals of Science* 29 (1972): 321–38.

53. William C. Redfield, "Remarks on the Prevailing Storms of the Atlantic Coast, of the Northeastern States," *American Journal of Science* 20 (1831): 17–51.

54. William C. Redfield, "Observations on the Storm of December 15, 1839," *Transactions of the American Philosophical Society* 8 (1843): 77–80.

55. Quoted in ibid., 77.

56. Dove and Redfield, both of whom rejected the centripetal models of Brandes and Espy, were correspondents; see Fleming, *Meteorology in America, 1800–1870*, 64.

57. Redfield, "Observations on the Storm of December 15, 1839," 80.

58. Ibid.

59. James P. Espy, "Essays on Meteorology no. III: Examination of Hutton's, Redfield's and Olmstead's Theories," *Journal of the Franklin Institute* n.s. 18 (1836): 100–8; quotation on 107–8.

60. For accounts of Espy's activities, see Fleming, *Meteorology in America, 1800–1870*, 23–8, 43–53, 57–9, 67–81.

61. Ibid., 25-6.

62. "Third Report of the Joint Committee on Meteorology, of the American Philosophical Society and the Franklin Institute of the State of Pennsylvania, for the Promotion of the Mechanic Arts," *Journal of the Franklin Institute* n.s. 19 (1837): 17–21; map on p. 19. Spitz suggested the map was "one of the first birds'-eye-view weather maps of a specific weather phenomenon to be published in this country, if not the first." Fleming later called it the first American weather map to cover a wide area. See Fleming, *Meteorology in America, 1800–1870*, 59; and Armand N. Spitz, "Meteorology in the Franklin Institute," *Journal of the Franklin Institute* 237 (1944): 271–87, 331–57, esp. 281.

63. James P. Espy, *The Philosophy of Storms* (Boston: Charles C. Little and James Brown, 1841), 105.

64. James P. Espy, *First Report on Meteorology to the Surgeon General of the United States Army* (Washington, D.C., 1843). Although the stated publication date is 1843, the report was not printed until 1845; see Fleming, *Meteorology in America, 1800–1870*, 70.

65. James P. Espy, *Fourth Report on Meteorology* (Washington, D.C.: 1857), U.S. Senate, Ex. Doc., 34th Cong., 3d sess., no. 65. The maps engraved for the report were a small fraction of the 1,100 charts—one a day for more than three

years—prepared for the second and third reports and the 700 maps produced for the fourth report. Espy noted, almost apologetically, that "it would be very expensive to print them all." Ibid., 9.

66. Espy used color not only to differentiate barometric minimums and maximums but also to distinguish the quantity of rain at the time of observation (indicated in red on some maps, but not the example in figure 2.9) from the total quantity of rain before or after the observation (shown, as here, in black).

67. Elias Loomis, "On the Storm Which Was Experienced throughout the United States about the 20th of December, 1836," *Transactions of the American Philosophical Society* 7 (1841): 125–63. A single map, bound ahead of the article as plate II, uses arrows to show wind direction on the morning of December 21 and slightly parabolic lines to describe the barometric minimums at 14 different times, between 6 A.M. December 20 and noon December 23. Plate I, which follows the article, consists of two time-series graphs describing changes in temperature and barometric pressure at various cities.

68. In his closing paragraph, Loomis apologizes for this apparent omission: "I trust it will not be inferred from my silence with respect to the labours of others in this important field, that I am wholly ignorant of them, or am insensible of their value. I have availed myself of the labours of others as far as was in my power. To have credited every suggestion to its original author would have been inconvenient, and generally superfluous, being found in most treatises on meteorology. I am happy, however, to express my particular obligations to the labours of Messrs. Redfield, Espy, and Col. Reid, and shall esteem myself well repaid if the present communication shall contribute something to the progress of that science which they have done so much to promote." Ibid., 163.

69. Loomis read his paper to the American Philosophical Society in 1843. Elias Loomis, "On Two Storms Which Were Experienced throughout the United States, in the Month of February, 1842," *Transactions of the American Philosophical Society* 9 (1846): 161–84. In the copy at the NOAA Library, all maps but chart 12 are colored. In copies held by the libraries of Syracuse University and the University of Wisconsin (Madison) none of the maps are colored.

70. Ibid., 165.

71. Scholars who recognized the significance of Loomis's maps include Gustav Hellmann and Eric Miller. See Hellmann, *Meteorologische Karten*, 6–7; and Eric R. Miller, "American Pioneers in Meteorology," *Monthly Weather Review* 61 (1933): 189–93, esp. 191.

72. Loomis, "On Two Storms," 164.

73. Ibid., 164.

74. Ibid., 184.

CHAPTER THREE

1. W. C. Redfield, "On Three Several Hurricanes of the American Seas and Their Relations to the Northers, So Called, of the Gulf of Mexico and the Bay of Honduras, with Charts Illustrating the Same," *American Journal of Science,* 2d ser., 2 (1846): 311–34, esp. 334. See also Cleveland Abbe, "An Item in the Early History of Weather Telegraphy," *Monthly Weather Review* 23 (1895): 215–6. Not the first to suggest sending messages by wire, Morse proposed his version of the electric telegraph in 1837. His 44-mile government-financed experiment in 1842

was the first successful demonstration of the telegraph's commercial potential. See Alvin F. Harlow, *Old Wires and New Waves: The History of the Telegraph, Telephone, and Wireless* (New York: D. Appleton–Century, 1936), 77–99; and Robert Luther Thompson, *Wiring a Continent: The History of the Telegraph Industry in the United States, 1832–1866* (Princeton, N.J.: Princeton University Press, 1947), 21–34. Electronic transmission was not the first relatively rapid method for sending messages over long distances: in 1794, for instance, Claude Chappe used a visual telegraph, with messages transmitted by large coded signs between relay stations equipped with telescopes and spaced roughly six miles apart, to link Paris with Lille, a distance of about 130 miles. See Geoffrey Wilson, *The Old Telegraphs* (London: Phillimore, 1976), 122–3. Aware of Chappe's project, Lavoisier had suggested making forecasts based on rapidly collected weather data. Although details are sketchy, Karl Kreil of Prague apparently scooped Redfield in 1842, by noting that the electric telegraph would be vastly superior to the visual telegraph for gathering weather data. See H. Hildebrand Hildebrandsson and Léon Teisserenc de Bort, *Les bases de la météorologie dynamique: historique-état de nos connaissances* (Paris: Gauthier-Villars et Fils, 1907), 1: 63–4; and Mark Harrington, "History of the Weather Map," *Report of the International Meteorological Congress, Chicago, Ill., August 21–24, 1893*, U.S. Weather Bureau Bulletin no. 11, pt. 2 (Washington, D.C., 1894), 327–35, esp. 328.

2. Henry published letters from Loomis and Espy in the *Second Annual Report of the Board of Regents of the Smithsonian Institution* (Washington, D.C., 1848), U.S. Senate, Misc. Doc., 30th Cong., 1st sess., no. 23, 193–208, noted in Abbe, "An Item in the Early History of Weather Telegraphy," 216. See also Marcus Benjamin, "Meteorology," in George Brown Goode, ed., *The Smithsonian Institution, 1846–1896: The History of Its First Half Century* (Washington, D.C.: Smithsonian Institution, 1897), 647–78.

3. See Joseph Henry, "Report of the Secretary of the Smithsonian Institution to the Board of Regents, December 8, 1847," reprinted in *Eighth Annual Report of the Board of Regents of the Smithsonian Institution, up to January 1, 1854, and the Proceedings of the Board up to July 8, 1854* (Washington, D.C., 1854), House of Representatives, Misc. Doc., 33d Cong., 1st sess., no. 97, 119–39; quotations on 139.

4. Nathan Reingold, "Joseph Henry," in Charles Coulston Gillispie, ed., *Dictionary of Scientific Biography* (New York: Charles Scribner's Sons, 1972), 6: 277–81. See also Joseph Henry, "Application of the Telegraph to the Prediction of Changes of Weather, Particularly in the City of Boston and Its Vicinity," *Proceedings of the American Academy of Arts and Sciences* 4 (1857–60): 271–5.

5. See James Rodger Fleming, *Meteorology in America, 1800–1870* (Baltimore: Johns Hopkins University Press, 1990), 10–20. For an earlier proposal to collect and map climatological data, see Simeon DeWitt, "Respecting a Plan of a Meteorological Chart, for Exhibiting a Comparative View of the Climates of North America, and the Progress of Vegetation," *Transactions of the Society for the Promotion of Agriculture, Arts, and Manufactures, Instituted by the State of New York* 1 (1801): 88–92.

6. Fleming, *Meteorology in America*, 20–1.

7. Loomis estimated the cost as three thousand dollars. Henry hoped to obtain additional funds from the Smithsonian in 1848 and 1849. See Joseph Henry,

"Second Annual Report of the Secretary of the Smithsonian Institution, Giving an Account of the Operations of the Year 1848," reprinted in *Eighth Annual Report of the Smithsonian Institution*, 148–59, esp. 152–3.

8. Henry commissioned the Swiss meteorologist Arnold Guyot to prepare an observer's manual. See, for example, Arnold Guyot, *Directions for Meteorological Observations, Intended for the First Class of Observers* (Washington, D.C.: Smithsonian Institution, 1850). The first instructions called for observations at "sunrise," 9:00 A.M., 3:00 P.M., and 9:00 P.M., local time, when army medical officers were required to record climatological data. The Smithsonian revised its observation times in 1853, 1855, and 1860. See Fleming, *Meteorology in America*, 81–2.

9. Joseph Henry, "Fourth Annual Report of the Secretary of the Smithsonian Institution, for the Year 1850," reprinted in *Eighth Annual Report of the Smithsonian Institution*, 187–91, quotation on 191.

10. The number of observers fluctuated markedly. A disagreement with a disgruntled former employee precipitated a drop to slightly over 200 in 1855, cooperation by the U.S. Patent Office promoted a vigorous rebound to more than 500 around 1859, and the Civil War caused the number to plunge again, to fewer than 300 in 1863. See Fleming, *Meteorology in America*, 82.

11. *Annual Report of the Board of Regents of the Smithsonian Institution, for the Year 1857* (Washington, D.C., 1858), U.S. Senate, Misc. Doc., 35th Cong., 1st sess., no. 272, 27–8.

12. *Annual Report of the Board of Regents of the Smithsonian Institution, for the Year 1858* (Washington, D.C., 1859), House of Representatives, Misc. Doc., 35th Cong., 2d sess., no. 57, 84.

13. Ibid., 26, 73.

14. *Annual Report of the Board of Regents of the Smithsonian Institution, for the Year 1860* (Washington, D.C., 1861), House of Representatives, Misc. Doc., 36th Cong., 2d sess., [no number], 101.

15. Benjamin, "Meteorology," 659; and Eric R. Miller, "The Evolution of Meteorological Institutions in the United States," *Monthly Weather Review* 59 (1931): 1–6, esp. 3. Although most sources indicated that the map was installed in 1856, one suggested that the map was put in place in 1850; see Anon., "Meteorological Work of the Smithsonian Institution," *Annual Report of the Board of Regents of the Smithsonian Institution, to July, 1892* (Washington, D.C., 1893), 89–93. Moreover, Cleveland Abbe suggested that the first American weather map based on telegraphic data appeared in 1849, when Alexander Jones, an Associated Press reporter in New York, sent Henry samples of experimental maps based on weather data for July 19 and 20. Henry was apparently not interested in Jones's proposal for a commercial weather data scheme based in New York. See Cleveland Abbe, "Weather Telegraphy in England and America," *Monthly Weather Review* 25 (1897): 205–6; and Fleming, *Meteorology in America*, 142–3.

16. *Annual Report of the Smithsonian Institution for the Year 1858*, 52.

17. Abbe, "Lectures on Meteorology," *Monthly Weather Review* 28 (1900): 208–10, esp. 210.

18. S. P. Langley, "The Meteorological Work of the Smithsonian Institution," *Report of the International Meteorological Congress*, 216–20.

19. Fleming, *Meteorology in America*, 146–8; and Webster Prentiss True, *The Smithsonian Institution* (Washington, D.C., 1929), 306–8.

20. *Annual Report of the Board of Regents of the Smithsonian Institution, for the Year 1870* (Washington, D.C., 1871), House of Representatives, Ex. Doc., 42d Cong., 1st sess., no. 20, 43–4.

21. John Ball, "On Rendering the Electric Telegraph Subservient to Meteorology," *Report of the Meeting of the British Association for the Advancement of Science* 18 (1848): 12–3; quotation on 13. See also Napier Shaw, *Manual of Meteorology*, vol. 1, *Meteorology in History* (Cambridge: Cambridge University Press, 1926), 302.

22. The first synchronous daily report was published on June 14, 1849. See R. G. K. Lempfert, "British Weather Forecasts: Past and Present," *Quarterly Journal of the Royal Meteorological Society* 39 (1913): 173–84. For a detailed description of the *Daily News*'s weather bulletin, see William Marriott, "The Earliest Telegraphic Daily Meteorological Reports and Weather Maps," *Quarterly Journal of the Royal Meteorological Society* 29 (1903): 123–31. According to Marriott, the *Daily News* had published an earlier daily weather report, based on data delivered to London by telegraph, between August 31 and October 30, 1848, the year before Glaisher's bulletin. Suggesting that the 1848 observations were not consistently and reliably taken at the same time each day, and noting that the earlier reports did not include wind speed, Marriott designated August 31, 1848, as the date of "Publication in the *Daily News* of the first telegraphic Daily Weather Report," and June 14, 1849, as the date of "Publication in the *Daily News* of the first synchronous Meteorological Observations." Ibid., 131. Glaisher described his project in a July 8, 1850, letter to Joseph Henry, reprinted in Abbe, "Weather Telegraphy," 206. According to Abbe, the reports were delivered by train, not telegraph, which would have been more expensive. See also William Marriott, "Some Account of the Meteorological Work of the Late James Glaisher, F.R.S.," *Quarterly Journal of the Royal Meteorological Society* 30 (1904): 1–28.

23. Maps portraying wind direction, barometric pressure, and general weather conditions were sold between August 8 and October 11, 1851. See William Marriott, "An Account of the Bequest of George James Symons, F.R.S., to the Royal Meteorological Society," *Quarterly Journal of the Royal Meteorological Society* 27 (1901): 241–60. The library that Symons willed to the society included a set of maps from August 11 to October 11, 1851. Symons had reproduced the first map, dated August 8, 1851, in the *Meteorological Magazine* for September 1896. See also Lempfert, "British Weather Forecasts," 173–4.

24. The prospectus was reprinted in Marriott, "An Account of the Bequest of George James Symons," 258–9. See also Marriott, "The Earliest Telegraphic Daily Meteorological Reports and Maps"; and Shaw, *Manual of Meteorology*, 305, 308–9. According to Marriott, the firm was founded in 1861, but because companies with limited liability, as described in the prospectus, were not authorized until August 1863, Shaw gave the date as 1863.

25. Galton, who seems to have had wider horizons than Glaisher, appealed for telegraphic weather reports in his "Circular Letter to Meteorological Observers," dated July 1861, and sent translations in French and German to meteorologists on the continent. The printed circular included a sample two-color weather map of England describing wind, cloud cover, temperature, and barometric pressure observations collected at 9 A.M. on January 16, 1861, by volun-

teers and mailed to London. Galton used data from his network of volunteer observers to produce numerous synoptic maps. See Francis Galton, *Meteorologica, or Methods of Mapping the Weather, Illustrated by Upwards of 600 Printed and Lithographed Diagrams Referring to the Weather of a Large Part of Europe during the Month of December 1861* (London: Macmillan, 1863); and Shaw, *Manual of Meteorology*, 304–7.

26. H. Landsberg, "Storm of Balaklava and the Daily Weather Forecast," *Scientific Monthly* 79 (1954): 347–52; and Josef W. Konvitz, *Cartography in France, 1660–1848: Science, Engineering, and Statecraft* (Chicago: University of Chicago Press, 1987), 145. For discussion of the destruction caused by the November 14, 1854, storm, see George MacMunn, *The Crimea in Perspective* (London: G. Bell & Sons, 1935), 144–8.

27. According to Herve Faye, president of the Bureau of Longitudes, Le Verrier apparently was unaware of the work of Redfield, Espy, and Loomis on the development and movement of storms; see H. Faye, "Accessory Phenomena of Cyclones," *American Meteorological Journal* 7 (1890): 295–302.

28. For examples, see G. Hellmann, *Meteorologische Karten*, Neudrucke von Schriften und Karten über Meteorologie und Erdmagnetismus, no. 8 (1897; reprint, Nendeln/Liechtenstein: Kraus Reprint, 1969), 8; and Hildebrandsson and de Bort, *Les bases de la météorologie dynamique*, 1: 65–7. Edgar Woolard attributed the "first isobaric chart based on observational data" to Renou; see Edgar W. Woolard, "Historical Note on Charts of the Distribution of Temperature, Pressure, Winds over the Surface of the Earth," *Monthly Weather Review* 48 (1920): 408–11; quotation on 410.

29. Harrington, "History of the Weather Map," 330.

30. A. Kh. Khrgian, *Meteorology: A Historical Survey*, 2d ed., rev., trans. Kh. P. Pogosyan (Jerusalem: Israel Program for Scientific Translation, 1970), 1: 116–7; and Shaw, *Manual of Meteorology*, 303–5.

31. Since 1852, Buys Ballot had published daily weather maps in the *Yearbook* of the Netherlands Meteorological Institute. For a concise summary of Buys Ballot's law, see Professor [C.] Buys-Ballot, "On the System of Forecasting the Weather Pursued in Holland," *Report of the British Association for the Advancement of Science* [Transactions section] 33 (1864): 20–1.

32. Landsberg, "Storm of Balaklava," 351; and R. P. W. Lewis, "The Founding of the Meteorological Office, 1854–55," *Meteorological Magazine* 110 (1981): 221–7. A captain at the time of his initial appointment, FitzRoy was promoted to rear admiral in 1857 and vice admiral in 1863. See George Basalla, "Robert FitzRoy," in Charles Coulston Gillispie, ed., *Dictionary of Scientific Biography* (New York: Charles Scribner's Sons, 1972), 5: 16–8.

33. Lempfert, "British Weather Forecasts," 175–6.

34. P. G. Parkhurst, "Ocean Meteorology: A Century of Progress (Part I)," *Marine Observer* 25 (1955): 16–21.

35. Robert FitzRoy, Letter to the editor, *Times* (London), February 12, 1861, 10.

36. Patrick Hughes, "FitzRoy the Forecaster: Prophet without Honor," *Weatherwise* 41 (1988): 200–4.

37. David Brunt, "A Hundred Years of Meteorology (1851–1951)," *Advancement of Science* 8, no. 30 (1951): 114–24; and Shaw, *Manual of Meteorology*, 311.

38. R. F. M. Hay, "Some Landmarks in Meteorological Progress, 1855–1955," *Marine Observer* 25 (1955): 11–16.

39. Although the Meteorological Office produced manuscript maps as early as March 11, 1872, maps did not appear in the printed bulletin until March 23. See Harrington, "History of the Weather Map," 331; and R. H. Scott, "Weather Charts in Newspapers," *Journal of the Society of Arts* 23 (1875): 776–82.

40. The maps issued for March 11–22 were probably only manuscript maps.

41. For concise summaries of the role of the Cincinnati Astronomical Observatory, see Fleming, *Meteorology in America*, 150–2. For Abbe's own account and recollection, see Cleveland Abbe, "Historical Notes on the Systems of Weather Telegraphy and Especially Their Development in the United States," *American Journal of Science* 3d ser. 2 (1871): 81–8; and Cleveland Abbe, "How the United States Weather Bureau Was Started," *Scientific American* 114 (1916): 529.

42. Truman Abbe, *Professor Abbe and the Isobars: The Story of Cleveland Abbe, America's First Weatherman* (New York: Vantage Press, 1955), 1–5.

43. Abbe, "How the United States Weather Bureau Was Started," 529.

44. The first map, measuring 7 by 10 inches, was apparently used on February 22, 1870. The larger, 14-by-21-inch map was introduced two days later. For an image of the map and a description of the reproduction process by Abbe's son, see Abbe, *Professor Abbe and the Isobars*, 115–6. Facing page 116, a facsimile of a "specimen" map, dated April 19, 1870, includes point symbols for 37 stations, the most distant of which are Cheyenne, Wyoming; Havana, Cuba; Halifax, Nova Scotia; and Marquette, Michigan. Next to each point symbol is the temperature, printed as a number. The map does not indicate time of day and contains neither isobars nor isotherms.

45. Eric R. Miller, "New Light on the Beginning of the Weather Bureau from the Papers of Increase A. Lapham," *Monthly Weather Review* 59 (1931): 65–70.

46. Ibid., 67.

47. Roy Popkin, *The Environmental Science Services Administration* (New York: Frederick A. Praeger, 1967), 55–6.

48. Fleming, *Meteorology in America*, 155–6; and Donald R. Whitnah, *A History of the United States Weather Bureau* (Urbana: University of Illinois Press, 1961), 19. For discussion of earlier meteorological programs in the American military, see Charles C. Bates and John F. Fuller, *America's Weather Warriors* (College Station: Texas A&M University Press, 1986); W. H. Beehler, "The Origin and Work of the Division of Marine Meteorology," *Report of the International Meteorological Congress*, 221–32; and Charles Smart, "The Connection of the Army Medical Department with the Development of Meteorology in the United States," *Report of the International Meteorological Congress*, 207–16.

49. Paine to Duane Mowry, 8 October 1903, quoted in Miller, "New Light," 68.

50. Popkin, *Environmental Science Services Administration*, 56. See also Fleming, *Meteorology in America*, 152–6.

51. Miller, "New Light," 69–70.

52. For a map of the 25 original Signal Office weather stations, see Fleming, *Meteorology in America*, 158.

53. For a concise account of Abbe's work with the weather service and his contributions to meteorology, see Alfred Judson Henry, "Memoir of Cleveland

Abbe," *Annals of the Association of American Geographers* 7 (1917): 61–66.

54. Whitnah, *History of the Weather Bureau*, 21. Obviously Congress's add-on appropriation of $15,000, for the year ending June 30, 1871, could never sustain 233 full-time workers, most of whom, I suspect, were soldiers recently reassigned to Myer's command. The appropriation for 1873 is from Popkin, *The Environmental Science Services Administration*, 59.

55. Office of the Chief Signal Officer, *The Practical Use of Meteorological Reports and Weather-Maps*, War Department circular (Washington, D.C., 1871), 5; and *Report of the Chief Signal Officer for 1880*, vol. 4 of the Report of the Secretary of War (Washington, D.C., 1881) House of Representatives, Exec. Doc., 46th Cong., 3d sess., no. 1, part 2, p. 10. The total for 1880, which includes military bases as well as stations reporting only once a day, is not comparable with station counts for the U.S. Weather Bureau, a civilian agency established in 1891. Annual lists of weather stations reveal an ongoing reevaluation of locations, with many new stations opened and others closed each year, especially in smaller cities.

56. Cleveland Abbe, "The Meteorological Work of the U.S. Signal Service, 1870 to 1891," *Report of the International Meteorological Congress*, 232–85, esp. 251. These outwardly unusual times were chosen so that the morning and afternoon cartographic snapshots might represent minimum and maximum disturbance by solar heating. The Signal Office changed the late evening observation to 11 P.M., Washington time, on August 25, 1872, and the morning and afternoon observations to 7 A.M., and 3 P.M., Washington time, on November 1, 1879. The bureau adopted the standard, zone time for Washington, based on the 75th meridian, and changed the evening observation to 10 P.M., on January 1, 1887. See Gustavus A. Weber, *The Weather Bureau: Its History, Activities, and Organization* (New York: D. Appleton, 1922), 7. For information on the "manifold" process used to print the Washington map, see Edgar B. Calvert, "Development of the Daily Weather Map," *Proceedings of the Convention of Weather Bureau Officials, Held at Omaha, Nebr., October 13–14, 1898*, U.S. Weather Bureau Bulletin no. 24 (Washington, D.C., 1899), 144–50.

57. Office of the Chief Signal Officer, *The Practical Use of Meteorological Reports and Weather-Maps*, War Department circular (Washington, D.C., 1871), 6.

58. Ibid., 10.

59. Ibid., 11.

60. Works Projects Administration Writers' Program, *Utah: A Guide to the State* (New York: Hastings House, 1941), 361–3. Corinne's prominence as a business center soon faded, along with its weather station, which was relocated to Salt Lake City. The telegraph through Corinne was not the first transcontinental telegraph line, which opened in 1861 and followed the route of the Pony Express rather than the lines of the Union Pacific and Central Pacific. For discussion of the symbiotic relationship of railway and telegraph companies, see Thompson, *Wiring a Continent*, 212–6.

61. In 1901, a weather service official noted that "the growth of the Weather Bureau system of reports and warnings has been coextensive with the growth of the electric telegraph." See J. H. Robinson, "The Telegraph and the Weather Service," *Proceedings of the Second Convention of Weather Bureau Officials, Held at Milwaukee, Wisc., August 27, 28, 29, 1901*, U.S. Weather Bureau Bulletin no. 31

(Washington, D.C., 1902), 145–6; quotation on 146. For a table reporting the number of weather stations in operation for each year from 1870 to 1888, see *Report of the Chief Signal Officer for 1888* (Washington, D.C., 1889), 171.

62. Thomas Russell, "Prediction of Cold Waves from Signal Service Weather Maps," *Report of the Chief Signal Officer for 1890*, 13–184, plus numerous charts on unnumbered pages; quotation on 88.

63. *Report of the Chief Signal Officer for 1882* (Washington, D.C., 1882), 62.

64. The estimate of approximately 175 telegraphic weather stations includes 26 in Canada and 1 in Havana. See J. Warren Smith, "The U.S. Signal Service Weather Forecasts," *American Meteorological Journal* 8 (1891): 49–52.

65. Weber, *Weather Bureau*, 7. For discussion of Abbe's role in promoting standard time, see Ian R. Bartky, "The Adoption of Standard Time," *Technology and Culture* 30 (1989): 25–56.

66. Whitnah, *History of the United States Weather Bureau*, 43–60.

67. Ibid., 27.

68. Weber, *Weather Bureau*, 8–9; and James Kenealy, "Weather Bureau Stations and Their Duties," *Yearbook of the Department of Agriculture: 1903* (Washington, D.C., 1904), 109–20.

69. *Report of the Chief Signal Officer for 1891* (Washington, D.C., 1892), 12; Harrington, "History of the Weather Map," table 2; and "Report of the Chief of the Weather Bureau for 1903," *Monthly Weather Review* 31 (1903): 627. Estimates are for a 12-month period ending June 30 of the year indicated. Values for 1887 and 1891 were estimated by dividing the reported yearly totals (52,248 and 1,007,156) by 365.

70. "Report of the Chief of the Weather Bureau for 1903," 627.

71. Harrington, "History of the Weather Map."

72. For a belated announcement of the new map as well as quotations from meteorologists acknowledging complimentary copies, see "New Daily Weather Map," *Monthly Weather Review* 42 (1914): 356. For a comment from the February 26, 1914, issue of *Nature*, see W. M. Shaw, "Daily Synoptic Charts of the Northern Hemisphere and Absolute Units," *Monthly Weather Review* 42 (1914): 100. For additional information on the history and use of the map and the collection of data, see E. B. Calvert, "Weather-Charts for the Northern Hemisphere," *Transactions of the American Geophysical Union* 13 (1932): 136-40.

73. For a reproduction of FitzRoy's synoptic chart for October 26, 1859, see "Early Synoptic Charts," *Marine Observer* 25 (1955): 9–10; the foldout map faces p. 24. For a reproduction of one of Buchan's charts, see Shaw, *Manual of Meteorology*, 310.

74. Harrington, "History of the Weather Map," 334; and Lempfert, "British Weather Forecasts," 178–80. The Signal Office, which began a daily northern hemisphere map in 1887, had published a monthly northern hemisphere map since 1879, using data a year or two old. See "The International Weather Chart," *Natural History Journal* 4 (1880): 134–7; and "Weather Charts for the Northern Hemisphere," *Nature* 20 (21 August 1879): 381–3.

75. "New Daily Weather Map," 35.

76. Ibid.

77. Cleveland Abbe, "The Weather Map on the Polar Projection," *Monthly Weather Review* 42 (1914): 36–8; quotation on 37.

78. "Northern Hemisphere Map Interrupted," *Monthly Weather Review* 42 (1914): 457.

79. Calvert, "Weather-Charts for the Northern Hemisphere," 139.

CHAPTER FOUR

1. For discussion of Weather Bureau resistance to the Norwegian concepts, see Charles C. Bates, "The Formative Rossby-Reichelderfer Period in American Meteorology, 1926–40," *Weather and Forecasting* 4 (1989): 593–603; Jerome Namias, "The History of Polar Front and Air Mass Concepts in the United States: An Eyewitness Account," *Bulletin of the American Meteorological Society* 64 (1983): 734–55; Richard J. Reed, "The Development and Status of Modern Weather Prediction," *Bulletin of the American Meteorological Society* 58 (1977): 390–400; and Donald R. Whitnah, *A History of the United States Weather Bureau* (Urbana: University of Illinois Press, 1961), 159–62.

2. See Ralph Jewell, "The Bergen School of Meteorology: The Cradle of Modern Weather-Forecasting," *Bulletin of the American Meteorological Society* 62 (1981): 824–30.

3. For discussion of related ideas that preceded Bjerknes's concept of a front, see A. Kh. Khrgian, *Meteorology: A Historical Survey*, 2d ed., rev., trans. Kh. P. Pogosyan (Jerusalem: Israel Program for Scientific Translation, 1970), 1: 204–12.

4. J. Bjerknes, "On the Structure of Moving Cyclones," *Monthly Weather Review* 49 (1919): 95–9. Written in English, Bjerknes's paper was also published in Norway, with the same title, shortly before or shortly afterward, as the second issue of volume 1 of the Norwegian Geophysics Commission's new monograph series *Geofysiske Publikationer*.

5. Written in 1918, the "Structure" paper described a preliminary version of the cyclone model, which Bjerknes did not develop into a three-dimensional weather front until 1919. See Robert Marc Friedman, *Appropriating the Weather: Vilhelm Bjerknes and the Construction of a Modern Meteorology* (Ithaca, N.Y.: Cornell University Press, 1989), 128–30.

6. Ibid., 186–8. Citing Friedman, culture critic Andrew Ross interpreted this use of "front" as a "metaphoric extension" of World War I. See Andrew Ross, *Strange Weather: Culture, Science, and Technology in the Age of Limits* (New York: Verso, 1991), 228.

7. J. Bjerknes and H. Solberg, "Life Cycle of Cyclones and the Polar Front Theory of Atmospheric Circulation," *Geofysiske Publikationer* 3, no. 1 (1922): 3–18. For earlier theories on the distribution of weather within a cyclone, see William Napier Shaw and R. G. K. Lempfert, *The Life History of Surface Air Currents* (London: Meteorological Office, 1906); and Ralph Abercromby, *Principles of Forecasting by Means of Weather Charts*, 2d ed. (London: Meteorological Office, 1885).

8. Bergeron described occlusions in 1928, in his doctoral dissertation; see T. Bergeron, "Über die dreidimensional verknüpfende Wetteranalyse, Part I: Prinzipielle Einführung in das Problem der Luftmassen- und Frontenbildung" (Three-Dimensionally Combining Synoptic Analysis, Part I: The Formation of Air Masses and Fronts), *Geofysiske Publikationer* 5, no. 6 (1928). According to Jewell, Bergeron first suggested the process of occlusion in November 1919; see Jewell, "The Bergen School," 826, 828.

9. Bjerknes, "Structure of Cyclones," 98–99.

10. J. Bjerknes and H. Solberg, "Life Cycle of Cyclones and the Polar Front Theory of Atmospheric Circulation," *Geofysiske Publikationer* 3, no. 1 (1922): 1–18. The U.S. Weather Bureau published a detailed abstract of the "Life Cycle" article, including many of the original graphics, as Alfred J. Henry, "J. Bjerknes and H. Solberg on the Life Cycle of Cyclones and the Polar Front Theory of Atmospheric Circulation," *Monthly Weather Review* 50 (1922): 468–73. Modern treatments of cyclogenesis differ markedly from the Bergen model in their reliance on maps of the upper-air pressure surface; for contemporary cartographic illustrations of the process Bjerknes and Solberg portrayed in figure 4.3, see Toby N. Carlson, *Mid-Latitude Weather Systems* (New York: Harper-Collins Academic Press, 1991), 102–5; and Frederick K. Lutgens and Edward J. Tarbuck, *The Atmosphere: An Introduction to Meteorology,* 6th ed. (Englewood Cliffs, N.J.: Prentice-Hall, 1995), 230–8.

11. For Ferrel's original paper, see William Ferrel, "An Essay on the Winds and Currents of the Ocean," *Nashville Journal of Medicine and Surgery* 11 (1856): 287–301, 375–89. Ferrel derived his idea from a two-cell (per hemisphere) model published in 1855 by Matthew Maury (1806–1873). For a concise summary of Ferrel's concept and later modifications, see William Herbert Hobbs, *The Glacial Anticyclones: The Poles of the Atmospheric Circulation* (New York: Macmillan, 1926), 16–9.

12. Bjerknes, "Structure of Cyclones," 99.

13. Halvor Solberg's letter to Theodore Hesselberg, March 28, 1920, quoted in Ralph Jewell, "Tor Bergeron's First Year in the Bergen School," in *Weather and Weather Maps: A Volume Dedicated to the Memory of Tor Bergeron,* Contributions of Current Research in Geophysics, vol. 10 (Basel: Birkhäuser Verlag, 1981), 474–90; quotation on 489.

14. The sketch is also reproduced in ibid., plate opposite p. 488.

15. Bjerknes and Solberg, "Life Cycle of Cyclones."

16. Jewell, "The Bergen School," 827. According to Jewell, Vilhelm Bjerknes's assistants used red pencils for cold fronts and blue pencils for warm fronts in their first few years of drawing weather maps, but later adopted the widely accepted color convention of red as warm and blue as cold.

17. The postcard is also reproduced in Ralph Jewell, "Tor Bergeron's First Year in the Bergen School," plate between pp. 488 and 489.

18. For a concise description of air-mass classification, see H. C. Willett, "Characteristic Properties of North American Air Masses," in Jerome Namias, ed., *An Introduction to the Study of Air Mass and Isentropic Analysis,* 5th ed. (Milton, Mass.: American Meteorological Society, 1940), 73–108.

19. For a detailed, insightful examination of Bjerknes's motives and self-promoting shrewdness, see Friedman, *Appropriating the Weather.*

20. V. Bjerknes, "Weather Forecasting," *Monthly Weather Review* 49 (1919): 90–5.

21. Ibid., 90.

22. V. Bjerknes, "Possible Improvements in Weather Forecasting, with Special Reference to the United States," *Monthly Weather Review* 49 (1919): 99–100; quotation on 99.

23. Alfred J. Henry, "Discussion" [of "The Earth's Atmosphere as a Circular

Vortex," by Anne Louise Beck], *Monthly Weather Review* 50 (1922): 401.

24. Bjerknes, "Possible Improvements," 99–100. Although an editorial footnote in the 1919 essay corrected Bjerknes's "about 3,000" to "about 4,500," Henry's predecessor registered no objection to the author's "about 300 telegraphic stations." However eager to inflate the percentage increase called for by Bjerknes, Henry accurately estimated the numbers of climatic stations (4,500) and telegraphic stations (slightly more than 200); see Gustavus A. Weber, *The Weather Bureau: Its History, Organization, and Activities* (New York: D. Appleton, 1922), 44.

25. Anne Louise Beck, "The Earth's Atmosphere as a Circular Vortex," *Monthly Weather Review* 50 (1922): 393–401; quotations on 400.

26. Henry, "Discussion."

27. Alfred J. Henry and others, *Weather Forecasting in the United States*, Weather Bureau publication no. 583 (Washington, D.C., 1916). The list of illustrations enumerates a frontispiece map of forecast districts and 199 separate figures, all but 16 of which are maps. Although some of the maps describe average conditions, storm tracks, or short-term change in pressure or temperature, the majority are cartographic snapshots of pressure and temperature.

28. Ibid., 5.

29. Willis L. Moore, *Weather Forecasting: Some Facts Historical, Practical, and Theoretical*, U.S. Weather Bureau Bulletin no. 24 (Washington, D.C., 1899), 15.

30. *Report of the Chief of the Weather Bureau, 1898–99* (Washington, D.C., 1900), 2: 439–40.

31. Frank H. Bigelow, "The Structures of Cyclones and Anticyclones on the 3500-foot and 10,000-foot Planes for the United States," *Monthly Weather Review* 31 (1903): 26–9.

32. Frank H. Bigelow, "IV. The Mechanism of Countercurrents of Different Temperatures in Cyclones and Anticyclones," *Monthly Weather Review* 31 (1903): 72–84.

33. Frank H. Bigelow, "VI. The Circulation in Cyclones and Anticyclones, with Precepts for Forecasting by Auxiliary Charts of the 3500-foot and the 10,000-foot Planes," *Monthly Weather Review* 32 (1904): 212–6.

34. For descriptions of the Weather Bureau's kite project, see W. R. Blair, "The Methods and Apparatus Used in Obtaining Upper Air Observations at Mount Weather, Va.," *Bulletin of the Mount Weather Observatory* 1 (1908): 12–9; and C. F. Marvin, "A Weather Bureau Kite," *Monthly Weather Bureau* 23 (1895): 418–20. Cleveland Abbe also described the kites and kite-collected data in *Encyclopaedia Britannica*, 11th ed, s.v. "meteorology." For an overview of kites, balloons, and other methods of sampling the upper air, see S. P. Fergusson, "The Early History of Aerology in the United States," *Bulletin of the American Meteorological Society* 14 (1933): 252–6; Khrgian, *Meteorology: A Historical Survey*, 1: 254–98; and W. E. Knowles Middleton, *Invention of the Meteorological Instruments* (Baltimore: Johns Hopkins Press, 1969), 265–85.

35. See, for example, C. F. Marvin, "Kite Experiments at the Weather Bureau," *Monthly Weather Review* 24 (1896): 199–206.

36. "Preliminary Results of Weather Bureau Kite Observations in 1898," *Monthly Weather Review* 27 (1899): 413–5.

37. Willis Ray Gregg, "History of the Application of Meteorology to Aero-

nautics with Special Reference to the United States," *Monthly Weather Review* 61 (1933): 165–9. Also see Willis Ray Gregg, "Aerological Investigations of the Weather Bureau during the War," *Monthly Weather Review* 47 (1919): 205–10; and Willis Ray Gregg and others, *Aeronautical Meteorology*, 2d ed. (New York: Ronald Press, 1930), esp. 246–51.

38. For examples of the use of pilot balloons to collect wind and cloud-ceiling data, see Middleton, *Invention of Meteorological Instruments*, 287–315; and L. W. Pollak, "A Graphical Method of Evaluating Pilot-Balloon Ascents," *Bulletin of the American Meteorological Society* 22 (1941): 345–8. For an early proposal to collect upper-air data with pilot balloons, see A. de Quervain, "A Proposal That Pilot Balloons Be More Generally Used in Making Meteorological Observations," *Monthly Weather Review* 35 (1907): 454–6. In addition, the Weather Bureau had also used captive (tethered) balloons, to take measurements when winds were too gentle to support kites; see "Annual Report by Willis L. Moore, Chief of the Weather Bureau, for the Fiscal Year Ending June 30, 1908," *Monthly Weather Review* 36 (1908): 454–65, esp. 454.

39. Charles Joseph Maguire, *Aerology: A Ground School Manual in Aeronautical Meteorology* (New York: McGraw-Hill, 1931), 48.

40. See Frederik Nebeker, *Calculating the Weather: Meteorology in the Twentieth Century* (San Diego: Academic Press, 1995), 114, 121; and R. F. M. Hay, "Some Landmarks in Meteorological Progress, 1855–1955," *Marine Observer* 25 (1955): 11–6, 79–83, esp. 80–1.

41. See, for example, "Aviation Weather Problems," *Science* 66 (October 28, 1927, supp.): xii; Donald Duke, *Airports and Airways: Cost, Operation, and Maintenance* (New York: Ronald Press, 1927), 92; and "The Weather," *Fortune* 21 (April 1940): 58–63, 137–44.

42. A more sensational motive for investigating the Weather Bureau was the rumor that a faulty weather forecast had led to the wreck of the navy airship *Akron;* see Bates, "Formative Rossby-Reichelderfer Period," 597.

43. For the full text of the preliminary report, see "The Work of the Weather Bureau," *Science* 78 (1933): 582–5, 604–7. Also see "Urges Air Service in Weather Bureau," *New York Times*, December 1, 1933. For the special committee's full report, see *Report of the Science Advisory Board, 1933–34* (Washington, D.C., 1934), 45–58. A subsequent report, comparatively brief, noted that the Weather Bureau had adopted air-mass analysis and several other recommendations; see *Second Report of the Science Advisory Board, 1934–35* (Washington, D.C., 1935), 40–2.

44. Ibid., 604.

45. "Urged as Weather Chief," *New York Times*, January 12, 1934; and "Dr. C. F. Marvin Retires," *New York Times*, August 26, 1934.

46. Gregg, *Aeronautical Meteorology*.

47. Willis Ray Gregg, "Progress in Development of the U.S. Weather Service in Line with the Recommendations of the Science Advisory Board," *Science* 80 (1934): 349–51.

48. *Report of the Chief of the Weather Bureau, 1938*, 5.

49. J. B. Lippincott, "Improvements Proposed in U.S. Weather Bureau," *Civil Engineering* 5 (1935): 654–5; *Report of the Chief of the Weather Bureau, 1935*, 9; and Bates, "Formative Rossby-Reichelderfer Period," 599–600.

50. Namias, "History of Polar Front and Air Mass Concepts"; and Bates, "Formative Rossby-Reichelderfer Period."

51. Hurd C. Willett, "Routine Daily Preparation and Use of Atmospheric Cross Sections," *Monthly Weather Review* 63 (1935): 4–7.

52. "New Weather Map Symbols Recall Indian Picture Writing," *Science News Letter* 40 (August 23, 1941): 117; *Report of the Chief of the Weather Bureau, 1941*, esp. 142; and Frank Thone, "A New Type of Weather Map," *Science* 94 (August 1, 1941, supp.): 10.

53. Quoted in Reed, "Development and Status of Modern Weather Prediction," 393.

54. Richard A. Kerr, "A Frontal Attack on a Paradigm of Meteorology," *Science* 254 (1991): 1591–2; and Richard J. Reed, "Advances in Knowledge and Understanding of Extratropical Cyclones during the Past Quarter Century: An Overview," in *Palmén Memorial Symposium on Extratropical Cyclones, Helsinki, Finland, 29 August–2 September 1988*, preprints (Boston: American Meteorological Society, 1988), 6–9. For examples of alternative interpretations, see Keith A. Browning, "Organization of Clouds and Precipitation in Extratropical Cyclones," in Chester W. Newton and Eero O. Holopainen, eds., *Extratropical Cyclones: The Erik Palmén Memorial Volume* (Boston: American Meteorological Society, 1990), 129–53; Peter V. Hobbs and others, "The Mesoscale and Microscale Structure and Organization of Clouds and Precipitation and Midlatitude Cyclones. I: A Case Study of a Cold Front," *Journal of the Atmospheric Sciences* 37 (1980): 568–96; Clifford F. Mass and David M. Schultz, "The Structure and Evolution of a Simulated Midlatitude Cyclone over Land," *Monthly Weather Review* 121 (1993): 889–917. Of particular relevance is the conveyor-belt model, described in K. A. Browning, "Conceptual Models of Precipitation Systems," *Meteorological Magazine* 114 (1985): 293–319.

55. Browning, "Conceptual Models of Precipitation Systems." For a much earlier indication of the ambiguity inherent in delineating some fronts, see C. E. Lamoureax, "On Quantitative Evaluation of Fronts," *Bulletin of the American Meteorological Society* 21 (1940): 125–6.

CHAPTER FIVE

1. The example cited is one of several rules for forecasting the course of a West Indian hurricane; see Alexander McAdie, *The Principles of Aërography* (Chicago: Rand McNally and Co., 1917), 88–9.

2. Ralph Abercromby, *Weather: A Popular Exposition on the Nature of Weather Changes from Day to Day* (New York: D. Appleton, 1887), 25–61. Abercromby provided detailed diagrams for all situations except the col.

3. See, for example, Napier Shaw, *The Drama of Weather* (Cambridge: Cambridge University Press, 1934), 227–33; and Alfred J. Henry and others, *Weather Forecasting in the United States*, Weather Bureau publication no. 583 (Washington, D.C., 1916), 74–5.

4. A 24-hour difference was preferable because of the difficulty of comparing morning and evening (or day and night) pressures. Even so, forecasters occasionally used 12-hour pressure-change charts, which could identify new trends and sudden shifts. Henry and others, *Weather Forecasting in the United States*, 72.

5. Ibid., 88–9.

6. See, for example, Victor P. Starr, *Basic Principles of Weather Forecasting* (New York: Harper and Brothers, 1942), esp. 16–17.

7. R. Hanson Weightman, "Advances and Developments in Weather Forecasting," *Journal of the Franklin Institute* 222 (1936): 527–49.

8. For discussion of upper-air weather charts, see John M. Wallace and Peter V. Hobbs, *Atmospheric Science: An Introductory Survey* (New York: Academic Press, 1977), 108–11, 128–39.

9. C. L. Mitchell and H. Wexler, "How the Daily Forecast Is Made," in *Climate and Man*, Yearbook of the Department of Agriculture, 1941 (Washington, D.C., 1941), 579–98.

10. Wind direction at the surface is strongly influenced by the pressure-gradient force, which moves air from an area of higher pressure to an area of lower pressure. Wind thus blows somewhat inward toward a low and somewhat outward from a high because friction with the land surface slows air movement and thereby lessens the Coriolis effect, which is proportional to velocity. In the upper air, also called the "free air," surface friction is negligible, so that the Coriolis force balances the effect of difference in pressure and the air no longer moves down the pressure gradient.

11. C.-G. Rossby and others, "Relation between Variations in the Intensity of the Zonal Circulation of the Atmosphere and the Displacements of the Semi-permanent Centers of Action," *Journal of Marine Research* 2 (1939): 38–55; and C.-G. Rossby, "The Scientific Basis of Modern Meteorology," in *Climate and Man*, Yearbook of the Department of Agriculture, 1941 (Washington, D.C., 1941), 599–655, esp. 613–21.

12. Elmar R. Reiter, *Jet-Stream Meteorology* (Chicago: University of Chicago Press, 1961), 2, 103–8.

13. C.-G. Rossby and others, "Forecasting of Flow Patterns in the Free Atmosphere by a Trajectory Method," appendix in Starr, *Basic Principles of Weather Forecasting*, 268–84; and H. Wexler, "Use of Rossby's Trough Formula on the Mean 5-Day 3-Km Pressure Charts," *Bulletin of the American Meteorological Society* 23 (1942): 32–4. Although Rossby's algorithm was new, the concept of periodic weather phenomena progressing in waves was not; see, for example, Henry Helm Clayton, "A Proposed New Method of Weather Forecasting by Analysis of Atmospheric Conditions into Waves of Different Lengths," *Monthly Weather Review* 35 (1907): 161–7.

14. See, for example, Sverre Petterssen, *Weather Analysis and Forecasting* (New York: McGraw-Hill, 1940), 378–425. An important part of this calculus was the isentropic chart, a thermodynamic tool that addressed the flow of moisture in the upper air. See Jerome Namias, "Isentropic Analysis," in ibid., 351–77.

15. Lewis F. Richardson, *Weather Prediction by Numerical Process* (Cambridge: Cambridge University Press, 1922), 181–213. For suggestions that the failure of Richardson's forecast might have set back computational forecasting, see Jack Fishman and Robert Kalish, *The Weather Revolution: Innovation and Imminent Breakthroughs in Accurate Forecasting* (New York: Plenum Press, 1994), 62; and Richard J. Reed, "The Development and Status of Modern Weather Prediction," *Bulletin of the American Meteorological Society* 58 (1977): 390–400, esp. 394. Fishman and Kalish unfairly assert that "the failure of Richardson's first numerical forecast resulted in the complete halt of further numerical experiments for

more than two decades"—as if other researchers were pursuing numerical forecasting in the early 1920s.

16. For a critical evaluation of Richardson's model and its significance, see Frederik Nebeker, *Calculating the Weather: Meteorology in the Twentieth Century* (San Diego: Academic Press, 1995), 58–82; and George W. Platzman, "A Retrospective View of Richardson's Book on Weather Prediction," *Bulletin of the American Meteorological Society* 48 (1967): 514–50.

17. Philip D. Thompson, *Numerical Weather Analysis and Prediction* (New York: Macmillan, 1961), 44–7.

18. For General Circulation Models (GCMs), used in long-term forecasting and studies of global change, the distinction between grid cells and grid points can have an important effect on the choice of an appropriate contouring algorithm. See William C. Skelly and Anne Henderson-Sellers, "Grid Box or Grid Point: What Type of Data Do GCMs Deliver to Climate Impacts Researchers?" *International Journal of Climatology* 16 (1996): 1079–86. This distinction can also be important for the short-term forecast models discussed in this chapter. For an overview of GCMs, see Anthony R. Hansen and Alfonso Sutera, "Weather Regimes in a General Circulation Model," *Journal of the Atmospheric Sciences* 47 (1990): 380–91; and Xiaoqing Wu and Mitchell W. Moncrieff, "Collective Effects of Organized Convection and Their Approximation in General Circulation Models," *Journal of the Atmospheric Sciences* 53 (1996): 1477–95.

19. At high latitudes, Richardson doubled the cells' longitudinal extent to counteract the shrinkage caused by converging meridians. He began doubling at 63°, and doubled the width several times again, as seemed appropriate, so that just below the north pole the cells were spherical triangles covering 90° in longitude. Richardson, *Weather Prediction*, 155.

20. Ibid., 219–20.

21. Edgar W. Woolard, "L. F. Richardson on Weather Prediction by Numerical Process," *Monthly Weather Review* 49 (1922): 72–4.

22. Platzman, "Retrospective View of Richardson's Book," 533–7.

23. Thompson, *Numerical Weather Analysis*, 16–8.

24. R. Courant, K. Friedrichs, and H. Lewy, "Über die partiellen Differenzengleichungen der mathematishcen Physik," *Mathematische Annalen* 100 (1928): 32–74; and Fishman and Kalish, *Weather Revolution*, 62–3.

25. Even if high-speed computing had been available in the 1920s, it is doubtful that the eccentric Richardson would have gained access—as a Quaker, he had been a conscientious objector during World War I, and consequently was denied a university professorship despite having driven a battlefield ambulance. The first computers were large, expensive, and scarce, and the technological bureaucracy that controlled them was wary of freethinkers.

26. Michael R. Williams, *A History of Computing Technology* (Englewood Cliffs, N.J.: Prentice-Hall, 1985), 302–3.

27. Nebeker, *Calculating the Weather*, 136–8. For a concise overview of numerical weather forecasting, see Frederick G. Shuman, "Numerical Weather Prediction," *Bulletin of the American Meteorological Society* 59 (1978): 5–17; and Philip Duncan Thompson, "The Mathematics of Meteorology," in Lynn Arthur Steen, ed., *Mathematics Today: Twelve Informal Essays* (New York: Springer-Verlag, 1978), 127–52.

28. For a concise analysis of Charney's one-layer model, see Thompson, "The Mathematics of Meteorology," 145–8.

29. J. G. Charney, R. Fjörtoft, and J. von Neumann, "Numerical Integration of the Barotropic Vorticity Equation," *Tellus* 2 (1950): 237–54; quotation on 245.

30. Frederick G. Shuman, "History of Numerical Weather Prediction at the National Meteorological Center," *Weather and Forecasting* 4 (1989): 286–96; quotation on 287. Also see Philip Duncan Thompson, "A History of Numerical Weather Prediction in the United States," *Bulletin of the American Meteorological Society* 64 (1983): 755–69.

31. Philip D. Thompson, "The Role of Computers in the Development of Numerical Weather Prediction," *Atmospheric Technology* [National Center for Atmospheric Research], no. 3 (September 1973): 13–6.

32. Sidney Teweles, Jr., and Hermann B. Wobus, "Verification of Prognostic Charts," *Bulletin of the American Meteorological Society* 35 (1954): 455–63. For discussion of improvements in skill scores during the 1950s, 1960s, and 1970s, see George J. Haltiner and Roger Terry Williams, *Numerical Prediction and Dynamic Meteorology*, 2d ed. (New York: John Wiley and Sons, 1980), 422–38.

33. Shuman, "History of Numerical Weather Prediction," 288; and David Laskin, *Braving the Elements: The Stormy History of American Weather* (New York: Doubleday, 1996), 158.

34. For discussion of numerical modeling as a unifying force in meteorology, see Nebeker, *Calculating the Weather*, 173–87.

35. Peter R. Chaston, *Weather Maps: How to Read and Interpret All the Basic Weather Charts* (Kearney, Mo.: Chaston Scientific, 1995), 105–8.

36. A third level might have a still finer grid, perhaps with a spacing of 5 km (3 miles). For examples of nested-grid models with two or three levels, see David M. Mocko and William R. Cotton, "Evaluation of Fractional Cloudiness Parameterizations for Use in a Mesoscale Model," *Journal of the Atmospheric Sciences* 52 (1995): 2884–901. For an overview of the National Meteorological Center's ETA model, see Richard A. Kerr, "Hurricane Forecasting Shows Promise," *Science* 247 (1990): 917. Also see Stanley L. Barnes, Fernando Caracena, and Adrian Marroquin, "Extracting Synoptic-Scale Diagnostic Information from Mesoscale Models: The Eta Model, Gravity Waves, and Quasigeostrophic Diagnostics," *Bulletin of the American Meteorological Society* 77 (1996): 519–28. Another promising short-term forecast model is the RUC (Rapid Update Cycle) model; see, for example, Tatiana G. Smirnova, John M. Brown, and Stanley G. Benjamin, "Performance of Different Soil Model Configurations in Simulating Ground Surface Temperature Fluxes," *Monthly Weather Review* 125 (1997): 1870–84. For a definitions of "mesoscale" and other scale divisions used on meteorology, see B. W. Atkinson, *Meso-scale Atmospheric Circulations* (London: Academic Press, 1981); and Isidoro Orlanski, "A Rational Subdivision of Scales for Atmospheric Processes," *Bulletin of the American Meteorological Society* 56 (1975): 527–30. For discussion of improvements resulting from blending (averaging) nested-grid and limited fine-mesh models, see Robert L. Vislocky and J. Michael Fritsch, "Improved Model Output Statistics Forecasts through Model Consensus," *Bulletin of the American Meteorological Society* 76 (1995): 1157–64.

37. See Vincent Kiernam, "Hurricane Forecasters Predict Real Storms

Ahead," *New Scientist* 147 (July 1,1995): 10; and Tim Vasquez, "Forecast Models for Hurricanes," *Weatherwise* 47 (June/July 1994): 30–1.

38. Chaston, *Weather Maps*, 113–41; and Joseph J. Tribbia and Richard A. Anthes, "Scientific Basis of Modern Weather Prediction," *Science* 237 (1987): 493–7. The mix of models changes frequently; for discussion of the array of models available in the mid-1990s, see Laskin, *Braving the Elements*, 158–63.

39. Tribbia and Anthes, "Scientific Basis of Modern Weather Prediction," 497.

40. Edward N. Lorenz, "Large-Scale Motions of the Atmosphere," in P. M. Hurley, ed., *Advances in Earth Science* (Cambridge, Mass.: MIT Press, 1964), 95–109; quotation on 108.

41. For a readable introduction to chaos theory, see James Gleick, *Chaos: Making a New Science* (New York: Viking Penguin, 1987).

42. For discussion of Lorenz's contributions to meteorology and chaos theory, see Stanley David Gedzelman, "Chaos Rules: Edward Lorenz Capped a Century of Progress in Forecasting by Explaining Unpredictability," *Weatherwise* 47 (August/September 1994): 21–6; Gleick, *Chaos*, 11–23; and Nebeker, *Calculating the Weather*, 190–4.

43. Ronald D. McPherson, "The National Centers for Environmental Prediction: Operational Climate, Ocean, and Weather Prediction for the Twenty-first Century," *Bulletin of the Meteorological Society of America* 75 (1994): 363–73.

44. For an introduction to the ensemble approach, see Zoltan Toth and Eugenia Kalnay, "Ensemble Forecasting at NCEP and the Breeding Method," *Monthly Weather Review* 125 (1997): 3297–319; Zoltan Toth and Eugenia Kalnay, "Ensemble Forecasting at NMC: The Generation of Perturbations," *Bulletin of the American Meteorological Society* 74 (1993): 2317–30; and M. Steven Tracton and Eugenia Kalnay, "Operational Ensemble Prediction at the National Meteorological Center: Practical Aspects," *Weather and Forecasting* 8 (1993): 379–98.

45. Stephen Lord, telephone conversation with author, August 28, 1996.

CHAPTER SIX

1. I emphasize this point a bit stridently because two meteorologists who read an earlier version of the manuscript thought that this chapter didn't belong in the book. Although the material discussed here is, I agree, meteorologically peripheral, the chapter is cartographically essential.

2. A fundamental distinction between point-source and non-point-source pollution underlies marked differences in modeling strategy and map scale. For general discussion of atmospheric pollution models, see Richard W. Boubel and others, *Fundamentals of Air Pollution*, 3d ed. (San Diego, Calif.: Academic Press, 1994), 291–363; Norman E. Bowne, "Atmospheric Dispersion," in Seymour Calvert and Harold M. Englund, eds., *Handbook of Air Pollution Technology* (New York: John Wiley and Sons, 1984), 859–91; and James F. Lape, "Air Dispersion and Deposition Models," in David R. Patrick, ed., *Toxic Air Pollution Handbook* (New York: Van Nostrand Reinhold, 1994), 226–40. For a glimpse of the widely varied results produced by different multiple-source models, see Robin L. Dennis and others, *Evaluation of Regional Acidic Deposition Models (Part I)*, NAPAP report no. 5 (Washington, D.C.: National Acid Precipitation Assessment Program, 1990), esp. 5-49 to 5-52.

3. Meaghan Boice, *How to Create a Toxic Plume Map*, CEC Fact Sheet no. 2 (Albany, N.Y.: Citizens' Environmental Coalition, 1992).

4. Ibid., 22.

5. Ibid.

6. For discussion of the Bhopal accident, see William Bogard, *The Bhopal Tragedy: Language, Logic, and Politics in the Production of a Hazard* (Boulder, Colo.: Westview Press, 1989); and Dan Kurzman, *A Killing Wind: Inside Union Carbide and the Bhopal Catastrophe* (New York: McGraw-Hill, 1987). For discussion of SARA Title III, see Susan G. Hadden, *A Citizen's Right to Know: Risk Communication and Public Policy* (Boulder, Colo.: Westview Press, 1989); and Rosemary O'Leary, *Emergency Planning: Local Government and the Community Right-to-Know Act* (Washington, D.C.: International City/County Management Association, 1993).

7. For a fuller discussion of the cartographic display and analysis of atmospheric hazards, see Mark Monmonier, *Cartographies of Danger: Mapping Hazards in America* (Chicago: University of Chicago Press, 1997), 148–72 and 216–38.

8. Daniel S. Davis and others, *Accidental Releases of Air Toxics: Prevention, Control, and Mitigation* (Park Ridge, N.J.: Noyes Data Corporation, 1989), 506–16.

9. William Peat, Jr., New York State Emergency Management Office, letter to author, 6 June 1995. The graphs in figures 6.2 and 6.3 were plotted originally by EIS (Emergency Information System) using the ALOHA (Areal Locations of Hazardous Atmospheres) plume modeling package. In addition to the conditions described, the model plume represents a release at 3 A.M. at 25 percent relative humidity under only 10 percent cloud cover without an inversion layer, which would retard upward movement of air.

10. U.S. Environmental Protection Agency, Federal Emergency Management Agency, and U.S. Department of Transportation, *Technical Guidance for Hazards Analysis: Emergency Planning for Extremely Hazardous Substances*, EPA publication no. OSWER-88-0001 (Washington, D.C., 1987), D-11. The IDLH level is defined as "the maximum level to which a healthy worker can be exposed for 30 minutes and escape without suffering irreversible health effects or escape-impairing symptoms." Ibid., A-6.

11. For a demonstration of the value of integrating model plumes with maps of demographic data, see Jayajit Chakraborty and Marc P. Armstrong, "Using Geographic Plume Analysis to Assess Community Vulnerability to Hazardous Accidents," *Computing, Environment, and Urban Systems* 19 (1996): 341–56.

12. Rudolf J. Engelmann and William F. Wolff, "Role of Meteorology in Emergency Response," in Mark L. Kramer and William M. Porch, eds., *Meteorological Aspects of Emergency Response* (Boston: American Meteorological Society, 1990), 5–14.

13. Ibid., 2-29 to 2-30.

14. Daniel J. McNaughton, Gary G. Worley, and Paul M. Bodner, "Evaluating Emergency Response Models for the Chemical Industry," *Chemical Engineering Process* 83 (January 1987): 46–51.

15. Michael K. Miller and Kau-Fui Vincent Wong, "Prediction of Vulnerable Zones for Reactive Substances," *Journal of Environmental Health* 56 (October 1993): 17–9.

16. For examples, see Jean-Michel Guldermann and Daniel Shefer, *Industrial Location and Air Quality Control: A Planning Approach* (New York: John Wiley and Sons, 1980), 63–75; and Gary S. Samuelsen, "Air Quality Impact Analysis," in John G. Rau and David C. Wooten, eds., *Environmental Impact Analysis Handbook* (New York: McGraw-Hill Book Co., 1980), 3-1 to 3-165.

17. For examples, see D. Bruce Turner, *Workbook of Atmospheric Dispersion Estimates: An Introduction to Dispersion Modeling*, 2d ed. (Boca Raton, Fla.: Lewis Publishers, 1994); U.S. Environmental Protection Agency, Office of Air Quality Planning and Standards, *Screening Procedures for Estimating the Air Quality Impact of Stationary Sources*, EPA publication no. 450/5-88-010 (Research Triangle Park, N.C., 1988); and Paolo Zannetti, *Air Pollution Modeling: Theories, Computational Methods, and Available Software* (New York: Van Nostrand Reinhold, 1990), 355–425. For a pioneering application of an air-diffusion model, see D. Bruce Turner, "A Diffusion Model for an Urban Area," *Journal of Applied Meteorology* 3 (1964): 83–91.

18. Calvin R. Brunner, *Hazardous Air Emissions from Incineration* (New York: Chapman and Hall, 1985), 1–7, 27–37; and David A. Tillman, *Trace Metals in Combustion Systems* (San Diego, Calif.: Academic Press, 1994), 121–50.

19. The air-quality analysis is reported in County of Onondaga, New York, Solid Waste Management Program, *Waste-to-Energy Facility: Draft Supplemental Environmental Impact Statement* (June 1988), chap. 5 and appendix J.

20. This approach is called *nesting*. After identifying high-impact areas with the coarse grid, the consultants used the intermediate grid to establish a new, local network of receptors (estimation points) 0.25 km apart in areas with the highest simulated fallout. The resulting calculations usually identified new maximums, with even higher concentrations, at intermediate positions. The engineers then established a smaller, finer network of points 0.1 km apart around these new high points and repeated the process of refinement. Nesting thus uses considerably less computational effort to find the maximum estimated concentration that would have been detected by applying the fine, 0.1-km grid throughout the entire 20-by-20-km study region.

21. Perry W. Fisher, John A. Foster, and James W. Sumner, "Comparison of the ISCST Model with Two Alternative U.S. EPA Models in Complex Terrain in Hamilton County, Ohio," *Journal of the Air and Waste Management Association* 44 (1994): 418–27; and U.S. Environmental Protection Agency, Office of Air Quality Planning and Standards, *Guideline* [sic] *on Air Quality Models (Revised)*, EPA publication no. 450/2-78-027R (Research Triangle Park, N.C., 1986), A-21 to A-24.

22. For discussion of the COMPLEX I model, see Paul D. Gutfreund and others, "COMPLEX I and II Model Performance Evaluation in Nevada and New Mexico," *Journal of the Air Pollution Control Association* 33 (1983): 864–71.

23. U.S. General Accounting Office, *Air Pollution: Reliability and Adequacy of Air Quality Dispersion Models*, report no. RCED-88-192 (August 1988). GAO officials, who did not validate the models themselves, reviewed EPA and independent data and criticized the agency for not testing all models under the same, comparatively rigorous standards.

24. Bruce A. Kunkel and Yutaka Izumi, *WADOCT: An Atmospheric Dispersion Model for Complex Terrain*, Environmental Research Paper no. 1062 (Hanscom AFB, Mass.: U.S. Air Force Geophysics Laboratory, 1990).

25. Richard M. Eckman and Ronald J. Dobosy, *The Suitability of Diffusion and Wind-Field Techniques for an Emergency Response Dispersion System*, NOAA Technical Memorandum no. ERL ARL-191 (Silver Spring, Md.: NOAA Air Resources Laboratory, 1989); and Steven R. Hanna and David G. Strimaitis, "Rugged Terrain Effects on Diffusion," in William Blumen, ed., *Atmospheric Processes over Complex Terrain*, Meteorological Monographs no. 45 (Boston: American Meteorological Society, 1990), 109–43.

26. For discussion of the evaluation methods, see Marvin H. Dickerson and Donald L. Ermak, "The Evaluation of Emergency Response Trace Gas and Dense Gas Dispersion Models," in Mark L. Kramer and William M. Porch, eds., *Meteorological Aspects of Emergency Response* (Boston: American Meteorological Society, 1990), 71–115.

CHAPTER SEVEN

1. Information on the design and operation of TIROS I is from Abraham Schnapf and others, "A History of Civilian Weather Satellites," in P. Krishna Rao and others, eds., *Weather Satellites: Systems, Data, and Environmental Applications* (Boston: American Meteorological Society, 1990), 7–19; J. Gordon Vaeth, *Weather Eyes in the Sky: America's Meteorological Satellites* (New York: Ronald Press, 1965), 23–87; and William K. Widger, Jr., *Meteorological Satellites* (New York: Holt, Rinehart and Winston, 1966), 50–91. For a concise summary of the satellite's characteristics, see W. G. Stroud, "Initial Results of the TIROS I Meteorological Satellite," *Journal of Geophysical Research* 65 (1960): 1643–4.

2. Subject to small but nonnegligible amounts of friction, TIROS gradually slowed down over several weeks from an initial spin of 12 rpm. To maintain the satellite's gyroscopic stability, NASA ground control fired an opposing pair of tiny spin-up rockets when the rotation rate dropped below 9 rpm. See Vaeth, *Weather Eyes in the Sky*, 31.

3. Russell C. Doolittle, Laurence I. Miller, and Irwin S. Ruff, "Geographic Location of Cloud Features," in Goddard Space Flight Center, *Final Report on the TIROS I Meteorological Satellite System*, NASA Technical Report R-131 (Washington, D.C.: National Aeronautics and Space Administration, 1962), 335–55.

4. Jay S. Winston, "Satellite Pictures of a Cut-Off Cyclone over the Eastern Pacific," *Monthly Weather Review* 88 (1960): 295–314.

5. "Session on TIROS Added to Washington Meeting," *Bulletin of the American Meteorological Society* 41 (1960): 522–3; quotation on 522.

6. Quoted in "Tiros I Concludes Mission: Tiros II Readied," *Weatherwise* 13 (1960): 158–61, 180; quotation on 159.

7. See Harry Wexler, "Observing the Weather from a Satellite Vehicle," *Journal of the British Interplanetary Society* 13 (1954): 269–76. Wexler is quoted in John W. Finney, "U.S. Will Share TIROS I Pictures," *New York Times*, April 5, 1960.

8. Richard Witkin, "U.S. Orbits Weather Satellite; It Televises Earth and Storms; New Era in Meteorology Seen," *New York Times*, April 2, 1960.

9. David Lawrence, "'TIROS I' Is Called a Lesson to Critics of U.S. Program," *New York Herald Tribune*, April 6, 1960.

10. Witkin, "U.S. Orbits Weather Satellite."

11. "Outer Space and Man," *New York Times*, April 3, 1960, Editorial section.

12. See, for example, Widger, *Meteorological Satellites*, 104.

13. For a description of the TIROS I vidicon camera and its recording system, see ibid., 71–6.

14. For a description of the TIROS II infrared radiometer, see W. R. Bandeen and others, "Infrared and Reflected Solar Radiation Measurements from the Tiros II Meteorological Satellite," *Journal of Geophysical Research* 66 (1961): 3169–85.

15. Temperatures are for a hypothetical "blackbody"—a completely dark object that reflects nothing and thus both absorbs and reemits all radiation received. According to the Stefan-Boltzman law, the peak wavelength of radiation from a blackbody is a function of its absolute temperature. See Bandeen and others, "Infrared and Reflected Solar Radiation Measurements," 3182.

16. TIROS IX and X, both launched in 1965, were substantially different in their stabilization system and orbit, respectively. See Widger, *Meteorological Satellites*, 91.

17. With orbits inclined at 48° to the equator, TIROS I through IV provided useful images only between 55°N and 55°S. With orbits inclined at 58°, TIROS V through VIII extended the imaged zone to within 25° of the poles. See Robert M. Rados, "The Evolution of the TIROS Meteorological Satellite Operational System," *Bulletin of the American Meteorological Society* 48 (1967): 326–37.

18. William K. Widger, Jr., "An Explanation of the Limitations to the Coverage Provided by TIROS," *Weatherwise* 14 (1961): 230–7.

19. For further information on contemporary polar-orbiting weather satellites, about which I say little, see Donald Blersch, "The National Polar-orbiting Operational Environmental Satellite System (NPOESS)," *Earth System Monitor* 7 (March 1997): 5–7, 16.

20. Information on the design and operation of Nimbus 2 is from Janice Hill, *Weather from Above: America's Meteorological Satellites* (Washington, D.C.: Smithsonian Institution Press, 1991), 23–6; W. Nordberg and others, "Preliminary Results from Nimbus II," *Bulletin of the American Meteorological Society* 47 (1966): 857–71; and Schnapf and others, "A History of Civilian Weather Satellites."

21. Stanley O. Kidder and Thomas H. Vonder Haar, *Satellite Meteorology: An Introduction* (San Diego: Academic Press, 1995), 6.

22. Hill, *Weather from Above*, 29–31; and Robert H. McQuain, "ATS-I Camera Experiment Successful," *Bulletin of the American Meteorological Society* 48 (1967): 74–9.

23. For examples, see L. F. Hubert and L. F. Whitney, Jr., "Wind Estimation from Geostationary Satellite Pictures," *Monthly Weather Review* 99 (1971): 665–72; and Frederick R. Mosher, "Cloud Drift Winds from Geostationary Satellites," *Atmospheric Technology* no. 10 (Winter 1978–79): 53–60.

24. Hill, *Weather from Above*, 31.

25. Schnapf and others, "A History of Civilian Weather Satellites," 16.

26. "A New Breed of Weather Satellite: The Fixed Stare," *Science News* 105 (1974): 332–3; "New Synchronous Satellite to Aid Storm Forecasting," *Aviation Week & Space Technology* 100 (May 13, 1974): 20; and Walter Sullivan, "Weather Station Due to Be Orbited Today," *New York Times*, May 16, 1974.

27. For additional information, see M. J. Bader and others, *Images in Weather*

Forecasting (Cambridge: Cambridge University Press, 1995), 9–11; and World Meteorological Organization, *World Weather Watch: Information on Meteorological and Other Environmental Satellites*, WMO publication no. 411 (Geneva, Switzerland, 1994).

28. Robert C. Cowen, "Hurricane Tracking Goes on without One GOES Satellite," *Christian Science Monitor*, June 12, 1989; and "Weather Satellite GOES Blind," *Science News* 135 (1989): 77. A similar strategy had proved effective in 1984, when the failure of GOES-5 left the east exposed. NOAA not only moved GOES-6 eastward to a central position for two years but shifted it with the seasons: farther east in the summer to watch for hurricanes and westward in winter to cover Pacific storms. See J. Eberhart and S. Weisburd, "The Road to Space Gets Steeper Still," *Science News* 129 (1986): 292.

29. William J. Broad, "U.S. Asking Europe to Lend Satellite," *New York Times*, September 1, 1991. In addition to providing better substitute coverage of west Africa, EUMETSAT's gesture returned an earlier favor: in 1985, when an earlier Meteosat had failed, the United States temporarily repositioned one of its satellites to provide fuller coverage of the eastern Atlantic. See "EUMETSAT Supports the Global Satellite System," *WMO Bulletin* 38 (1989): 302–4. Intelligible only from a receiving station in Germany, Meteosat-3 made the journey to 75°W in two steps, interrupted by a one-and-a-half-year pause at 50°W while NESDIS built an appropriate ground relay station at Wallops Island, Virginia. See Warren E. Leary, "Europeans Lend Satellite to U.S. for the Weather," *New York Times*, February 25, 1993.

30. Geostationary satellites regularly need to adjust their orbits, to avoid wobbling well out of the equatorial plane into a more inclined orbit. Although failure of electronic systems can result in premature death, satellites die of old age by running out of the hydrazine fuel used to power their thrusters.

31. For information on the attempted sale, see U.S. Congress, House Committee on Science and Technology, *Commercialization of Land and Weather Satellites: Report Prepared by the Congressional Research Service, Library of Congress, for the Subcommittee on Space Science and Applications and the Subcommittee on Natural Resources, Agricultural Research and Environment*, 98th Cong., 1st sess., 1983.

32. The satellite destroyed along with its Delta launch vehicle was GOES-G. (NASA identifies satellites by letter before launch and by number once in orbit.) GOES-H, intended to become GOES-8, thus became GOES-7. See Eberhart and Weisburd, "The Road to Space Gets Steeper Still."

33. U.S. General Accounting Office, *Weather Satellites: Action Needed to Resolve Status of the U.S. Geostationary Satellite Program*, report no. NSIAD-91-252 (July 1991).

34. Ibid., 4.

35. Ibid., 5. Representative Howard Wolpe, who attributed the difficulties of GOES-Next to the "incompetence of NASA and its aerospace contractors," apparently considered NOAA more a victim than a culprit. See Warren E. Leary, "Mismanagement Found on U.S. Weather Satellite," *New York Times*, July 26, 1991. For a chronology of GOES-Next and its technical problems, see U.S. Congress, Senate Committee on Appropriations, *GOES Weather Satellite Problems: Special Hearing*, 102d Cong., 1st sess., 1991.

36. Warren E. Leary, "Craft in Orbit: A Major Gain for Forecasts," *New York Times*, April 12, 1994; "Satellite Launched to Help in Scouting for Rough Weather," *New York Times*, May 24, 1995.

37. Internet postings by NESDIS officials Gary Davis and Jamie Hawkins described the history of GOES-7, which was

> launched on February 26, 1987, and went operational on March 25, 1987, at 75°W. It was moved to 108°W on February 1989 as "GOES Prime," a one-GOES system following the January 1989 failure of GOES-6. It was then moved to 98° in July 1989 for the hurricane season. It was moved back to 108° in November 1989 for the Pacific storm season. It was moved back to 98° in July 1990, then to 108° in November 1990, back to 98° in July 1991, to 112° in April 1992, when Meteosat-3 was brought to 75°, and finally to 135° in January 1995, following the commissioning of GOES-8. ... operational for over 8 years 9.5 months, [the spacecraft] was truly the GOES workhorse.

With its fuel supply running low, GOES-7 had started to wobble out of a strictly geosynchronous orbit, but it could be reactivated if needed. According to Hawkins, NOAA's newer satellites produce far better cloud shots, but pictures from GOES-7 would be "better than no imagery at all." Jamison Hawkins, telephone conversation, December 12, 1996.

38. See, for example, "Mission Overview: Geostationary Operational Environmental Satellite," unnumbered joint NOAA-NASA publication (ca. 1995). For descriptions of the satellites and their imaging systems, see W. Paul Munzel and James F. W. Purdom, "Introducing GOES-I: The First of a New Generation of Geostationary Operational Environmental Satellites," *Bulletin of the American Meteorological Society* 75 (1994): 757–81.

39. NASA plans to launch GOES-L and -M, with additional refinements, in 2001 and 1999, respectively. See Warren E. Leary, "New Weather Satellite to Boost Ailing Fleet with Better Imagery," *New York Times*, April 12, 1994; and National Oceanic and Atmospheric Administration, *National Environmental Satellite, Data, and Information Service: 1995* (Washington, D.C., ca. 1996), p. 7. In addition to carrying the Imager and the Sounder, the satellites collect environmental data, send facsimiles of satellite images and conventional weather maps, and support the Space Environment Monitor System, which measures solar activity and the earth's magnetic field. For an overview and critique of NOAA's plans to assure continuous coverage, see U.S. General Accounting Office, *Weather Satellites: Planning for the Geostationary Satellite Program Needs More Attention*, report no. AIMD-97-37 (March 1997). The GOES program has been on the GAO's list of high-risk government programs. Because of deferred planning, NOAA faces a gap in coverage as early as 2002; see U.S. General Accounting Office, *National Oceanic and Atmospheric Administration: Weather Service and NOAA Corps Issues* (testimony of Joel C. Willemssen before the Subcommittee on Energy and Environment, Committee on Science, House of Representatives), report no. T-AIMD-97-63 (March 13, 1997), 5–7.

40. Scanning is controlled by a program specifying where the Imager is to look and when. There are three scan scenarios: routine, rapid, and super-rapid. Super-rapid scans replace some of the routine images with more frequent sub-scans of small areas. The scenarios as well as modifications for testing, recalibra-

tion, and other housekeeping operations are described at the *NESDIS Office of Satellite Operations* Web site (Appendix: NESDIS Satellite). For an example of storm monitoring in rapid-scan mode, see James R. Asker, "GOES-Next Improves Hurricane Monitoring," *Aviation Week and Space Technology* 143 (August 14, 1995): 62.

41. With different beats, GOES-East and GOES-West follow different scenarios, which NESDIS modifies as needed. For example, GOES-East provides frequent "CONUS" images of the conterminous United States whereas GOES-West provides equivalent "PACUS" images of the western U.S. and eastern Pacific Ocean. In its routine scenario, GOES-East provides CONUS coverage twice every half hour: once as a tightly framed CONUS image and about a quarter hour later as part of a wider Canada-to-Brazil image. Both images are interrupted every three hours by the full-disk scan, of which the conterminous states are a subset.

42. Jamison Hawkins, "NOAA's New Space Sentinels," *Earth System Monitor* 6 (September 1995): 1–2; quotation on 2.

43. GOES-4 through -7 provided less vertically precise soundings with an instrument that could not operate independently of the imaging radiometer. For an example of the use of soundings from GOES-7, see Randall J. Alliss and Sethu Raman, "A Comparison of Saturation Pressure Differences and GOES VAS Estimates to Surface Observations of Cloudiness," *Journal of Applied Meteorology* 35 (1996): 521–31. For details on the independent GOES-Next Sounder and its operation, see Raymond J. Komajda, *An Introduction to the GOES I–M Imager and Sounder Instruments and the GVAR Retransmission Format,* NOAA Technical Report no. NESDIS-82 (National Oceanic and Atmospheric Administration, 1994). For an overview of principles and techniques of atmospheric sounding, see J. T. Houghton, F. W. Taylor, and C. D. Rodgers, *Remote Sounding of Atmospheres* (Cambridge: Cambridge University Press, 1984).

44. Donald G. Gray, Christopher M. Hayden, and W. Paul Menzel, "Review of Quantitative Satellite Products Derived from GOES-8/9 Imager and Sounder Instrument Data," ca. 1995, <http://orbit7i.nesdis.noaa.gov:8080/gray95.html>.

CHAPTER EIGHT

1. For a comprehensive general history of radar, see Robert Buderi, *The Invention That Changed the World: How a Small Group of Radar Pioneers Won the Second World War and Launched a Technological Revolution* (New York: Simon and Schuster, 1996).

2. For a history of meteorological radar, see Stuart G. Bigler, "Radar: A Short History," *Weatherwise* 34 (1981): 158–63; and Jack Fishman and Robert Kalish, *The Weather Revolution* (New York: Plenum Press, 1994), 180–93.

3. James W. Wilson and Kenneth E. Wilk, "Nowcasting Applications of Doppler Radar," in K. A. Browning, ed., *Nowcasting* (London: Academic Press, 1982), 87–105.

4. For general information on the underlying principles and operation of weather radar, see M. J. Bader and others, *Images in Weather Forecasting* (Cambridge: Cambridge University Press, 1995), 50–9; William R. Cotton, *Storms,* Geophysical Science Series vol. 1 (Fort Collins, Colo.: ASTeR Press, 1990), 11–9; Henri Sauvageot, *Radar Meteorology* (Norwood, Mass.: Artech House, 1992), esp. 1–4;

and Aaron Williams, Jr., *The Use of Radar Imagery in Climatological Research,* Commission on College Geography Resource Paper no. 21 (Washington, D.C.: Association of American Geographers, 1973).

5. For an early account, see S. S., "Radar Detection of the Florida Hurricane," *Bulletin of the American Meteorological Society* 26 (1945): 451.

6. For examples, see Ralph J. Donaldson, Jr., "Methods for Identifying Severe Thunderstorms by Radar: A Guide and Bibliography," *Bulletin of the American Meteorological Society* 46 (1965): 174–93.

7. Ibid., 175–7.

8. Ibid., 177–8; Robert Davies-Jones, "Tornadoes," *Scientific American* 273 (August 1995): 48–57; and J. S. Marshall and W. E. Gordon, "Radiometeorology," *Meteorological Monographs* 3 (July 1957): 73–113, esp. 90–1.

9. Also see Stuart G. Bigler, "An Analysis of Tornado and Severe Weather Echoes," *Proceedings of the Weather Radar Conference,* Signal Corps Engineering Laboratories, Fort Monmouth, N.J., September 12–15, 1955, 167–75, esp. 173-4; Keith A. Browning, "The Evolution of Tornadic Storms," *Journal of the Atmospheric Sciences* 2 (1965): 664–8; J. S. Marshall and W. E. Gordon, "Radiometeorology," *Meteorological Monographs* 3 (1957): 73–113, esp. 90; and John T. Snow, "The Tornado," *Scientific American* 250 (April 1984): 86–96.

10. Vaughn D. Rockney, "New Developments in Observations and Instrumentation in the Weather Bureau," *Bulletin of the American Meteorological Society* 40 (1959): 554–60.

11. Wesley Irvin, "Networks for Weather Radar and Upper Air Observing Systems," *Bulletin of the American Meteorological Society* 45 (1964): 388–91.

12. U.S. Weather Service, *Operations of the National Weather Service*, Department of Commerce publication no. 72–10470 (Washington, D.C., 1971), 114–5.

13. Ibid., 116–7.

14. Stuart G. Bigler, "A Comparison of Synoptic Analysis and Composite Radar Photographs of a Cold Front and Squall Line," *Proceedings of the Weather Radar Conference,* Signal Corps Engineering Laboratories, Fort Monmouth, N.J., September 12–15, 1955, 113–9. Also see Myron G. H. Ligda, *Study of the Synoptic Application of Weather Radar Data*, Final Report under U.S. Air Force Contract AF 19(604)-573 (College Station: Texas A&M University, Department of Oceanography, 1956).

15. Bigler, "Radar: A Short History," 158.

16. Louis J. Battan, *Radar Observation of the Atmosphere* (Chicago: University of Chicago Press, 1973), 97–101. For an example of the use of maps in validating radar-based rainfall measurements, see G. E. Stout and J. C. Neill, "Utility of Radar in Measuring Areal Rainfall," *Bulletin of the American Meteorological Society* 34 (1953): 21–7.

17. J. S. Marshall, "Peak Reading and Thresholding in Processing Radar Weather Data," *Journal of Applied Meteorology* 10 (1971): 1213–23.

18. Ibid., 104–13; and Clayton E. Jensen, "A Review of Federal Meteorological Programs for Fiscal Years 1965–1975," *Bulletin of the American Meteorological Society* 56 (1975): 208–24.

19. See "Federal Plan for Weather Radars," *Bulletin of the American Meteorological Society* 55 (1974): 465–6.

20. See A. E. Woodruff, "Doppler, Johann Christian," in *Dictionary of Scientific Biography* (New York: Scribner, 1971), 4: 167–8.

21. For an early examination of Doppler radar's potential value to forecasters, see James Wilson and others, "Forecasting Applications of Doppler Radar," *Atmospheric Technology* no. 13 (1981): 105–18. For a more recent appraisal, see Jack Williams, "Doppler Effects: New Radar Puts Forecasters a Step Ahead of the Weather," *Weatherwise* 47 (August-September 1994): 43–6.

22. Robert L. Smith and David W. Holmes, "Use of Doppler Radar in Meteorological Observations," *Monthly Weather Review* 89 (1961): 1–7.

23. See Robert E. Rinehart, *Radar for Meteorologists, Or You, Too, Can Be a Radar Meteorologist, Part III*, 2d ed. (Grand Forks, N.D.: University of North Dakota, Center for Atmospheric Sciences, 1991), 181–2; and U.S. General Accounting Office, *Weather Forecasting: Radar Availability Requirement Not Being Met*, report no. AIMD-95-132 (May 1995), 16.

24. "Maker Challenges Deal on U.S. Weather Radar," *New York Times*, December 22, 1990.

25. Noam S. Cohen, "Officials Divided on Troubled Weather Radar Plan," *New York Times*, August 13, 1991.

26. Richard A. Kerr, "Upgrade of Storm Warnings Paying Off," *Science* 262 (1993): 331–3; quotation on 331.

27. John Livingston, "NEXRAD Now!" *Mariner's Weather Log* 37 (Winter 1993): 12–5. NOAA made a more modest but nonetheless impressive claim to the General Accounting Office: an increase from 7 minutes in 1993 to 9 minutes in 1995 in the lead time for tornado predictions, as well as an increase in accuracy from 47 to 60 percent. Further improvements were planned for 1996: 10 minutes of lead time and 64 percent accuracy. See U.S. General Accounting Office, *Executive Guide: Effectively Implementing the Government Performance and Results Act*, report no. GGD-96-118 (June 1996), 26. Another GAO report called this improvement an "important success"; see U.S. General Accounting Office, *Weather Service Modernization: Risks Remain That Full Systems Potential Will Not Be Achieved* (testimony of Joel C. Willemssen before the Subcommittee on Oversight of Government Management, Restructuring, and the District of Columbia, Committee on Government Affairs, U.S. Senate), report no. T-AIMD-97-85 (April 24, 1997), 5.

28. Williams, "Doppler Effects."

29. Estimate includes costs of land acquisition and construction. See U.S. General Accounting Office, *Weather Forecasting: Radars Far Superior to Predecessors, but Location and Availability Questions Remain*, report no. T-AIMD-96-2 (October 17, 1995), 5. For discussion of factors considered in designing the NEXRAD network, see D. A. Leone and others, "Meteorological Considerations in Planning the NEXRAD Network," *Bulletin of the American Meteorological Society* 70 (1989): 4–13.

30. U.S. General Accounting Office, *Weather Forecasting: Radars Far Superior to Predecessors*, 3.

31. See National Weather Service Employees Organization, *Response to the Strategic Plan for the Modernization and Associated Restructuring of the National Weather Service* (1989), reproduced in U.S. Congress, House Committee on Science, Space, and Technology, *Tornado Warnings and Weather Service Moderniza-*

tion: Hearing before the Subcommittee on Natural Resources, Agricultural Research, and Environment, 101st Cong., 1st sess., 1989, 91–115.

32. NEXRAD Panel, National Weather Service Modernization Committee, Commission on Engineering and Technical Systems, National Research Council, *Toward a New National Weather Service: Assessment of NEXRAD Coverage and Associated Weather Services* (Washington, D.C.: National Academy Press, 1995). Private meteorologists and officials in numerous localities were also concerned about what they perceived as less than adequate coverage. See, for example, U.S. Congress, House Committee on Science, *Next Generation Weather Radar (NEXRAD): Are We Covered?: Hearing before the Subcommittee on Energy and Environment*, 104th Cong., 1st sess., 1995. For an official response from the Executive Branch, see U.S. Department of Commerce, *Secretary's Report to Congress on Adequacy of NEXRAD Coverage and Degradation of Weather Services under National Weather Service Modernization for Thirty-two Areas of Concern* (October 12, 1995), vol. 1 [government document no. C 55.012: AD 3/v.1].

33. The ranges reported here are generalized from a table compiled by the NRC panel from 23 separate sources. See NEXRAD Panel, National Weather Service Modernization Committee, Commission on Engineering and Technical Systems, National Research Council, *Toward a New National Weather Service*, 16–7. The table, which reports "experience" ranges and theoretical, calculated distances for NEXRAD and the two older generations of radars, lists medium maximum range as well as the range of typical experience values. In most instances I relied on the calculated values, converted from kilometers to statute miles, and rounded to a multiple of 5 or 10 in the direction of the experience range.

34. Ibid., 61.

35. For a concise summary of changes and goals by the nation's chief meteorologist, see Elbert W. Friday, Jr., "The Modernization and Associated Restructuring of the National Weather Service: An Overview," *Bulletin of the American Meteorological Society* 75 (1994): 43–52. Plans changed slightly as the weather service responded to complaints of inadequate coverage. As of early 1994, NWS plans called for 116 WFOs: 112 in the conterminous United States, two in Alaska, and one each in Hawaii and Puerto Rico. Other sources indicate a slight increase in the number of modernized forecast offices from 115 in 1992 to 119 in 1996. See, for example, the timeline at the *National Weather Service History* Web site (Appendix: Weather History); and U.S. General Accounting Office, *Weather Forecasting: NWS Has Not Demonstrated That New Processing System Will Improve Mission Effectiveness*, report no. AIMD-96-29 (February 1996), 3.

36. For a pre-NEXRAD listing of WSOs and WSFOs, see National Oceanic and Atmospheric Administration, *National Weather Service Offices and Stations*, 25th ed. (September 1990) [government document no. C 55.102: Of 2/990].

37. Principal critics were local air-traffic controllers, who complained that the 10-minute averages reported by ASOS occasionally indicate clear sky when a glance out the window reveals snow. Although the National Weather Service officially closed the WSO at Syracuse's Hancock International Airport in October 1995, as of early 1996, an NWS contractor with four employees monitored the ASOS unit 24 hours a day. But NEXRAD images from Binghamton addressed many of the air controllers' complaints. See Charles Hannagan, "Monitoring

Weather Not Automatic," *Syracuse Post-Standard*, February 21, 1995; and Jeff Theodore and Robert L. Smith, "Our Weather Looks Better at a Distance," *Syracuse Post-Standard*, March 11, 1996. In the early 1990s, private forecasters and other commercial users of weather data had reservations about ASOS; see Mark D. Powell, "Wind Measurement and Archival [*sic*] under the Automated Surface Observing System (ASOS): User Concerns and Opportunity for Improvement," *Bulletin of the American Meteorological Society* 74 (1993): 615–23.

38. Peter Ahnert, National Weather Service, interview by author, Binghamton, N.Y., January 28, 1997.

39. Jeff Waldstreicher, National Weather Service, interview by author, Binghamton, N.Y., January 28, 1997.

40. For a concise overview of NEXRAD's subsystems and products, see Timothy D. Crum and Ron L. Alberty, "The WSR-88D and the WSR-88D Operational Support Facility," *Bulletin of the American Meteorological Society* 74 (1993): 1669–87.

41. For a concise summary of analysis products, see Gerald E. Klazura and David A. Imy, "A Description of the Initial Set of Analysis Products Available from the NEXRAD WSR-88D System," *Bulletin of the American Meteorological Society* 74 (1993): 1293–1311.

42. For contemporary news accounts, see Steve Marshall, "Tornado Whips through Mass. Town," *USA Today*, May 30, 1995; and Ronald Sullivan, "Tornado Tears through Berkshires, Killing 3 in an Auto," *New York Times*, May 30, 1995.

43. Steven LaPointe, "The Great Barrington, Massachusetts, Tornado, May 29, 1995: The Case Account," *WRGB-TV* Website,<http://www.wrgb.com/weather/GBTRP.html> (as of March 11, 1997).

44. Timothy D. Crum, Ron L. Alberty, and Donald W. Burgess, "Recording, Archiving, and Using WSR-88D Data," *Bulletin of the American Meteorological Society* 74 (1993): 645–53.

45. Peter R. Ahnert, Jeff S. Waldstreicher, and John F. Chiaramonte, "Integrating New Data and Technology into Operations: Forecasting the Snowmelt Flood of January 19–20, 1996," paper presented at the American Meteorological Society's 13th International Conference on Interactive Information and Processing Systems for Meteorology, Oceanography, and Hydrology, February 2–7, 1997, Long Beach, Calif.

46. Friday, "Modernization and Associated Restructuring of the National Weather Service," esp. 47. AWIPS will also integrate detailed vertical sounding data from vertical radars called wind profilers; see F. M. Ralph, P. J. Neiman, and D. W. van de Kamp, "Using Spectral Moment Data from NOAA's 404-MHz Radar Wind Profilers to Observe Precipitation," *Bulletin of the American Meteorological Society* 76 (1995): 1717–39.

47. For descriptions of AWIPS capabilities and reports on implementation, see the National Weather Service's *AWIPS* Web sites (Appendix: AWIPS 1 and AWIPS 2).

48. See, for example, "NWS Computer System Gets Chilly Review," *Science* 271 (1996): 1351; U.S. General Accounting Office, *Weather Forecasting: NWS Has Not Demonstrated That New Processing System Will Improve Mission Effectiveness*; and U.S. General Accounting Office, *Weather Forecasting: Recommendations to Address New Weather Processing System Development Risks*, report no.

AIMD-96-74 (May 1996). AWIPS is the key element of the National Weather Service Modernization plan, which is on the GAO's list of high-risk government programs; see U.S. General Accounting Office, *National Oceanic and Atmospheric Administration: Weather Service and NOAA Corps Issues* (testimony of Joel C. Willemssen before the Subcommittee on Energy and Environment, Committee on Science, House of Representatives), report no. T-AIMD-97-63 (13 March 1997), 1–5.

CHAPTER NINE

1. Mark Monmonier, *Maps with the News: The Development of American Journalistic Cartography* (Chicago: University of Chicago Press, 1989).

2. William Marriott, "The Earliest Telegraphic Daily Meteorological Reports and Weather Maps," *Quarterly Journal of the Royal Meteorological Society* 29 (1903): 123–31.

3. For contemporary accounts of the *Times* weather map, see *Report of the Meteorological Committee of the Royal Society, for the Year Ending 31st December 1874* (London, 1875), 16–8, 63; R. H. Scott, "Weather Charts in Newspapers," *Journal of the Society of Arts* 23 (1875): 776–82; and "The 'Times' Weather Chart," *Nature* 11 (1875): 473–4. These accounts make clear that although the *Times* weather chart is, arguably, the first regular newspaper weather map, it was not the first regular weather-related newspaper map. Since 1871, the *Shipping Gazette*, a London trade newspaper, had run a less comprehensive daily wind chart drawn at the Meteorological Office.

4. Scott, "Weather Charts in Newspapers," 777.

5. A cousin of Charles Darwin, Galton encouraged meteorological work within the Royal Society and served on the oversight committee of the Meteorological Office. His meteorological writings include a richly illustrated monograph on the importance of weather maps. See Francis Galton, *Meteorographica, or Methods of Mapping the Weather; Illustrated by Upwards of 600 Printed and Lithographed Diagrams Referring to the Weather of a Large Part of Europe during the Month of December 1861* (London: Macmillan, 1863).

6. Karl Pearson, *The Life, Letters, and Labours of Francis Galton*, vol. 2, *The Middle Years* (Cambridge: Cambridge University Press, 1924), 41–7.

7. Scott, "Weather Charts in Newspapers," 781.

8. R. G. K. Lempfert, "British Weather Forecasts: Past and Present," *Quarterly Journal of the Royal Meteorological Society* 39 (1913): 173–84.

9. *Report of the Meteorological Committee of the Royal Society, for the Period of Seventeen Months, Ending 31st May 1877* (London, 1877), 16–7, 52; and Robert H. Scott, "The Publication of Daily Weather Maps and Bulletins," *Report of the International Meteorological Congress Held at Chicago, Ill., August 21–24, 1893*, U.S. Weather Bureau Bulletin no. 11 (Washington, D.C., 1894), 6–9.

10. Report of the Chief Signal Officer (1881), in U.S. Department of War, *Report of the Secretary of War*, House of Representatives, 47th Cong., 1st sess., Exec. doc. 1, pt. 2, 1881, vol. 4, esp. 26. Also see Edgar B. Calvert, "Development of the Daily Weather Map," *Proceedings of the Convention of Weather Bureau Officials, Held at Omaha, Nebr., October 13–14, 1898*, U.S. Weather Bureau Bulletin no. 24 (Washington, D.C., 1899), 144–50; Henry L. Heiskell, "The Commercial Weather

Map of the United States Weather Bureau," *Yearbook of the Department of Agriculture* (Washington, D.C., 1912), 537–9; and Robert De Courcy Ward, "The Newspaper Weather Maps of the United States," *American Meteorological Journal* 11 (1894): 96–107.

11. Stephen Henry Horgan, "The World's First Illustrated Newspaper," *Penrose Annual* 35 (1933): 23–4; and Allen Hutt, *The Changing Newspaper: Typographic Trends in Britain and America, 1622–1972* (London: Gordon Fraser, 1973), 83–4.

12. Report of the Chief Signal Officer (1881).

13. Report of the Chief Signal Officer (1883), in U.S. Department of War, *Report of the Secretary of War*, House of Representatives, 47th Cong., 2d sess., Exec. doc. 1, pt. 2, 1883, vol. 4, esp. 64–5.

14. Calvert, "Development of the Daily Weather Map," 145–6; quotation on 146.

15. Report of the Chief Signal Officer (1883).

16. Report of the Chief Signal Officer (1884), in U.S. Department of War, *Report of the Secretary of War*, House of Representatives, 48th Cong., 1st sess., Exec. doc. 1, pt. 2, 1884, vol. 4, esp. 17.

17. Ward, "Newspaper Weather Maps of the United States."

18. Ward's description of engraving processes is sketchy at best. For additional details on the use of chalk plates to make reproducible weather maps, see Calvert, "Development of the Daily Weather Map"; "Chalk-Plate Weather Maps," *Science* n.s. 5 (1897): 337–8; and "How a Weather Map Is Made," *Scientific American* 82 (January 20, 1900): 38.

19. Ward, "Newspaper Weather Maps of the United States," 106.

20. Calvert, "Development of the Daily Weather Map," 147. Yearly totals are by fiscal year. Quotation from *Report of the Chief of the Weather Bureau, 1896–1897* (Washington, D.C., 1898), 9.

21. Calvert, "Development of the Daily Weather Map," 147; and *Report of the Chief of the Weather Bureau, 1908–1909* (Washington, D.C., 1910), 31.

22. *Report of the Chief of the Weather Bureau, 1902–1903* (Washington, D.C., 1904), xii. Moore was also proud of the timeliness of the printed station maps, which were ready for distribution, by mail or messenger, "within three hours from the time observations were made." See "Report of the Chief of the Weather Bureau for the Fiscal Year Ending June 30, 1905," *Monthly Weather Review* 33 (1905): 572–82; quotation on 572.

23. Calvert, "Development of the Daily Weather Map," 147; and *Report of the Chief of the Weather Bureau, 1908–1909* (Washington, D.C., 1910), 31.

24. Calvert, "Development of the Daily Weather Map," 147; and "How a Weather Map Is Made."

25. Willis Isbister Milham, *Meteorology* (New York: Macmillan, 1912), 368.

26. However primitive, milliograph duplication was a substantial improvement over manifolding, which the Signal Office used widely until the mid-1880s. The manifold was a steel frame that could hold up to 15 tissue-paper outline maps interleaved with sheets of carbon paper. An original map placed on top of the stack was then reproduced by tracing over its lines with a stylus. Point symbols and labels were added by placing symbol dies and figure dies face down on the master and tapping gently but firmly with a mallet. For information on manifold printing,

see U.S. Signal Office, *Instructions to Observers of the Signal Service* (Washington, D.C., 1881), 45. Because the Weather Bureau's Washington office lithographed 19,460 outline maps for manifolding during the 1908–9 accounting year, the process apparently was still in use. See *Annual Reports of the Department of Agriculture for the Year Ended June 30, 1909* (Washington, D.C., 1910), 185. For information on the job titles and salaries of Weather Bureau personnel who helped prepare and reproduce weather maps, see U.S. Congress, House Committee on Expenditures, *Hearings on Estimates of Appropriations for the Fiscal Year Ending June 30, 1908*, 59th Cong., 2d sess., 1906–7, vol. 2, esp. 889, 893, 927–8.

27. "Report of the Chief of the Weather Bureau for the Fiscal Year Ending June 30, 1903," *Monthly Weather Review* 31 (1903): 626–44; quotations on 627.

28. "Statement of Prof. Willis L. Moore, Chief of the Weather Bureau, Department of Agriculture" (January 7, 1907), in U.S. Congress, House Committee on Expenditures, *Hearings*, 59th Cong., 2d sess., 1906–7, vol. 2, 893.

29. U.S. Congress, House Committee on Agriculture, *Hearings on the Estimates of Appropriations for the Fiscal Year Ending June 30, 1909*, 60th Cong., 1st sess., 1907–8, 37.

30. U.S. Congress, House Committee on Agriculture, *Hearings on the Estimates of Appropriations for the Fiscal Year Ending June 30, 1910*, 60th Cong., 2d sess., 1908–9, 37.

31. Donald R. Whitnah, *A History of the United States Weather Bureau* (Urbana: University of Illinois Press, 1961), 125.

32. *Report of the Chief of the Weather Bureau, 1909–1910* (Washington, D.C., 1911), 30.

33. Ibid.

34. Whitnah, *History of the Weather Bureau*, 125.

35. Heiskell, "The Commercial Weather Map of the United States Weather Bureau"; and Whitnah, *History of the Weather Bureau*, 98.

36. According to Whitnah, "Less than twenty-four newspapers provided readers with this service in 1897," but verification could not be found in the sources he lists. See Whitnah, *History of the Weather Bureau*, 98. Two sentences in a short article lauding the new commercial weather map in the 1912 *Yearbook of Agriculture* suggest that only 27 newspapers ever carried a weather map between 1881 and 1910: "In 1881 the Cincinnati Commercial published a daily weather map, including Sundays and holidays, which was continued until November, 1892, the longest period in which the weather map was published in any newspaper up to 1910. Twenty-six other papers published the map at various periods previous to 1910." See Heiskell, "The Commercial Weather Map of the United States Weather Bureau," 537. By contrast, Ward lists 27 American newspapers that had carried a daily weather map at some point between 1879 and 1894. See Ward, "The Newspaper Weather Maps of the United States," 98, 99, 102.

37. *Report of the Chief of the Weather Bureau, 1911–1912* (Washington, D.C., 1913), 24.

38. U.S. Congress, House Committee on Agriculture, *Hearings on the Agricultural Appropriation Bill for the Fiscal Year Ending June 30, 1913*, 62d Cong., 2d sess., 1911–12, 4.

39. Whitnah, *History of the Weather Bureau*, 127–30. In March 1913, Moore submitted a letter of resignation in which he agreed to step down on July 31. But

the president, citing irregularities and misconduct, removed him three and a half months before his planned departure.

40. Mark Monmonier, "Telegraphy, Iconography, and the Weather Map: Cartographic Weather Reports by the United States Weather Bureau, 1870–1935," *Imago Mundi* 40 (1988): 15–31; and *Report of the Chief of the Weather Bureau, 1917–1918* (Washington, D.C., 1919), 12.

41. See Monmonier, *Maps with the News,* 115–7. In the absence of tabulations by the Weather Bureau or anyone else, my evidence for the decline of newspaper weather maps is largely anecdotal. No major newspapers I have been able to check on microfilm carried a weather map in 1920, although several had run one earlier. Among newspapers in central New York, the *Syracuse Herald* carried a weather map from May 15, 1910, through July 13, 1919, when cuts in weather service personnel probably forced its cancelation.

42. *Report of the Chief of the Weather Bureau, 1919–1920* (Washington, D.C., 1921), 10. That Marvin was a less enthusiastic and effective advocate of the weather map than his predecessor might reflect the reduced role of intensively empirical map study in the training of Weather Bureau personnel. In 1927, for instance, the agency discontinued issuing bound copies of the Washington weather map to 45 field stations. "Unless such stations make specific request for continuance, with adequate reason therefor, hereafter bound volumes of the Washington weather map will be sent only to the forecast centers." See "Bound Weather Maps," *Weather Bureau Topics and Personnel* (April 1927): 207.

43. An uncataloged envelope contained a memorandum to the [Weather Bureau] Library from an official identified only by initials. Sent in response to a request from the library, the February 16, 1926, memo included sample maps and lists of stations using the Weather Bureau's three outline maps: Form DD for page-size [8.5 by 11 in.] printed maps, Form CM for half-size printed maps, and Form E for page-size milliograph maps. In addition, one station, San Francisco, used Form DD-Pacific, on which a map centered on the Pacific coast eliminated much of the East Coast but provided room for ad hoc reports from ships up to 1,500 miles off the coast.

44. *Report of the Chief of the Weather Bureau, 1923–24* (Washington, D.C., 1925), 6.

45. For contemporary accounts, see "A.P. Wirephotos Flash across Nation," *Editor and Publisher* 67 (January 5, 1935): 7; and John W. Perry, "Wirephoto Found Flexible and Speedy; Operates with Uncanny Precision," *Editor and Publisher* 67 (January 12, 1935): III and XI.

46. For a line-art reproduction of an early Wirephoto weather map, reproduced as a halftone in the *Syracuse Herald* for January 3, 1935, see Monmonier, *Maps with the News,* 117.

47. Whitnah, *History of the Weather Bureau,* 229–30. For examples of much earlier Weather Bureau experiments with facsimile transmission by radio, see B. Francis Dashiell, "Broadcasting Weather Maps by Radio," *Monthly Weather Review* 54 (1926): 419–20; B. Francis Dashiell, "Maps by Radio Accomplished," *Radio News* 8 (1927): 791-B; and S. R. Winters, "The Broadcasting of Weather," *Radio News* 8 (1927): 791-A. For an example of the National Weather Service's continued commitment to broadcasting facsimile weather maps, see Lee Ches-

neau, "Expanded High Seas Atlantic HF Radiofacsimile Broadcast," *Mariners Weather Log* 39 (fall 1995): 26–8.

48. For a concise history of photowire development between 1950 and 1985, see Monmonier, *Maps with the News*, 97–101.

49. The Weather Bureau suspended nationwide weather reports on December 16, 1945, a week after the Japanese attack on Pearl Harbor. See "Weather Conditions: Reports Suspended," *Syracuse Herald-Journal*, December 16, 1945, 25. Microfilm copies of the *Herald-Journal* suggest that the AP Wirephoto weather map was not resumed until February 4, 1946, almost six months after Japan surrendered. That the revived weather maps included fronts partly reflected the military utility of air mass concepts. The *Herald-Journal*, if not the AP, lagged more than a year behind the *New York Times*, the *Providence (R.I.) Journal*, and several other metropolitan dailies that resumed publication of a daily weather map in late 1944, while fighting continued in Europe and the Pacific. See C. F. B., "Newspaper Weather Maps Return," *Bulletin of the American Meteorological Society* 25 (1944): 410.

50. This change was an apparent response to a recommendation by the Photo and Graphics Committee of the Associated Press Managing Editors, an advisory group of executives of member newspapers, which asked the AP to substitute a forecast weather map for the slightly stale current-conditions map; see *Associated Press Managing Editors Red Book* 2 (1949): 109. Examination of microfilm copies of the *Syracuse Herald-Journal* suggest that the redesigned weather map with a forecast inset first appeared on March 8, 1950. The *Herald-Journal* ran the afternoon edition of the AP weather map, which described conditions as of 1:30 that morning, eastern time, on the main map and "low temperatures and areas of precipitation expected tonight" in the inset.

51. *Associated Press Managing Editors Red Book* 28 (1975): 183.

52. For the recommendation by the Photo and Graphics Committee, see *Associated Press Managing Editors Red Book* 27 (1974): 212.

53. See, for example, Debra Gersh, "Another Chapter 11 for UPI," *Editor and Publisher* 124 (August 31, 1991): 14, 35.

54. For an early discussion of the *USA Today* weather page, see Jim Norman, "How USA Today Does It," *Design: Journal of the Society of Newspaper Design* no. 12 (summer 1983): 4–6; and Edward F. Taylor, "Telling the Weather Story," *Weatherwise* 36 (April 1983): 52–9.

55. See, for example, the map for April 15, 1986, in Monmonier, *Maps with the News*, 120. For the map of May 9, 1989, and a humorous critique, see Ron Carlson, "Weathering Heights," *Harpers Magazine* 279 (July 1989): 40–1.

56. These graphics were pioneered by *USA Today*'s weather page editor, Jack Williams, who authored a lucid, well-illustrated explanation of North American weather. See Jack Williams, *The Weather Book: An Easy-to-Understand Guide to the USA's Weather* (New York: Vintage Books, 1992).

57. Mark Monmonier and Val Pipps, "Weather Maps and Newspaper Design: Response to USA Today?" *Newspaper Research Journal* 8 (Summer 1987): 31-42. For discussion of the contents and uses of newspaper weather pages, see Mike Mogil and Barbara G. Levine, "Gallery of Weather Pages," *Weatherwise* 48 (August/September 1995): 15–19.

58. Newspapers with color presses that chose not to run a color weather map did so because management believed color either added little useful information or reduced flexibility in the placement of section breaks and advertising. See Mark Fitzgerald, "Whither Weather Color?" *Editor and Publisher* 127 (September 24, 1994): 34C–35C, 38C.

59. For discussion of the use of Macintosh graphics in newspapers, see George Garneau, "Weather Graphics Via PCs," *Editor and Publisher* 119 (October 4, 1986): 62–4; Monmonier, *Maps with the News*, 119–24, 154–9; Michael Reddy, "Fit to Print at the New York Times: Macs Used to Produce Weather Maps, Charts, Graphics," *MacWeek* 1 (September 29, 1987): 10; and Stuart Silverstone, "Newsroom Graphics," *Macworld* 4 (February 1987): 130–5. For discussion of Accu Weather, one of the leading weather forecast services providing syndicated graphics, see Mary Ann Whitley, "Accu-Weather Forecast: Continuing Bright Future," *Penn Stater* 81 (September/October 1993): 40–4.

60. Both the AP and UPI photowires began carrying images from the ESSA-1 weather satellite shortly after its launch on January 2, 1966. See Roy Popkin, *The Environmental Science Services Administration* (New York: Frederick A. Praeger, 1967), 98. As early as 1961, though, the Weather Bureau transmitted TIROS II images over its own facsimile network. See "New Weather Facsimile Network," *Monthly Weather Review* 89 (1961): 38.

61. For a cogent statement on meteorological knowledge as an element of scientific literacy, see A. M. Lucas, "Who Needs to Know What about Weather," in J. M. Walker, ed., *Weather Education: Proceedings of the First International Conference on School and Popular Meteorological Education* (Bracknell, Berkshire, U.K.: Royal Meteorological Society, 1985), 233–47.

62. Mark Fitzgerald, "Graphics on the Gulf," *Editor and Publisher* 124 (March 9, 1991): 12–3.

63. Although difficult to estimate, the number of lay weather enthusiasts is most certainly much smaller than the number of sports fans. According to a 1967 survey in Victoria, British Columbia, for instance, only 14 percent of newspaper readers regularly examined the weather forecast. See W. J. Maunder, *The Value of the Weather* (London: Methuen, 1970), 279–80.

64. For further insights on blocking highs, see Lee Grenci, "Blocking High Ahead: Meet the Unassuming Upper-Air Anomaly That Was an Accomplice in the Flood of 1993," *Weatherwise* 47 (August/September 1994): 38–42; and D. F. Rex, "Blocking Action in the Middle Troposphere and Its Effect upon Regional Climate II: The Climatology of Blocking Action," *Tellus* 2 (1950): 275–301.

65. For an essay blaming television for poor weather coverage in newspapers, see Peter Rogers, "Coverage of Weather Forecasts in UK National Newspapers," *Weather* 43 (1988): 406–8.

Chapter Ten

1. Jan Gecsei, *The Architecture of Videotex Systems* (Englewood Cliffs, N.J.: Prentice-Hall, 1983), 46–50. Not all scan lines are used for the picture, though. A vertical blanking interval, which separates consecutive refresh scans of the screen, consumes 41 of the 525 scan lines in the North American (and Japanese)

television standard and 50 of the 625 scan lines of the European standard, leaving only 484 and 575 active scan lines, respectively.

2. Legibility depends not only on the height of letters but also their thickness, shape, and contrast with the background. Size is further constrained because designers try not to place labels near the edges of the screen, which are often distorted on older receivers. On all-text screens the rule is a maximum of ten lines of type. See Ralph Ayers, *Graphics for Television* (Englewood Cliffs, N.J.: Prentice-Hall, 1984), 11–26, 56–61. For discussion of other technological limitations on television cartography, see Patricia S. Caldwell, "Television News Maps: The Effects of the Medium on the Map," *Technical Papers, American Congress on Surveying and Mapping*, 41st Annual Meeting, Washington, D.C., February 22–27, 1981, 382–92.

3. Adopted by the Federal Communications Commission in late 1996, the new HDTV standard increases the width/height ratio from 12:9 (4:3) to 16:9 and raises the number of active scan lines from 484 to over 1,000. A highly flexible compromise designed to serve both broadcasting and the computer industry, the standard supports current formats as well as several new, high-resolution formats. Full conversion, an expensive hurdle for households and broadcasters alike, is likely to take more than a decade. For further discussion of high-definition television, see Joel Brinkley, "No Narrowing of 2 Industries' Rift on Advanced TV," *New York Times*, April 14, 1997; and Raymond Sokolov, "The Best TV Picture You've Never Seen," *Wall Street Journal*, March 20, 1997.

4. Robert Henson, *Television Weathercasting: A History* (Jefferson, N.C.: McFarland, 1990), 5–6.

5. For a short account as well as a picture of "the world's first television weather chart," see P. F. McAllen, "A Brief History of Weather Broadcasting," *Weather* 34 (1979): 436–41.

6. For examples, see Robert Henson, "Show and Tell," *Weatherwise* 46 (October/November 1993): 12–9; Henson, *Television Weathercasting*, 33–46; and David Laskin, "A Change in the Weather," *New York Times*, February 18, 1996.

7. "Seal of Approval Program for Radio and Television," *Bulletin of the American Meteorological Society* 77 (1996): 1821–34; quotations from 1822. The *AMS* home page on the World Wide Web includes a detailed description of the requirements and certification process (Appendix: AMS Certificate).

8. Revocation occurs only after a careful investigation by a fact-finding panel, which may also dispose of valid complaints with a "written informal admonition." An accused seal holder may cross-examine witnesses as well as appeal an unfavorable decision to the society's executive committee.

9. L. Michael Trapassc, Randy Bowman, and Laura Daniel, "TV Weather Forecasters," *RTNDA Communicator* 39 (December 1985): 16–8.

10. A more recent account confirms the importance of "advanced scientific training and back-slapping skills." See Rick Bragg, "A Good Forecaster Is Hard to Find," *New York Times*, September 15, 1997.

11. Departmental home pages describe college courses, programs, and facilities in broadcast meteorology. For example, I found a list of Mississippi State's course requirements at its Web site (Appendix: Mississippi State), and a description of Lyndon's meteorology laboratory at its Web site (Appendix: Lyndon

State). For information on how Penn State prepares its meteorology majors for work in television, see Steve Mirsky, "Turning Weather Wonks into TV Stars," *Technology Review* 99 (November/December 1996): 41–5.

12. For a revealing glimpse of a television weather reporter's hectic three hours preparing for a noon weathercast, see Alan Sealls, "Adventures in Weathercasting," *Weatherwise* 47 (December 1994/January 1995): 58, 62.

13. Trapassc, Bowman, and Daniel, "TV Weather Forecasters."

14. For examples of the trickle down of computer graphics and related technology, see Jack Fishman and Robert Kalish, *The Weather Revolution: Innovations and Imminent Breakthroughs in Accurate Forecasting* (New York: Plenum Press, 1994), 259–63; David Laskin, *Braving the Elements: The Stormy History of American Weather* (New York: Doubleday, 1996), esp. 184–6; and C. Catherine Winslow, "Virtually Real: New Weather Technology," *RTNDA Communicator* 50 (December 1996): 10–7.

15. The U.S. Weather Bureau was one of weathercasting's earliest pioneers. In 1947, an experimental broadcasting unit in the bureau's training section produced a daily "weather brief" aired on stations in Washington, Baltimore, Philadelphia, and New York. For discussion of the unit's use of blackboards and other visual aids, see James C. Fidler, "Weather via Television," *Bulletin of the American Meteorological Society* 29 (1948): 329–31.

16. Perry Saltzman, a pioneer meteorologist on Canadian television, included sliding maps in a description of his experiences as a weathercaster. See Perry Saltzman, "The Weather on Television," *Proceedings of the Royal Meteorological Society, Canadian Branch* 5, no. 2 (1954): 1–11.

17. Gary A. England, *Weathering the Storm: Tornadoes, Television, and Turmoil* (Norman: University of Oklahoma Press, 1966), 56.

18. Instead of bright blue, chroma key systems can use lime green. For a fuller treatment of the principles and operation of chroma key, see Gerald Millerson, *The Techniques of Television Production*, 12th ed. (Boston: Focal Press, 1990), 482–99.

19. Henson, *Television Weathercasting*, 8, 63, 98; quotation on 8. Also see Stanley David Gedzelman, *The Science and Wonders of the Atmosphere* (New York: John Wiley and Sons, 1980), 8. Gedzelman, who ignored issues of sequence and scale, identified four main elements of televised weather reports: high-pressure areas, low-pressure areas, cold fronts, and warm fronts.

20. J. R. Carter, "Weather Maps on Television in the USA," *Proceedings of the Sixteenth International Cartographic Conference*, Cologne, Germany, May 3–5, 1993, 244–54.

21. Ibid., 245.

22. Tal White, "Morning vs. Evening," *RTNDA Communicator* 50 (December 1996): 11.

23. Jim Carter and Robert Henson provide informative accounts of *A.M. Weather,* a detailed 15-minute weathercast aired by as many as 300 public television stations, usually before seven o'clock local time. An outgrowth of a televised briefing for pilots, the program integrated satellite and radar loops with about 20 detailed National Weather Service charts, simplified for television viewing. Among a variety of specialized graphics, *A.M. Weather* included upper-air maps deemed too technical for commercial television. The program went off the air in

February 1995, when Maryland Public Television, the producer, canceled the series. See Carter, "Weather Maps on Television," 248–50; Henson, *Television Weathercasting*, 97–9; and "Wake-Up Weather," *Weatherwise* 48 (October/November 1995): 9–10.

24. Steven Beschloss, "Praying for Rain," *RTNDA Communicator* 9 (April 1989): 18–20.

25. Carter, "Weather Maps on Television."

26. Laskin, "Change in the Weather"; Fleming Meeks, "What Brand Is Your Weather?" *Forbes* 156 (October 23, 1995): 320–1; and Leonard Ray Teel, "The Weather Channel Turns 10," *Weatherwise* 42 (April/May 1992): 9–15. Personnel counts vary a bit: Teel enumerated 30 "weathercasters," Meeks counted 60 "behind-the-scenes meteorologists," and Laskin reported 24 on-camera and 42 off-camera meteorologists, respectively.

27. For information on TWC's origins and growth, see Henson, *Television Weathercasting*, 99–103; Leonard Ray Teel, "The Weather Channel," *Weatherwise* 35 (August 1982): 156–63; and Teel, "The Weather Channel Turns 10." Unique in American television, the Weather Channel has been the butt of jokes by David Letterman and a postmodernist rant by cultural studies guru Andrew Ross, who chided (unfairly, I think) TWC's "fictional world" for offering "no maps of acid rain damage, deforestation, oil spill concentrations, toxic dump locations, or downwind nuclear zones." See Andrew Ross, *Strange Weather: Culture, Science, and Technology in the Age of Limits* (London: Verso, 1991), 238–45; quotation on 240–1.

28. According to the *Landmark Communications* Web site the Weather Channel was received by "more than 64 million households" and "can be seen … in more than 66% of all homes with TVs." <http://www.landmarkcom.com/land-credit.html> (as of May 2, 1997).

29. "The Weather Channel," *Advertising Age* 67 (May 20, 1996): A34.

30. Carter, "Weather Maps on Television," 251–2; and Henson, *Television Weathercasting*, 101–2.

31. Meeks, "What Brand Is Your Weather?"

32. Ibid.

33. Ty Burr, "Wired News Tonight," *Entertainment Weekly* 338 (August 2, 1996): 60–1; Marc Gunther, "CNN Envy," *Fortune* 134 (July 8, 1996): 120–6; and Elizabeth Lesly and Kathy Rebello, "Network Meets Net," *Business Week* no. 3484 (July 15, 1996): 68–70.

34. According to Tom Kierein, a broadcast meteorologist at WRC-TV, in Washington, D.C., "the most visible improvements in TV weathercasting came just prior to 1980, with the introduction of computer graphics." See Tom Kierein, "The HI-TECH World of TV Weathercasting," *Weatherwise* 41 (1988): 150–4; quotation on 150. By late 1985, officials of WSI Corporation, a leading supplier of weather graphics systems, estimated that 40 to 45 percent of television news departments were using computer graphics for their weather reports; see Bob Paulson, "Weather Graphics," *RTNDA Communicator* 39 (December 1985): 20–1. Among the earliest users of computer graphics was WTSP-TV, in Tampa–St. Petersburg, whose chief meteorologist Dick Fletcher claimed in 1990 to have gone through "seven generations of software and hardware since 1979." For an essay on weather reporting at Tampa–St. Petersburg television stations, see

Fred W. Wright, Jr., "The Hottest TV Weather Market in the Country," *Weatherwise* 43 (1990): 151–3; quotation on 152.

35. According to Jeffrey Smith, executive director of the Commercial Weather Services Association, an industry trade group, "about 85 percent of the weather information that reaches the public is now provided by the private sector." See Barnaby J. Feder, "Highs and Lows Are Their Business," *New York Times*, October 22, 1996.

36. Transmitting frequently updated bulletins prepared by a nearby Weather Forecast Office, NOAA weather radio is available in most large- and medium-size cities. Seven channels in a band from 162.400 to 162.550 MHz afford slightly overlapping coverage without interference by over 400 transmitters, each with a range of approximately 40 miles. See Ronald L. Wagner and Bill Adler, Jr., *The Weather Sourcebook* (Old Saybrook, Conn.: Globe Pequot Press, 1994), 163–5; and the *NOAA Weather Radio* Web page, at <http://www.nws.noaa.gov/nwr/nwr.htm> (as of May 27, 1997).

37. Vico E. Baer, "The Transition from the Present Radar Dissemination System to the NEXRAD Information Dissemination Service," *Bulletin of the American Meteorological Society* 72 (1991): 29–33.

38. Ibid.; and Vico Baer, telephone conversation, May 29, 1997. In 1990, to recover added personnel cost and related expenses, the government began charging data providers an initial fee of $780 for each new NEXRAD site and an annual fee of $1,395 for each site. NOAA adjusts the fee every two years. A lower fee of $1,354 in 1997 reflects the increased number of commissioned radar sites.

39. "Reagan Orders Satellite Sale," *Aviation Week and Space Technology* 118 (March 14, 1983): 263; Alton K. Marsh, "Reports Criticize Plan to Sell Satellites," *Aviation Week and Space Technology* 118 (April 4, 1983): 48–9; and M. Mitchell Waldrop, "The Satellite Sale: Another Dose of Reality," *Science* 221 (August 12, 1983): 632–3.

40. For discussion of computer weather graphics systems available in 1985, see Barry Lee Myers and Joel N. Myers, "New Technology and the Presentation of Weather," *RTNDA Communicator* 39 (December 1985): 12–4; Paulson, "Weather Graphics"; and Joe Russell, "Custom-Drawn Computer Graphics," *RTNDA Communicator* 39 (December 1985): 10. For a comparable assessment a decade later, see Winslow, "Virtually Real: New Weather Technology."

41. In the late 1990s, these firms became prime candidates for mergers and acquisitions. WSI, for example, is now a unit of TASC, itself a subsidiary of Primark Corporation. An Alden press release dated February 5, 1997, announced that the firm had sold its weather information business to Platinum Equity Holdings; see <http://www.alden.com/news.html> (May 22, 1997).

42. <http://www.wsicorp.com/market.html> (May 23, 1997).

43. For general discussions of three-dimensional weather animation for television, see "Fly-Through Weathermen," *Popular Science* 247 (September 1995): 48; Paula T. McCaslin and Philip A. McDonald, "A Promising Forecast: Colorado Researchers Add Dimension to the Weather," *Geo Info Systems* 6 (October 1996): 30–3; and Winslow, "Virtually Real: New Weather Technology." For more technical insights, see Richard Grotjahn and Robert M. Chervin, "Animated Graphics in Meteorological Research and Presentations," *Bulletin of the American Meteorological Society* 65 (1984): 1201–8; A. F. Hasler and others, "Meteoro-

logical Data Fields 'in Perspective,'" *Bulletin of the American Meteorological Society* 66 (1985): 795–801; William L. Hibbard, "Computer-Generated Imagery for 4-D Meteorological Data," *Bulletin of the American Meteorological Society* 67 (1986): 1362–9; Florian Schröder, "Visualizing Meteorological Data for a Lay Audience," *IEEE Computer Graphics and Applications* 13 (September 1993): 12–4; and James A. Schiavone and Thomas V. Papathomas, "Visualizing Meteorological Data," *Bulletin of the American Meteorological Society* 71 (1990): 1012–20.

44. William LaRue, "When It Rains, It Soars," *Syracuse Post-Standard*, November 13, 1995. For an earlier, more detailed examination of television weathercasting in Syracuse, see Bob Niedt, "Syracuse's Wise Men of Weather," *Syracuse Herald American*, January 5, 1992, Stars section.

45. Author's tape of weather segments on WIXT's 11 P.M. newscast for November 30, 1995.

46. For information on lightning-detection systems and their uses, see E. P. Krider and others, "Lightning Direction-Finding Systems for Forest Fire Detection," *Bulletin of the American Meteorological Association* 61 (1980): 980–6; Richard Orville and Herbert Songster, "The East Coast Lightning Detection Network," *IEEE Transactions on Power Delivery* 2 (1987): 899–903; and Bruce F. Whitney and Haukur Asgeirsson, "Lightning Location and Storm Severity Display System," *IEEE Transactions on Power Delivery* 6 (1991): 1715–9. For a concise summary of lightning-detection charts used by the National Weather Service, see Peter R. Chaston, *Weather Maps: How to Read and Interpret All the Basic Weather Charts* (Kearny, Mo.: Chaston Scientific, 1995), 45.

47. Fred Ludwick, Niagara Mohawk research department, telephone conversation, May 30, 1997. Outages are identified when a circuit breaker trips at the same time a detector in the area senses a lightning strike.

48. John Bobel, "Fill-in Meteorologists," *RTNDA Communicator* 49 (April 1995): 34–7.

49. For information on the origin and ramifications of the Internet, the World Wide Web, and related network technology, see Tim Berners-Lee, "WWW: Past, Present, and Future," *Computer* 29 (October 1996): 69–77; and Rob Shields, ed., *Cultures of Internet: Virtual Spaces, Real Histories, Living Bodies* (Thousand Oaks, Calif.: Sage Publications, 1996). For insights on Web delivery of maps and geographic information, see Jeremy Crampton, "Cartography Resources on the World Wide Web," *Cartographic Perspectives* no. 22 (Fall 1995): 3–11; and Michael P. Peterson, "Cartography and the Internet: Introduction and Research Agenda," *Cartographic Perspectives* no. 26 (Winter 1997): 3–12. For discussion of the early use of telecommunications networks to retrieve weather information, see Richard N. Aarons, "Microchips and Isobars," *Business and Commercial Aviation* 54 (January 1984): 56–67.

50. The acronym "http" stands for "hypertext transfer protocol." Because they are subject to change, all URLs that are not primarily a "literature" citation have been assigned a brief title, referenced with a parethetical note of the form "(Appendix: USA Today)," and listed in the Appendix.

51. Content indexes are compiled by automatic indexing software called Web crawlers, which systematically visit site after site, cataloging what they find. For a critique of Web crawlers, see Gary Taubes, "Indexing the Internet," *Science* 269 (September 8, 1995): 1354–6.

52. For a more complete list of links to weathercam sites, see the *WeatherNet WeatherCams* page at <http://cirrus.sprl.umich.edu/wxnet/wxcam.html>.

53. For reviews of North American weather pages, see Doug Le Comte, "Heavy Hitters on the Web," *Weatherwise* 48 (December 1995/January 1996): 38–39; and Douglas Le Comte, "Home Page, Sweet Home Page," *Weatherwise* 48 (April/May 1995): 36–37. For similar articles on weather pages in Britain, see R. Brugge, "Computer Networks and Meteorological Information," *Weather* 49 (September 1994): 298–306; and Trevor Lawson, "Whatever the Weather," *Geographical Magazine* 68 (May 1996): 22–4. For an essay on the distribution of meteorological software and technical documentation over the Internet, see Jon Ahlquist, "Free Software and Information via Computer Network," *Bulletin of the American Meteorological Society* 74 (1993): 377–86; and Thomas J. Phillips, "Documentation of the AMIP Models on the World Wide Web," *Bulletin of the American Meteorological Society* 77 (June 1996): 1191–6. For a concise history of weather reporting in Canada, including on-line forecasts, see David Phillips, "Winds of Change: Past, Present, and Future of the Forecast," *Canadian Geographic* 115 (November/December 1995): 33–4.

54. The Weather Visualizer site is maintained by the Collaborative Visualization (Appendix: CoVis Project) project of the University of Illinois, Northwestern University, San Francisco's Exploratorium, and Bellcore. For a comparison of several university weather sites, see Greg R. Notess, "The Internet Weather Channel," *Database* 17 (October-November 1994): 95–8.

55. For concise discussions of VRML's structure and promise, see Joe Flower, "How to Build a Metaverse," *New Scientist* 148 (October 14, 1995): 36–40; G. Jason Mathews, "Visualizing Space Science Data in 3D," *IEEE Computer Graphics and Applications* 16 (November 1996): 6–9; and Karen Whitehouse, "VRML Adds a New Dimension to Web Browsing," *IEEE Computer Graphics and Applications* 16 (July 1996): 7–9.

56. Laurence Zuckerman, "Pushing the Envelope on Delivery of Customized Internet Data," *New York Times*, December 9, 1996.

57. And so apparently did the cable channel's owner, Landmark Communications, which not only invested heavily in an impressive Web site (Appendix: Weather Channel) but managed to sell ad space to many of TWC's regular sponsors. What's more, TWC's advertisers seem to know what they're doing. According to PC Meter, a service that rates Web sites like Nielsen rates television programs, the Weather Channel is among the top five news/information/entertainment sites. See Richard Tedesco, "Weather Channel, Disney, Warner Gain in PC Meter Ratings," *Broadcasting and Cable* 127 (February 17, 1997): 47.

CHAPTER ELEVEN

1. For a concise overview of the scope and objectives of climatology, see Reid A. Bryson, "The Paradigm of Climatology: An Essay," *Bulletin of the American Meteorological Society* 78 (1977): 449–55.

2. Lorin Blodget, *Climatology of the United States and of the Temperate Latitudes of the North American Continent* (Philadelphia: J. B. Lippincott and Co., 1857).

3. James K. McGuire, "The Father of American Climatology," *Weatherwise* 10 (1957): 93–4, 97.

4. James Rodger Fleming, *Meteorology in America, 1800–1870* (Baltimore: Johns Hopkins University Press, 1990), 110–5.

5. William Jackson Humphreys, "Lorin Blodget," in Allen Johnson, ed., *Dictionary of American Biography* (New York: Charles Scribner's Sons, 1929), 2: 379. However peeved at his former assistant's unauthorized use of Smithsonian data, Henry was probably rankled even more by *Climatology*'s preface, in which Blodget inflated his own role in compiling the charts, mentioned "a Letter of strong approval from the illustrious Baron Humboldt," and conspicuously ignored Henry in acknowledging the contributions of the Smithsonian Institution, the army, the Patent Office, several learned societies, General Lawson, and various federal officials and prominent scientists. See Blodget, *Climatology of the United States*, v–x.

6. Cleveland Abbe, "Lorin Blodget," *Monthly Weather Review* 29 (1901): 174. Contradicting Lincoln's assertion that "the evil that men do lives after them, while the good is oft interred with their bones," Abbe eulogized Blodget as an "eminent statistician" who "labored unselfishly to promote the public interest." Apparently aware of Henry's difficulties with his assistant, Abbe noted tactfully that "owing to a difference of opinion as to [Blodget's] right to use these official records for his own publications, this arrangement was terminated."

7. Robert De C. Ward, "Lorin Blodget's 'Climatology of the United States': An Appreciation," *Monthly Weather Review* 42 (1914): 23–7.

8. Ibid., 23.

9. Ibid., 26.

10. Ibid., 23.

11. H. E. Landsberg, "Early Stages of Climatology in the United States," *Bulletin of the American Meteorological Society* 45 (1964): 268–75; quotation on 272.

12. Blodget's maps of mean precipitation are even more difficult to reproduce for a wholly different reason: between the isohyets (lines of equal precipitation) are finely textured brownish gray tints varying from light for low values to medium for comparatively high values. Although effective in portraying variation in rainfall, these drab tints largely obscure boundary lines, rivers, and tiny labels identifying places.

13. U.S. Department of Agriculture, *Climate and Man*, Yearbook of Agriculture (Washington, D.C., 1941), esp. 701–47.

14. U.S. Department of Commerce, Environmental Data Service, *Climatic Atlas of the United States* (Asheville, N.C.: National Climatic Data Center, 1968), 1. An identical caveat appears on all four January temperature maps on page 1, as well as on temperature maps for other months, and similar caveats appear elsewhere throughout the atlas.

15. Notable exceptions are maps in sections with a lengthy introduction that includes an even longer caveat and maps on which tiny wind roses, histograms, and similar symbols describing conditions at specific places are less amenable to precise guessing for intermediate locations.

16. For a critique of the use of "normal," as in "above normal" and "below normal," see Helmut Landsberg, *Physical Climatology*, 2d ed. (DuBois, Pa.: Gray Printing Co., 1958), 68–80.

17. A hundredth of an inch makes a relevant threshold because "0.01 inch or more of precipitation" is the event for which the National Weather Service esti-

mates and reports the probability of precipitation. See J. Michael Fritsch and Robert L. Vislocky, "Reply to comments on 'Operational Omission and Misuse of Numerical Precipitation Probability Expressions,'" *Bulletin of the American Meteorological Society* 76 (1995): 1813–4.

18. For a catalog of products and services, write to National Climatic Data Center, Federal Plaza, Asheville, NC 28801-5001.

19. For discussion of these centers' role, see Stanley A. Changnon, Peter J. Lamb, and Kenneth G. Hubbard, "Regional Climate Centers: New Institutions for Climate Services and Climate-Impact Research," *Bulletin of the American Meteorological Society* 71 (1990): 527–37.

20. William M. Alley, "The Palmer Drought Severity Index: Limitations and Assumptions," *Journal of Climate and Applied Meteorology* 23 (1984): 1100–9.

21. There are 344 climate divisions in the conterminous United States. For an essay on the evolution and standardization of division boundaries and the uses and limitations of the divisions, see Nathaniel B. Guttman and Robert G. Quayle, "A Historical Perspective of U.S. Climate Divisions," *Bulletin of the American Meteorological Society* 77 (1996): 293–303.

22. Karl Schneider-Carius, *Weather Science, Weather Research: History of Their Problems and Findings from Documents during Three Thousand Years* (New Delhi: Indian National Scientific Documentation Centre, 1975), 344.

23. Ibid., 347. For Supin's original description of his climatic provinces, see Alexander Supin, *Grundzüge der physischen Erdkunde* (Leipzig: Veit, 1884). For examples of Thornthwaite's work, see C. Warren Thornthwaite and Benjamin Holzman, "The Determination of Evaporation from Land and Water Surfaces," *Monthly Weather Review* 67 (1939): 4–11; and C. W. Thornthwaite and J. R. Mather, "The Water Budget and Its Use in Irrigation," in U.S. Department of Agriculture, *Water*, Yearbook of Agriculture (Washington, D.C., 1955), 346–58. For a concise summary of Thornthwaite's contributions, see Joe R. Eagleman, *The Visualization of Climate* (Lexington, Mass.: D. C. Heath, 1976), 15–29. For an overview of climatology's classical era and early stages, see H. E. Landsberg, "Roots of Modern Climatology," *Journal of the Washington Academy of Science* 54 (1964): 130–41.

24. For a list in English of Supin's 35 provinces, see Robert De Courcy Ward, *Climate, Considered Especially in Relation to Man* (New York: G. P. Putnam's Sons, 1908), 55–60. For a map of Supin's climatic provinces, see Talcott Williams, "The Link Relations of South-Western Asia," *National Geographic Magazine* 12 (1901): 249–65, map on 259. For Thornthwaite's classification, introduced in 1931 and modified in 1948, see C. Warren Thornthwaite, "The Climates of North America According to a New Classification," *Geographical Review* 21 (1931): 633–55; and C. Warren Thornthwaite, "An Approach toward a Rational Classification of Climate," *Geographical Review* 38 (1948): 55–94. Although Köppen wrote widely on climate and its classification, he presented the key descriptions of his system in W. Köppen, "Versuch einer Klassifikation der Klimate vorzugsweise nach ihren Beziehungen zur Pflanzenwelt" (An attempt at a classification of climates primarily according to their relations to the plant world), *Geographischen Zeitschrift* 6 (1900): 593–611 and 657–79; W. Köppen, "Klassifikation der Klimate nach Temperatur, Niederschlag, und Jahreslauf" (Classification of climate according to temperature, precipitation, and the annual cycle), *Petermann's Mit-*

teilungen 64 (1918): 193–203; and W. Köppen, "Das geographische System der Klimate," in *Handbuch der Klimatologie* (Berlin: Gebrüder Borntraeger, 1936), 1 (pt. C): 190–5; map on 194. For a critical evaluation, see Edward A. Ackerman, "The Köppen Classification of Climates in North America," *Geographical Review* 31 (1941): 105–11.

25. For a convenient method of applying the Köppen system, see Peirce F. Lewis, "Dichotomous Keys to the Köppen System," *Professional Geographer* 13 (September 1961): 25–31.

26. For discussion of nineteenth-century vegetation maps that provided Köppen with valuable data, see David J. de Laubenfels, *Mapping the World's Vegetation: Regionalization of Formations and Flora* (Syracuse, N.Y.: Syracuse University Press, 1975), esp. 335.

27. For an example, see Werner H. Terjung, "Physiologic Climates of the Conterminous United States: A Bioclimatic Classification Based on Man," *Annals of the Association of American Geographers* 56 (1966): 141–79.

28. Comfort-region maps are conspicuously absent from tour guides, including the cartographically noteworthy *Traveling Weatherwise in the U.S.A.,* which illustrates an overview of the country with 35 separate small-scale maps addressing topics ranging from hours of sunshine to smog and hay fever. Even so, the book contains a map of climatic regions with highly generalized boundary lines faintly similar to the boundaries between Köppen's B, C, and D divisions. See Edward Powers and James Witt, *Traveling Weatherwise in the U.S.A.* (New York: Dodd, Mead, 1972), 9.

29. Matt Damsker, "New Hardiness Zone Map," *Organic Gardening* 37 (March 1990): 83–4; and "New Map to Help Plants Find Good Homes," *Garden* 14 (May/June 1990): 14–5.

30. James D. Laver, interview with author, November 14, 1996. In 1997, 6-to-10-day outlooks were issued three times a week, on Monday, Wednesday, and Friday.

31. Cliff Nielsen, "Take a Long Look," *Weatherwise* 47 (December 1994/January 1995): 9–10; and "Climate Prediction Center Issues Long-Lead Climate Forecasts," *Bulletin of the American Meteorological Society* 75 (1994): 2193–4. For discussion of the usefulness of seasonal outlooks by electric companies, see Stanley A. Changnon, Joyce M. Changnon, and David Changnon, "Uses and Applications of Climate Forecasts for Power Utilities," *Bulletin of the American Meteorological Society* 76 (1995): 711–20.

32. Edward O'Lenic, telephone conversation, June 16, 1997.

33. Because a seasonal mean can fall outside the range of 30 means recorded for the reference period, "above-average" might include values greater than the 10 means in the highest third of the distribution, whereas "below average" might include values lower than the 10 means in lowest third.

34. Occasionally the forecast models indicate an increased likelihood of conditions in the middle, "normal" range of the distribution. For these situations, the outlook map identifies anomalous regions with the letter "N."

35. Anthony G. Barnston and others, "Long-Lead Seasonal Forecasts: Where Do We Stand?" *Bulletin of the American Meteorological Society* 75 (1994): 2097–2114.

36. "Frequently Asked Questions [concerning the Climate Prediction Cen-

ter's 90-day outlooks]," <http://met-www.cit.cornell.edu/cpc/faq.hml>, as of June 12, 1997.

37. Edward O'Lenic, telephone conversation, June 16, 1997.

38. Barnston and others, "Long-Lead Seasonal Forecasts."

39. For a comprehensive overview, see S. George Philander, *El Niño, La Niña, and the Southern Oscillation* (San Diego: Academic Press, 1990). The strong El Niño of 1997–98 received considerable attention in the media. For a cartographically illustrated, unhyped discussion of the phenomenon's human impact, see Barnaby J. Feder, "El Niño versus El Nonsense," *New York Times*, November 8, 1997.

40. G. T. Walker, "Correlation in Seasonal Variations of Weather VIII: A Preliminary Study of World Weather," *Memoirs of the Indian Meteorological Department* 24 (1924): 75–131. For a concise history of scientific study of the El Niño–Southern Oscillation phenomenon, see Philander, *El Niño*, 1–8. According to Philander, Federic Alfonso Pezet mentioned the warm current associated with El Niño in an address to the Sixth International Geographical Congress, held in Lima, Peru, in 1895.

41. Walker refined his description of the Southern Oscillation in a number of subsequent papers; see, for example, G. T. Walker and E. W. Bliss, "World Weather VI," *Memoirs of the Royal Meteorological Society* 4 (1937): 119–39.

42. J. Bjerknes, "Atmospheric Teleconnections from the Equatorial Pacific," *Monthly Weather Review* 97 (1969): 163–72. Also see J. Bjerknes, "A Possible Response of the Atmospheric Hadley Circulation to Equatorial Anomalies of Ocean Temperature," *Tellus* 18 (1966): 820–9.

43. For a concise summary, see Subcommittee on Global Change Research, Committee on Environment and Natural Resources Research of the National Science and Technology Council, *Our Changing Planet: The FY 1996 U.S. Global Change Research Program* (Washington, D.C.: Executive Office of the President, Office of Science and Technology Policy, 1996), 14–5, 69–70.

44. <http://nic.fb4.noaa.gov:80/products/predictions/multi_season/13_seasonal_outlooks/fxus05.doc> (as of June 12, 1997).

45. Advisory Panel for the Tropical Oceans and Global Atmosphere Program, Climate Research Committee, Board on Atmospheric Sciences and Climate, Commission on Geosciences, Environment, and Resources, National Research Council, *Learning to Predict Climate Variations Associated with El Niño and the Southern Oscillation* (Washington, D.C.: National Academy Press, 1996), 132–40. Other coupled phenomena that might lead to improved models include the Asian monsoons and interconnections involving sea ice, snow, and land.

46. Especially intriguing is the likelihood that warm nighttime temperatures, the result of increased air pollution, will enhance plant growth; see Ian Strangeways, "Gaps in the Climate Map," *New Scientist* 132 (November 23, 1991): 46–9.

47. For discussion of greenhouse gasses, their production, and plausible effects, see Donald J. Wuebbles and Jae Edmonds, *Primer on Greenhouse Gases* (Chelsea, Mich.: Lewis Publishers, 1991).

48. For a concise media account of the findings and limitations of computer models of global change, see William K. Stevens, "Computers Model World's Climate, But How Well?" *New York Times*, November 4, 1997. Also see James Ben-

net, "Clinton Nudges TV Forecasters on Warming," *New York Times*, October 2, 1997; William K. Stevens, "Doubts on Cost Are Bedeviling Climate Policy," *New York Times*, October 6, 1997; William K. Stevens, "Experts on Climate Change Ponder: How Urgent Is It?" *New York Times*, September 9, 1997; and William K. Stevens, "In Kyoto, the Subject Is Climate; the Forecast Is for Storms," *New York Times*, December 1, 1997.

49. While skepticism and uncertainty about global warming are real, widespread concern about global change seems too easily interpreted as a consensus accepting global warming as a fact. For example, the Intergovernmental Panel on Climate Change declared in 1990 that it is "certain that there is a natural greenhouse effect that already keeps the Earth warmer than it would otherwise be [and] emissions resulting from human activities are substantially increasing the atmosphere concentrations of the greenhouse gases." See U.S. Congress, Office of Technology Assessment, *Changing by Degrees: Steps to Reduce Greenhouse Gases*, report no. OTA-O-482 (Washington, D.C., 1991), 46. If modified with appropriate caveats, this statement would probably reflect the opinion of a majority of scientists, as well as this writer. But reasoned critiques reflect a healthy skepticism about the certitude of warming from both ends of the political spectrum. For examples, see Patrick J. Michaels, *Sound and Fury: The Science and Politics of Global Warming* (Washington, D.C., Cato Institute, 1992); and Andrew Ross, *Strange Weather: Culture, Science, and Technology in the Age of Limits* (London: Verso, 1991), esp. 193–249. For the moment at least, I subscribe to the view articulated by meteorological historian David Laskin in *Braving the Elements: The Stormy History of American Weather* (New York: Doubleday, 1996). In his chapter titled "What's Happening to Our Weather," Laskin writes:

My own feeling is that the whole greenhouse issue has been so hyped since 1988 that it has assumed an independent life of its own: the concept has become part of the way we experience and talk about the weather. We don't have storms or heat waves or droughts anymore—we have *mounting evidence* or *ominous signs* or *inescapable indications* or *the onset*. Climate change is in the air, and we don't need the media any longer to point out the global, even cosmic implications of unusual weather patterns. The fact that the major climate story shifted in the course of a decade from global cooling to global warming doesn't bother us at all. The truth is, we've grown accustomed to the idea of climate change—even comfortable with it. . . . Both cooling and warming theories have the ultimate effect of demystifying the weather, of placing it within our grasp. Both theories put humankind, for good or ill, at the center of the atmosphere. Since we caused the climate change in the first place, it stands to reason that we can change it back again. (223)

50. Thomas R. Karl and others, "Indices of Climate Change for the United States," *Bulletin of the American Meteorological Society* 77 (1996): 279–92.

51. Wuebbles and Edmonds, *Primer on Greenhouse Gases*, 47–51, 89–90, 122–7. Ozone is also a greenhouse gas that, like water vapor and carbon dioxide, contributes to global warming, at least below 30 km (19 miles). At higher altitudes, adding ozone would decrease temperature by reducing the amount of solar radiation reaching the troposphere, where warming occurs.

52. Stephen S. Hall, *Mapping the Next Millennium: How Computer-Driven*

Cartography Is Revolutionizing the Face of Science (New York: Random House, 1992), 132.

53. Ibid., 127–38.

54. J. C. Farman, B. G. Gardiner, and J. D. Shanklin, "Large Losses of Total Ozone in Antarctica Reveal Seasonal ClOx/NOx Interaction," *Nature* 315 (1985): 207–10.

55. W. Wayt Gibbs, "The Treaty That Worked—Almost," *Scientific American* 273 (September 1995): 18. In 1995, an extensive black market in CFCs threatened to undermine the accord.

56. Hall, *Mapping the Next Millennium*, 135.

57. Ibid. Hall illustrated his narrative with an array of eight TOMS images showing December ozone concentrations over Antarctica for 1979 through 1982 and 1987 through 1990. For an array of twelve October maps, for 1979–90, see the *Public Affairs Office* Web site (Appendix: Goddard PAO) of the Goddard Space Flight Center.

58. The *Distributed Active Archive Center* Web site (Appendix: Goddard DAAC) at NASA's Goddard Space Flight Center offers samples of TOMS images and data sold on CD-ROMs.

CHAPTER TWELVE

1. For an insightful exception, see James R. Carter, "Cartographic Visualization in the Atmospheric Sciences," *Proceedings of the Seventeenth International Cartographic Conference*, Barcelona, Spain, September 3–9, 1995, 1631–9. In addition to noting meteorology's contributions to the visualization of geographic data, Carter identifies weather forecasters and atmospheric scientists as a distinct community of map users.

2. Many newly trained meteorologists acquired their perceptions of what a weather map should look like from WXP, one of several mapping and display packages supported by Unidata. For discussion of the mapping capabilities of WXP, McIDAS, and GEMPAK, see Gregory Byrd and others, "Integrating Technology into the Meteorology Classroom: a Summary of the 1993 Northeast Regional Unidata Workshop," *Bulletin of the American Meteorological Society* 75 (1994): 1677–83. The NCAR Graphics package, developed at the National Center for Atmospheric Research, in Boulder, Colorado, has also been influential. For examples, see R. Grotjahn and R. M. Chervin, "Animated Graphics in Meteorological Research and Presentations," *Bulletin of the American Meteorological Society* 65 (1984): 1201–8; and M. G. Morselli and others, "ICARO: A Package for Wind Field Studies over Complex Terrain," *Environmental Software* 7 (1992): 241–54.

3. Although historians of cartography have recently challenged the uncritical acceptance of notions that maps have grown better and better over time, the progressive evolution of meteorological cartography is difficult to dispute—assuming one ignores the increased confusion of naïve users, the inevitable fits and starts of technological change, and the obvious aesthetic flaws of most maps distributed by fax or the World Wide Web. For insightful critiques of cartographic progression, among other issues, see M. J. Blakemore and J. B. Harley, "Concepts in the History of Cartography," *Cartographica* 17 (monograph 26; Winter 1980): 1–120 (esp. 17–23); Matthew H. Edney, "Cartography without 'Progress': Rein-

terpreting the Nature and Historical Development of Mapmaking," *Cartographica* 30 (Summer/Autumn 1993): 54–68; and Matthew H. Edney, "Theory and the History of Cartography," *Imago Mundi* 48 (1996): 185–91.

4. Jim Carter notes the importance of unique viewing situations like television weathercasts and weather sites on the World Wide Web. Viewing environments like these pose the challenge of shared constraints, like poor resolution, as well as the experience and knowledge of a user community accustomed to seeing particular kinds of maps. See James R. Carter, "The Map Viewing Environment: A Significant Factor in Cartographic Design," *American Cartographer* 15 (1988): 379–85.

5. Typical examples are the retrospective Daily Weather Maps, which the National Weather Service publishes once a week as an eight-page booklet. A page with four maps summarizes each day's weather for the conterminous United States, northern Mexico, and the southern part of Canada. In the upper three-fifths of the page a map at approximately 1:25,000,000 describes surface weather at 7:00 A.M., eastern time; in addition to isobars, fronts, areas of precipitation, and isotherms for 0° and 32°F, the map reports conditions at nearly 180 weather stations and offshore locations. In the lower left, a continental map at approximately 1:60,000,000 reaches as far north as Alaska and Greenland and uses height contours (in decameters) and wind arrows to describe the 500-mb surface at 7:00 A.M., while to its right a pair of 1:60,000,000 maps focused on the conterminous 48 states report areas and amounts of precipitation during the preceding 24 hours and highest and lowest temperatures for selected United States weather stations.

6. For discussion of the division of atmospheric motions by scale, see Frederick K. Lutgens and Edward J. Tarbuck, *The Atmosphere: An Introduction to Meteorology*, 4th ed. (Englewood Cliffs, N.J.: Prentice Hall, 1989), 188–9. According to Lutgens and Tarbuck, "macroscale" phenomena include mesoscale features as well as planetary scale motions, like waves in the westerlies, with lengths as great as the earth's circumference.

7. Microscale phenomena are of considerable interest around airports, but wind-profiler systems that detect these short-lived features are concerned with vertical, not horizontal resolution. For discussion of a wind-profiler, see B. E. Martner and others, "An Evaluation of Wind Profiler, RASS, and Microwave Radiometer Performance," *Bulletin of the American Meteorological Society* 74 (1993): 599–613. For examples of large-scale maps used in microclimatology, see M. P. Coutts and J. Grace, eds., *Winds and Trees* (Cambridge: Cambridge University Press, 1995); and Rudolf Geiger, *The Climate Near the Ground*, trans. Milroy N. Stewart and others (Cambridge, Mass.: Harvard University Press, 1950), esp. 197, 381.

8. For discussion of the construction and use of the Daily Weather Map, Weekly Series, see Paul J. Kocin and others, "Surface Weather Analysis at the National Meteorological Center: Current Procedures and Future Plans," *Weather and Forecasting* 6 (1991): 289–98.

9. C. L. Godske and others, *Dynamic Meteorology and Weather Forecasting* (Boston: American Meteorological Society; Washington, D.C.: Carnegie Institution of Washington, 1957), 128–33. Flattened somewhat at the poles, the earth is not a perfect sphere. Although national mapping agencies responsible for large-scale topographic maps compensate for deviations from a perfect sphere by treat-

ing the earth as an ellipsoid, these deviations are negligible at scales used to map atmospheric data.

10. See, for example, Cleveland Abbe, "Comprehensive Maps and Models of the Globe for Special Meteorological Studies," *Monthly Weather Review* 35 (1907): 559–64, esp. 559.

11. For further information on the polar stereographic projection, see John P. Snyder, *Map Projections: A Working Manual*, U.S. Geological Survey Professional Paper no. 1395 (Washington, D.C., 1987), 141–2, 154–63. Secant constructions of the polar stereographic projection are not common; most presentations are based on a projection plane tangent at the north or the south pole.

12. W. R. Gregg and I. R. Tannehill, "International Standard Projections for Meteorological Charts," *Monthly Weather Review* 65 (1937): 411–5. Bjerknes, who made his recommendations in Paris in 1919 at the Fourth International Conference of Directors of Meteorological Institutes and Observatories, published a more detailed proposal the following year; see V. Bjerknes, "Sur les projections et les échelles à choisir pour les cartes géophysiques," *Geografiska Annaler* 2 (1920): 1–12.

13. Armond L. Griggs, "The Background and Development of Weather Charts," *Bulletin*, Geography and Map Division, Special Libraries Association, no. 21 (October 1955): 10–3. The United States Weather Bureau, which had been plotting its weather charts on a polyconic projection, accepted the Salzburg resolutions in 1938 and adopted the Lambert conformal conic projection secant at 30° and 60°N. It continued to use a 1:30,000,000 polar stereographic projection for its hemispherical maps.

14. Aside from the juxtaposition of two polar stereographic projections, one for each hemisphere, the Salzburg resolutions offered no recommendation for a world map projection. Since 1870, though, textbook authors and atlas editors had found Gall's stereographic projection a useful compromise between Mercator's conformal projection, which greatly distorted area and the separation of meridians toward the poles, and the plane-chart projection, on which evenly spaced parallels severely distorted angles and shape. For discussion of meteorological applications of Gall's projection, which was neither conformal nor equal-area, see I. R. Tannehill and Edgar W. Woolard, "Gall's Projection for World Maps," *Monthly Weather Review* 64 (1936): 294–7.

15. Kocin and others, "Surface Weather Analysis," 290–1; and Jamie Kousky, Climate Prediction Center, telephone communication, July 8, 1997. The Daily Weather Map is derived from a North American surface-analysis map, which meteorologists at the National Centers for Environmental Prediction prepare every three hours. Every second surface-analysis map is copied for use as part of the northern hemisphere surface-analysis chart. After copying, the map is expanded further, through the addition of data from Europe and Asia. Essential for the hemispherical chart, the polar stereographic projection tangent at 60°N is satisfactory but not optimal for the North American map.

16. For the complete text of the Salzburg resolutions, see Gregg and Tannehill, "International Standard Projections," 412.

17. C. Fitzhugh Talman, "Meteorological Symbols," *Monthly Weather Review* 44 (1916): 265–74; and World Meteorological Organization, *One Hundred Years of International Co-operation in Meteorology (1873–1973): A Historical Review* (Geneva, Switzerland, 1973), 7.

18. Except for the symbol for thunderstorms, Lambert (or his printer) constructed all of his symbols from type. See M. Lambert, "Exposé de quelques Observations qu'on pourroit faire pour répandre du jour sur la Météorologie," *Nouveaux mémoires de l'Académie Royale des sciences et belles-lettres, année 1771* (Berlin: Chrétien Fréderic Voss, 1773), 60–5; esp. 63.

19. Continued evolution was particularly apparent in 1941, when the U.S. Weather Bureau announced the replacement of isotherms with symbols representing fronts, and a concurrent reduction in the number of isobars, more widely spaced because of a larger interval. See "New Weather Map Symbols Recall Indian Picture Writing," *Science News Letter* 40 (August 23, 1941): 117; and Frank Thone, "A New Type of Weather Map," *Science* 94 (August 1, 1941): supp., 10.

20. Ibid., 265.

21. Founded in 1947 and affiliated with the United Nations since 1951, the World Meteorological Organization coordinates international atmospheric research and the exchange of data among 185 member nations. With headquarters in Geneva, the WMO administers complex international agreements on codes, measurements, and communications standards. See World Meteorological Organization, *One Hundred Years of International Co-operation*; and the *WMO* Web site (Appendix: WMO).

22. For a concise summary of the diversity among weather maps, see Peter R. Chaston, *Weather Maps: How to Read and Interpret All the Basic Weather Charts*, 2d ed. (Kearny, Mo.: Chaston Scientific, 1997).

23. Despite its small type, "Explanation of the Daily Weather Map," NOAA Public Affairs publication no. 85009 (March 1988), is a poster, nearly 2 feet by 3 feet. Further evolution is apparent in differences between several elements in the current symbol set and the 1940s standard, illustrated in figure 12.6 and described in George R. Jenkins, "Transmission and Plotting of Meteorological Data," in F. A. Berry, E. Bollay, and Norman R. Beers, eds., *Handbook of Meteorology* (New York: McGraw-Hill Book Co., 1945), 574–600.

24. Jamie Kousky, telephone communication, July 8, 1997. Also see the *METAR/TAF Information* Web site of the National Weather Service (Appendix: METAR). The official authority for United States agencies is the *Federal Meteorological Handbook*, maintained by the Office of the Federal Coordinator for Meteorological Services and Supporting Research, a collaborative endeavor of the departments of Commerce, Defense, Energy, and Transportation. For a list of departures from WMO practice, see the *U.S. National Coding Practices* Web site (Appendix: Coding Practices).

25. The station plotting model used in the United States is much closer to the international standard now than in the early 1940s; for an illustration of the earlier difference between the U.S. and international plotting models, see Sverre Petterssen, *Introduction to Meteorology* (New York: McGraw-Hill Book Co., 1941), 166–7. For examples of symbols not part of the international code, see the elements marked with an asterisk in Meteorological Office (United Kingdom), *Observer's Handbook*, 4th ed. (London: Her Majesty's Stationery Office, 1982), 72–5.

26. In 1771, J. H. Lambert recommended varying the number of symbol elements to portray force and duration. See Lambert, "Exposé de quelques Observations."

27. Bertin's most influential work is *Sémiologie Graphique*, published in

French in 1967. Although American cartographic textbooks adopted his logic soon thereafter, an English-language version did not appear until 1981. For a translation of the second edition of *Sémiologie Graphique*, see Jacques Bertin, *Semiology of Graphics: Diagrams, Networks, Maps*, trans. William J. Berg (Madison: University of Wisconsin Press, 1983). For additional discussion of visual variables, see Mark Monmonier, *Mapping It Out: Expository Cartography for the Humanities and Social Sciences* (Chicago: University of Chicago Press, 1993), 58–76; J. C. Muller, "Bertin's Theory of Graphics: A Challenge to North American Thematic Cartography," *Cartographica* 18 (Autumn 1981): 1–8; and Arthur H. Robinson and others, *Elements of Cartography*, 6th ed. (New York: John Wiley and Sons, 1995), 319–21.

28. Bertin's two locational and six retinal variables are not the only visual variables. For a lucid, well-illustrated discussion of static and dynamic visual variables as well as auditory variables relevant to cartographic communication, see Alan M. MacEachren, *How Maps Work: Representation, Visualization, and Design* (New York: Guilford Press, 1995), 270–90.

29. For an early essay on the role of color in radar displays, see G. R. Gray and others, "Real-time Color Doppler Radar Display," *Bulletin of the American Meteorological Society* 56 (1975): 580–8.

30. "Guidelines for Using Color to Depict Meteorological Information: IIPS Subcommittee for Color Guidelines," *Bulletin of the American Meteorological Society* 74 (1993): 1709–13.

31. Ibid., 1709–10. According to Jim Schiavone, who chaired the AMS Interactive Information and Processing Systems Subcommittee on Color Guidelines, the goal was to "put a stake in the ground since nothing else existed." The AMS endorsed the guidelines, which Schiavone then sent to the World Meteorological Organization. Although several WMO members responded—typically with only minor changes and other suggestions—the organization has neither endorsed nor rejected the AMS proposal. James A. Schiavone, telephone conversation, July 25, 1997.

32. Ibid., 1710.

33. Robert R. Hoffman, "Human Factors Psychology in the Support of Forecasting: The Design of Advanced Meteorological Workstations," *Weather and Forecasting* 6 (1991): 98–110.

34. Robert R. Hoffman and others, "Some Considerations in Using Color in Meteorological Displays," *Weather and Forecasting* 8 (1993): 505–18; quotation on 509.

35. Among the varied reasons for allowing meteorologists to depart from standard colors is the apparent color impairment of as many as a quarter of potential workstation users. See Hoffman, "Human Factors Psychology," 103.

36. For insights on automated interpolation, see O. Amble, "A Smoothing Technique for Pressure Maps," *Bulletin of the American Meteorological Society* 34 (1953): 293–7; Alfred Blackadar, "Computer Map Analysis: Drawing Contours," *Weatherwise* 42 (1989): 109–13; Alfred Blackadar, "Computer Weather Maps: Contouring Observed Data," *Weatherwise* 43 (1990): 154–8; P. Menmuir, "The Automatic Production of Weather Maps," *Cartographic Journal* 11 (1974): 12–7; and Thomas J. Wright, "Utility Plotting Programs at NCAR," *Atmospheric Technology* no. 3 (September 1973): 51–7. The typical approach to contouring

has been to use distance-weighted interpolation to estimate values at intersection points on a uniform grid, thread isolines between these grid points, and apply a smoothing algorithm to remove kinks in the threaded contours. For a concise examination of interpolation procedures used for computer models, see Kiran Alapaty, Rohit Mathur, and Talat Odman, "Intercomparison of Spatial Interpolation Schemes for Use in Nested Grid Models," *Monthly Weather Review* 126 (1998): 243–9.

37. For an intriguing study of the effects of map projection on contours plotted for small-scale climate maps, see Cort J. Willmott, Clinton M. Rowe, and William D. Philpot, "Small-Scale Climate Maps: A Sensitivity Analysis of Some Common Assumptions Associated with Grid-Point Interpolation and Contouring," *American Cartographer* 12 (1985): 5–16. Also see Scott M. Robeson, "Spherical Methods for Spatial Interpolation," *Cartography and Geographic Information Systems* 24 (1997): 3-20.

38. Alan M. MacEachren, lecture at Syracuse University, April 21, 1995. Adding to uncertainty is the failure of many small-scale computer-contoured plots to show the density of the sampling network. For insights on the effects of sampling density on small-scale climatic maps, see Scott M. Robeson, "A Spatial Resampling Perspective on the Depiction of Global Air Temperatures," *Bulletin of the American Meteorological Society* 76 (1995): 1179–83.

39. Clifford F. Mass, "Synoptic Frontal Analysis: Time for a Reassessment?" *Bulletin of the American Meteorological Society* 72 (1991): 348–63. For a critique of the kinematic explanation of frontogenesis, see Toby N. Carlson, *Mid-Latitude Weather Systems* (New York: Harper-Collins Academic Press, 1991), 361–2.

40. Ibid., 355.

41. A 1991 article by National Weather Service staff noted "an over-reliance" on the Norwegian Cyclone and Frontal model in the surface analysis presented in the Daily Weather Map. See Kocin and others, "Surface Weather Analysis," 297. For examples of alternative treatments of air-mass boundaries, especially the behavior of upper-level fronts, see K. A. Browning, S. P. Ballard, and C. S. A. Davitt, "High-Resolution Analysis of Frontal Fracture," *Monthly Weather Review* 125 (1997): 1212–30; John D. Locatelli, Jonathan E. Martin, and Peter V. Hobbs, "Organization and Structure of Clouds and Precipitation on the Mid-Atlantic Coast of the United States: The Role of an Upper-Level Front in the Generation of a Rainband," *Journal of the Atmospheric Sciences* 49 (1992): 1293–1303; John D. Locatelli and others, "Surface Convergence Induced by Cold Fronts Aloft and Prefrontal Stages," *Monthly Weather Review* 125 (1997): 2808–20; and M. A. Shapiro and Daniel Keyser, "On the Structure and Dynamics of Fronts, Jet Streams, and the Tropopause," in *Palmén Memorial Symposium on Extratropical Cyclones, Helsinki, Finland, 29 August–2 September 1988*, preprints (Boston: American Meteorological Society, 1988), 177–81.

42. Napier Shaw, *The Drama of Weather* (Cambridge: Cambridge University Press, 1934), 245–53.

43. Ibid., 254.

44. See, for example, Petterssen, *Introduction to Meteorology*, 143. According to Petterssen, "The isobars at a front must be refracted in such a manner that the kink in the isobar points from low to high pressure." Also see World Meteorological Organization, *The Preparation and Use of Weather Maps by Mariners*,

WMO publication no. 179.TP.89 (Geneva, Switzerland, 1966), 30.

45. Ibid., 170.

46. For a storm chaser's view of the dryline, see Tim Marshall, "Dryline Magic," *Weatherwise* 45 (April/May 1992): 25–8.

47. J. Owen Rhea, "A Study of Thunderstorm Formation along Dry Lines," *Journal of Applied Meteorology* 5 (1966): 58–63.

48. Frederick Sanders and Charles A. Doswell III, "A Case for Detailed Surface Analysis," *Bulletin of the American Meteorological Society* 76 (1995): 505–21.

49. For an example, see J. Michael Fritsch and Robert L. Vislocky, "Enhanced Depiction of Surface Weather Features," *Bulletin of the American Meteorological Society* 77 (1996): 491–506; esp. 493 and 496.

50. Joseph T. Schaefer, "The Life Cycle of the Dryline," *Journal of Applied Meteorology* 13 (1974): 444–9. Time is central standard time. The original maps included barbed arrows showing wind direction.

51. Ibid., 444.

52. George W. Bomar, *Texas Weather* (Austin: University of Texas Press, 1983).

53. George W. Bomar, "A West Texas Phenomenon," *Weatherwise* 38 (1985): 142–6; quotation on 146.

54. Shaw, *Drama of Weather*, 254–6.

55. Ibid., 255.

56. Ibid., 256.

57. The concise synopsis is hardly a new idea; see N. R. Taylor, "The Importance of a Well Written Synopsis of Weather Conditions," *Monthly Weather Review* 33 (1905): 475–6. As Taylor noted, "No better way can be imagined of teaching the public at least some of the principles which are involved in making weather predictions than an intelligently written summary of meteorological conditions" (475). Automatic sequences and user profiles describing a viewer's interests and visual preferences are comparatively recent concepts; see, for example, Mark Monmonier, "Authoring Graphic Scripts: Experiences and Principles," *Cartography and Geographic Information Systems* 19 (1992): 247–60, 272.

58. Elizabeth Dirks Fauerbach, Robert M. Edsall, David Barnes, and Alan M. MacEachren, "Visualization of Uncertainty in Meteorological Forecast Models," *Spatial Data Handling '96: Advances in GIS Research II, Proceedings* (1996) 1: 18A17–8A28. MacEachren's Web site (Appendix: Geo Visualization) contains an electronic version of the paper, with a link to the software.

Index